Mrs. Lincoln

Mary Lincoln

Mrs. Lincoln

A Life

Catherine Clinton

HARPER LUXE

An Imprint of HarperCollinsPublishers

HarperCollins books may be purchased for educational, business, or sales promotional use. For information please write: Special Markets Department, HarperCollins Publishers, 10 East 53rd Street, New York, NY 10022.

FIRST HARPERLUXE EDITION

HarperLuxe™ is a trademark of HarperCollins Publishers

Library of Congress Cataloging-in-Publication Data is available upon request.

ISBN: 978-0-06-171974-5

09 10 11 12 13 ID/RRD 10 9 8 7 6 5 4 3 2 1

To N. G. H. and St. D.

Contents

Introduction

For most people, it was an unimaginable act—Abraham Lincoln was shot and would spend a long night dying. For his wife, Mary Lincoln, her worst nightmare had come true. She had spent the past four years fearing for her husband's safety, and now he was felled by an assassin's bullet. She stared in sorrow at Abraham's ashen face, praying for him to wake up.

Only hours before, she had been sitting proudly by the side of her husband, the man who saved the Union. When they entered the presidential box at Ford's Theatre, the audience had applauded wildly, welcoming the interruption for "Hail to the Chief." After Lee's surrender just days before, the Lincolns felt a personal emancipation from the terrible curse of war. The president had been more carefree and animated than his wife had seen him in years, at least since their son

Willie had died three years before. While watching the play, she had been clinging to her husband's arm and teasing him about what their companion in the box, Miss Harris, might think of them. Lincoln replied, "She won't think anything about it." These words were his last, as the evening dissolved into chaos, noise, and blood.

Mary kept recalling with horror the moment of the audience's laughter masking the report of the pistol. She at first could not understand what was happening, when the demon emerged from the shadows, slashing out with a knife before leaping onto the stage. Everything went out of focus, except the vivid image of her husband's head slumped forward on his chest—his limbs slack in his rocker. And the blood. Once she saw the blood, Mary began to shriek, "They have shot the President!"

During the ghastly scenes that unfolded—the mad rush into the box, a doctor finding no pulse, mouth-to-mouth resuscitation, Lincoln's revived breathing, soldiers lifting the president onto a litter to carry him across the street into a house, her long-legged husband laid diagonally across a bed—Mary stood by. She was not a mute witness, but wondered aloud why it was not *she* who was shot. Her grief was so extreme that doctors agreed to withhold the information that there was no expectation of Lincoln's recovery—to allow her to cling to the hope that he might regain consciousness. Attendants warmed his limbs, changed his dressings, and all except Mary recognized that the end was near.

Robert Lincoln, their oldest son, recently home from the military, was collected from his bedroom at the White House and escorted to the scene. He arrived at the Petersen house to find his father clinging tenuously to life, and his mother in a state of shock. Mary wanted her younger son, Tad, fetched as well, but everyone vetoed this idea. Robert recognized his mother's fragility and summoned her close friend, Elizabeth Dixon (the wife of Connecticut Senator James Dixon), to assist Clara Harris, the Lincolns' theater guest, to tend to his mother. As the cabinet gathered by Lincoln's bedside, Secretary of War Edwin Stanton issued orders and took charge, updating the outside world while trying to organize the largest national manhunt to apprehend these assassins. Stanton learned that this attempt on Lincoln's life was part of a larger conspiracy, as Secretary of State William Seward had barely survived an invader's attack at his home, with several others who tried to protect him injured as well.

When word of the president's shooting spread throughout the city, many rushed to Petersen's house, across from Ford's Theatre, to hear news of Lincoln's condition within. Lincoln's confidants crowded in and out of the small upstairs bedroom, and later, lithographs of the death scene expanded the size of the group encircling the bed to preposterous proportions. The preservation of the Petersen home by the National Park Service allows visitors today to imagine the scene as they enter the cramped bedroom at the rear of the house.

Although portraits always included Mrs. Lincoln, kneeling or sitting by the bed during Lincoln's final moments, the truth was that she had been banished from the room. Mary wanted to remain by her husband's side, but when she began to sob hysterically, Mrs. Dixon or Miss Harris would take her into a nearby parlor. After calming herself, Mrs. Lincoln would reenter the room where her husband lay, and doctors hid the rust-red stains on his pillow with fresh linen, trying to disguise evidence of his lifeblood seeping out. Near dawn, Mrs. Lincoln realized that her husband was not getting better, but worse. She collapsed onto the floor in a faint. Edwin Stanton barked, "Take that woman out and do not let her in again."

The deathbed of a loved one was perhaps the most hallowed of nineteenth-century ritual settings. Mary's ancestors were Irish women who might keen for hours,

THE ASSASSINATION OF PRESIDENT LINCOLN.

Lincoln's assassination at Ford's Theatre

if not days, over the body of a departing loved one. In Victorian America, attending a dying husband was a wife's most privileged obligation, to press into her memory his final moments, to be there at the very end. Everyone crowded in the room that night knew a wife's sacred duty. When Lincoln's breathing became halting and labored around 7:00 a.m., however, no one summoned Mary. Instead, Rev. Phineas Gurley suggested, "Let us pray. " All knelt or bowed their heads, as more than a dozen men encircled their beloved leader. At 7:22 a.m., Abraham Lincoln, the sixteenth president of the United States, was pronounced dead. Edwin

Stanton uttered his famous tribute: "Now he belongs to the ages."

But forgotten are the words of Mary Lincoln, the wife kept from her husband's bedside. When informed of the president's passing, she cried: "Oh, why did you not tell me he was dying?" Her crying could be heard throughout the house, and her wails of grief alerted those gathered outside—Abraham Lincoln was gone.

This was a moment she had fearfully imagined for so long, that she might lose this man who was the length and breadth of her, the very core of her. Mary's very marrow was indistinguishable from her marriage. And now this unbelievable loss plunged her into an abyss from the lofty heights she had dared to scale, giddy at the prospect of a future with her husband. She lost all that in a long and terrible night.

How could Mary Lincoln have been so cruelly treated at such a time? The trauma of her husband's murder while they sat together, her gown splashed with his blood, was terrible enough. But the way in which she was shoved aside at this critical moment compounded her trauma, deepened and widened her grief, foretelling indignities to come. Her subsequent maltreatment—both real and imagined—contributed to her steady decline. From the time of her husband's assassination in 1865 until her own death sixteen years later in 1882, Mrs. Lincoln failed to recover from the pain and loss of that long night in the tiny bedroom at Petersen's boardinghouse.

Mary Lincoln was a formidable nineteenth-century role model for women, yet her life also contains many cautionary tales. She invested her talents and energies in her marriage and her family, like most women of her era. But as a witty and intellectual woman ahead of her time, she channeled her ambitions into her husband's political career. Mary stepped outside the conventions to which she was expected to conform. But in both cases, her fate was bound up with that of Abraham Lincoln.

With his tragic death, she lost not just a husband, but as she later suggested, she lost everything. Robbed of his protection as well as his love, her own reputation caved in on itself during widowhood. Lincoln's legacy grew in the wake of his martyr's death and developing interpretations of his role as the country's "savior." While Abraham Lincoln became immortal in the American imagination, Mary Lincoln would become infamous. Perhaps to some an object of pity, she was generally consigned to the sidebars of her husband's narrative text.

Mrs. Lincoln did survive her husband's passing to see grandchildren born, to visit the European capitals of which they both had dreamed, to establish rhythms and patterns as the widow of a slain president. But she endured a series of crippling humiliations and losses during widowhood: There were claims by her husband's former partner that Lincoln's sweetheart from New Salem, Ann Rutledge, was his one true love. Then there were claims by Lincoln's former business associates that the president's widow was a blackmailing harridan besmirching her husband's memory by trying to pawn her jewelry and wardrobe, in the Old Clothes Scandal, as the press dubbed this episode. Political enemies suggested that she was disloyal or even criminal while in the White House, and that she did not deserve any sympathy or financial support.

On her own, she would mourn in 1871 the death of her youngest son, Tad, the one reason, she claimed, that she did not take her own life following her husband's death. Next, she suffered an embittering alienation from her only remaining son, Robert. In 1875 he committed her to an asylum against her will, ostensibly for her own safety, an act that caused incalculable damage to her self-esteem. This created a permanent stain on her historical reputation. "Was she crazy?" continues to be the most common query her biographers encounter.

Mary Lincoln's sanity as well as her character has become a matter of lively debate, as scholars continue to wrangle over her life and legacy. Even her name has become contested terrain. The majority who write about her characteristically refer to her as Mary Todd Lincoln. Yet she never used this name herself. Lincoln's wife was proud to have been born a Todd, and contemporaries would acerbically suggest she was a snob about her pedigree. Lincoln himself commented that although God managed with just one *d* the Todds required two.

Born into an aristocratic Kentucky family, she was given a superior education. Mary had nearly a decade more formal education than her husband, and went to school many more years than the average male in America. Her educational accomplishments outdistanced all but a minority of the women of her generation. As an alluring young belle with impeccable connections, there was every reason to believe Mary would have made a grand match like two of her siblings. Her older sister Elizabeth had married the son of the governor of Illinois, while a younger sister, Emilie, married the son of the governor of Kentucky. But, after a few false starts, Mary wed Springfield lawyer Abraham Lincoln, a man of humble origins whose main recommendation was his intellectual promise.

Once married, Mary and Abraham endured personal setbacks—including the death of their second child, Eddie—and many professional disappointments. Lincoln's one term in Congress and his two failed U.S. Senate races weighed heavily on the ambitious couple. They managed to triumph over circumstances when, shortly after their eighteenth wedding anniversary, Lincoln was elected president in 1860. From this pinnacle of achievement, Mary tumbled into despair when the country erupted into war within weeks of her husband's inauguration. She wept as her extended family fractured and divided, like much of the nation during the American Civil War, Lincoln's great crucible. She remained undivided in her loyalty to her husband and his goal to restore the Union. She was forced to run a gauntlet as First Lady, with scorching press attacks more prolonged than those made on any presidential wife before. She was called names in the press and hunted like the vulnerable prey that she was.

She survived these White House travails, but just barely—once even escaping an attempt on her husband's life and sustaining a severe head injury. But at long last, Mrs. Lincoln celebrated Lee's surrender on April 9, 1865—and an end to the row upon row of soldiers' graves scarring the nation.

By 1865, the death lists posted at telegraph offices and read at railroad stations had touched families throughout the land. More than three million soldiers had served, North and South. The embalming business had boomed and tin-lined coffins were hawked to bring bodies home before decay made them unrecognizable to loved ones. Small towns across the Midwest had lost as many as half their male population to the armed forces, and sizable numbers never returned, felled by combat and disease. Now at last, the black crepe that decked the halls of American homes and the black-bordered stationery that signaled shared losses might be put away to let the healing begin.

The promise of peace electrified the American people; bonfires were lit and bells rang, as delirium reigned throughout the North, but especially in Washington City, where Lincoln had remained steadfast throughout the Civil War.

With victory secured, the presidential couple enjoyed a carriage ride alone on Good Friday, April 14, discussing plans for foreign travel after Lincoln's second term ended. Lincoln wanted Mary to try to be happier, after all they had been through, both remembering the loss of their son Willie in 1862. Later that evening, Mrs. Lincoln had not felt like going out to the theater, hoping to stay home to avoid a headache. But she did not want to disappoint her husband, who intended to occupy the Presidential Box at Ford's Theatre. She and her husband left the White House that night with high hopes, for enjoying the comedy as well as the pomp and cheer that accompanied them now. But Mrs. Lincoln would return home without her husband the next morning, on Easter Saturday, a widow.

The nation would heal and be reborn, but Mrs. Lincoln would never recover from the trauma of her husband's assassination. Her story is bound up with the story of the nation, with the story of her husband's presidency, but her own story remains singular and remarkable. She was one of the first American women in the White House to capture the public imagination and to maintain a historical reputation into the present. It was Mrs. Lincoln who was first labeled First Lady, and it was by this role that she gained her lasting fame. When she whispered to her husband that night, wondering what would be thought of her, little did she realize what a defining moment this question would mark.

1.
Kentucky Homes

The rolling hills of Bluegrass Kentucky remain astonishingly beautiful, unfurling with promise and glory along the road from Hodgenville to Lexington. The lush countryside was marked with tobacco and horses, which brought the region its fame. The miles between the two towns can be measured, but the distance between them—and what it represents—is more difficult to calculate, especially in the lives of Mary and Abraham Lincoln.

In 1809 a rough-hewn log cabin carved out of the woods near Hodgenville sheltered the newborn son of Thomas Lincoln and Nancy Hanks. The hardscrabble roots of Abraham Lincoln have become legendary. His reputation has soared dramatically in the years since his presidency, and his role in American history has risen to mythic proportions. More than a century and a half after his death, the Lincoln birthplace has been turned into a shrine—with piles of marble dwarfing and literally engulfing the reconstructed cabin at the National Park site.

The contrast between this backwoods crossroads and the thriving metropolis of Lexington, dubbed the Athens of the West, is striking. When Mary Anne Todd, the daughter of Robert Smith Todd and Eliza Parker, grew up in an elegantly appointed mansion full of European imports and family mementos, she was connected by blood or intermarriage with nearly all the important political leaders of the day and, unlike her future husband, grew up with a sense of rank and privilege. Mary Lincoln's girlhood home is also maintained as an historic site—where the trappings of her family's pedigree and taste are on prominent display.

The Todd name carried great weight within elite circles of the early republic. James Madison, the Virginia aristocrat who became the nation's fourth president, married widow Dolley Todd. Madison's wife became a legendary Washington hostess and maintained warm relations with her Todd kin.

Robert S. Todd, Mary Lincoln's father, grew up just down the road from Henry Clay's plantation, Ashland. Clay was a dynamic figure within the region and the

Lincoln Birthplace cabin, Hodgenville, Kentucky

Mary Todd Lincoln home, Lexington, Kentucky

era—a dashing and handsome character. He had been born into a family of middling wealth in Virginia but had come out onto the Kentucky frontier and carved out a fortune for himself, possessing over sixty slaves on his impressive estate. His skills as a public speaker were renowned, and once he headed to Washington to represent his state, his patrician looks and oratorical flourishes won him plaudits throughout the slaveholding South. He provided a role model for the rising young men of Robert's generation.

Nearly six feet tall and extremely handsome, Robert Todd entered Transylvania College in 1805. He went on to study law and passed the state bar in 1811. With the outbreak of the War of 1812, Todd enlisted in the local military. When struck down with pneumonia only weeks after embarking on his military career, he was brought back home to Lexington to recover.

Robert's return had a fortuitous result: while recuperating, he renewed his courtship of his distant cousin, Elizabeth Parker. The couple had vastly different temperaments: "Eliza was a sprightly, attractive girl with a sunny disposition, in sharp contrast to her impetuous, high-strung sensitive cousin." Regardless, on November 26, 1812, the twenty-one-year-old Todd wed his teenage sweetheart at the home of the bride's widowed mother. The next day, Robert rejoined his regiment, the Fifth Regular Kentucky Volunteers, and returned to health and duty.

After his army stint, Robert Todd built a house on Short Street, on a lot adjoining the home of his mother-in-law. Soon his household was filled with a parade of babies. On December 13, 1818, Mary Anne Todd joined older siblings, Elizabeth, Frances, and Levi. By this time, Robert Todd was well established and in possession of a flourishing dry-goods business; he was also a clerk of the Kentucky House of Representatives and a member of the Fayette County Court. He became an up-and-coming voice within state politics.

Mary's mother, Eliza Parker Todd, who had a child every other year following her marriage—not an uncommon pattern for southern brides—died in 1825 following the birth of her seventh child (George, who survived). She was buried next to her sixth child, Robert, who had died at fourteen months. The thirty-four-year-old widower, Robert Todd, was left with a half-dozen offspring, including six-year-old Mary.

Robert Todd's unmarried sister, Ann Maria, moved in to supervise the household and slave staff: Jane Saunders, the housekeeper, Chany, the cook, Nelson, the coachman and valet, Sally, the nanny, and Judy, the nurse. But the burdens of family life were considerable, and Mary's father felt he could only cope by taking a new wife. While Robert was in the state capital, he wooed the daughter of a well-connected political ally, and on November 1, 1826, married Elizabeth (Betsy) Humphreys in her father's home in Frankfort, where John J. Crittenden, a former speaker of the Kentucky house and future governor of the state, stood as Todd's best man. Crittenden himself had recently been widowed and remarried Todd's kinswoman. At her marriage, the never-wed Elizabeth Todd became stepmother to six children, ranging in age from eighteen months to fourteen years.

Mary was nearly eight when her father remarried. This pattern of disruption and displacement was common for children of the era, but nevertheless painful. Abraham Lincoln, who lost his beloved mother at nine, ever after referred to her as his "angel mother." His loss was compounded when his father left both Abraham and his sister Sarah behind on their Indiana farm, while he went back to Kentucky to seek a new wife. Young Abraham and Sarah were left to the care of a cousin, Dennis Hanks, for a prolonged period, barely able to scrape by until Thomas Lincoln returned to Pigeon Creek with his new wife, Sarah Bush Johnston, a widow with three small children of her own.

Mary Todd never suffered this kind of neglect but nonetheless found the transitions within her childhood traumatizing. She might have welcomed a new mother nearly eighteen months after her own had died, but her Grandmother Parker strongly opposed anyone who sought to take her daughter's place. This friction stimulated a crisis. Despite his former mother-in-law's objections, six motherless children and Elizabeth Humphrey's charms were more than enough to convince Robert to take a new bride back to Lexington. Betsy also brought with her a good dower, no small matter in the face of economic challenges for Robert Todd.

The fortunes of Lexington were waning after the turn of the nineteenth century when river trade boomed, leaving the landlocked Athens of the West high and dry. Even before the panic of 1819 jeopardized his business interests, Robert Todd was forced to sell the family estate of Ellerslie in 1817.

Robert Todd's new father-in-law was a wealthy, well-connected physician, Dr. Alexander Humphreys. Several of Betsy's uncles were prominent, including one who served as one of Kentucky's first senators, and another who was elected senator from Louisiana before his appointment as U.S. minister to France. A maternal uncle, Dr. Samuel Brown, studied medicine at Edinburgh, so when smallpox broke out in 1802, he experimented with cowpox as a vaccine, inoculating over five hundred people in Kentucky and saving scores of lives. His success became a new strand of prideful lore woven into young Mary's sense of family heritage.

One of the most revered figures among Mary's new kin was her step-grandmother, Mary Brown Humphreys of Frankfort. She impressed her new granddaughter with her Old World charm, speaking French fluently and remaining stylish and fashionable, even at an advanced age.

Betsy Todd entered her husband's household with six young children looking to her for care, and their grandmother looking at her for faults. The looming and disapproving presence of the imperious Elizabeth Parker may have dampened Betsy's enthusiasm for the role of surrogate mother. Mrs. Parker deeply resented any interference with the Todd children—and antagonism developed, especially with Mrs. Parker living next door. Parker's money had purchased the Todd home, and several of the servants who remained in the Todd household felt loyalty to Mrs. Parker—a difficult situation for all three generations.

The Todd family Christmas in 1826 must have been full of unspoken tension, with a new bride of only two months sitting at the foot of the family table. Holidays the following year were seriously dampened by the fact that Betsy had lost her first child in 1827, a son named

Robert, who died only a few days after his birth. Yet within a dozen years, Betsy gave birth to eight more Todds.

After the birth of her stepsister Margaret in 1828 and stepbrother Sam in 1830, and with another stepsibling on the way (David, born in March 1832), Mary's older sister Elizabeth Todd decided to marry her suitor, Ninian Wirt Edwards, son of the governor of Illinois. The couple wed on February 14, 1832, while Edwards was still finishing his studies at Transylvania College. The newlyweds settled in his family home back in Springfield later in the spring. Although it was difficult to make her way to a new state, a new city, a whole new life, Elizabeth landed in a community where Kentuckians thrived, and made a great success of her marriage and family connections. Over the next few years, dozens of Todd kin followed her lead and transplanted themselves, settling in Springfield.

Even after Elizabeth's departure, the house on Short Street was still overcrowded. To accommodate his growing family, Robert Todd decided to move. He surely could have added on to the original house, but perhaps this gesture was intended to satisfy his second wife. In May 1832, Todd purchased a more expansive and impressive residence on West Main Street—only two blocks away from his former residence, but at least some distance from Mrs. Parker. The large double-brick house with a wide hall in the center was a definite improvement for Betsy Todd. The new estate boasted large formal flower gardens, as well as stables. Elkhorn Creek ran through the property, and a small conservatory stood to the left of the house. The grandeur did not go unnoticed by young Mary Todd, but this new, improved household had drawbacks as well. Young Mary would find that with each passing year, she felt herself being more and more out of place in her own home.

When Abraham Lincoln's stepmother, Sarah Lincoln, arrived at the rude cabin Thomas Lincoln possessed, she immediately sought improvements to the household—like scrubbing her stepchildren clean, so they might appear "more human." She brought "luxuries" like knives, forks, and spoons into their rustic domain, as well as a sense of domestic order. She required her new husband to install a floor, a window, a sleeping loft, as well as a new roof and door. The remodeled cabin allowed Sarah Lincoln to care for all her children, and the blended family thrived. Lincoln would refer nostalgically to these particular years during his boyhood.

Lincoln's fond memories perhaps stemmed from his first contact with formal schooling. Although his stepmother was illiterate, she sensed in young Abraham a hunger for learning and enrolled him in classes taught by a local schoolmaster—about a mile from the Lincoln cabin. When the school closed, the Lincoln children had to do without until the following year when another tutor set up shop. But this school was nearly four miles from his home, so young Abraham could not always attend, as his farm chores took precedence. Thus his formal education was sporadic and abbreviated. According to Lincoln himself, the aggregate "did not amount to one year."

In stark contrast, his future wife, Mary Todd, enjoyed consistent and superior education for at least a decade. Mary was certainly better educated than most women within her society, and also was better educated than most of the men during her era as well. Her several years of formal training were in contrast to what was enjoyed by other Todd sisters. Several of her brothers attended college and appreciated advanced training, but Mary seemed the lone female Todd to excel in her studies.

She began formal education at the school run by Dr. John Ward, an Episcopal minister from Connecticut who prided himself on his routines of rigid discipline. He was fond of early morning recitations, and so Mary might be found repeating memorized verse shortly after her five o'clock rising and before her breakfast. She attended Dr. Ward's with her stepmother's niece, Elizabeth Humphreys.

Elizabeth had moved in with the Todds, sharing a bedroom with Mary and becoming a devoted companion. Mary did not complain at having to get up to dress by candlelight, and then walk to school several city blocks, even in sleet and snow. She was strong willed, and her temper and tongue were acknowledged within the family circle. A theology student brought in to tutor her siblings found her a trial of his patience, and worried that she was always insulting him or laughing at him around the dining table.

Whatever her problems at the family table, she shone in the classroom. Her childhood intimate, Elizabeth Humphreys, recalled Mary's "retentive memory" and envied her quick mind. Mary excelled in domestic arts as well—required to knit ten rounds of sock per evening, and consistently finishing the task ahead of her cousin.

The sense of being unwanted at home was presumably palpable by this time. Mary and her older siblings began to spend weekends at Walnut Hill, the home of her aunt and uncle, Hannah and John Stuart. Stuart, a Presbyterian minister who taught languages at Transylvania College. Another of her father's siblings,

Elizabeth Carr, also lived at Walnut Hill. Mrs. Carr and Mrs. Stuart became confidantes of Mary Todd, and they remained close during her Kentucky youth. She also enjoyed spending time at her Grandmother Parker's, where she was indulged by Anne, Prudence, and Cyrus, Parkers' household slaves. And she enjoyed her occasional escapes to Frankfort, especially practicing French with her Grandmother Humphreys.

Mary Todd polished her French at an academy run by Augustus and Victoire Charlotte Mentelle, Parisian refugees who had fled the Terror in 1792. A few years after the Mentelles turned up in Lexington, Mary Owen Russell, a widow who had been born a Todd, donated an estate to the couple to found their school. They turned this estate into a very renowned academy for young ladies—with fees of $120 per year. When Lafayette paid a visit to Kentucky in 1825, making a stopover in Lexington, where he paid his respects at the home of Henry Clay on May 25, he also might have visited the Mentelle home, not far from Clay's estate at Ashland.

The Mentelles' academy at Rose Hill was not merely a finishing school, but a place of learning that shaped young women. Mary Lincoln found the experience bracing, and acknowledged Madame Mentelle's significant influence on her during her formative years.

This cost for female education would not have been negligible to Robert Todd, especially as he had only one daughter wed out of six when Mary, at age fourteen, enrolled at Madame Mentelle's. Mary's older sister Frances had already moved to Springfield by 1836, but Anne, Margaret, and baby Martha were still at home—and would be joined by sisters Emilie, Elodie, and Katherine. By this time, Robert Todd's fortunes were on the rise again, as he became president of the Branch Bank of Kentucky, which opened its doors in 1835. The profits from his cotton manufacturing business, Oldham Todd & Co., one of the largest houses in the region, added to his wealth.

Were these Mary's happiest childhood years, when she could shine at her boarding school and get the attention she could not command at home? While Mary might have been lost in the shuffle of her siblings, she found a way to make an outstanding impression as a star student. She later commented, "My early home was truly at a boarding school."

Certainly Mary blossomed during these years, as a classmate remembered her as merry and smiling. She was one of the most intellectual of her circle of girlhood friends: her cousins Elizabeth Humphreys and Margaret Stuart, and the daughters of her father's friends: Margaret and Mary Wickliffe, Catherine Cordelia Trotter, Sarah Shelby, and Julia Warfield. She always earned the highest marks and captured prizes, speaking French as fluently as Madame Mentelle. Her cousin Margaret Stuart Woodrow remembered that "Mary's love for poetry, which she was forever reciting, was the cause of many a jest among her friends." Finally, Mary had a flair for theatrics and became the star actress of the academy.

Several childhood stories emphasize Mary's precocious charm and penchant for melodrama. She enjoyed being the center of attention, and would insinuate herself into adult situations from a very young age, as one observer recalled: "She heard politics discussed by eminent men who had patted her on the head and who sometimes, much to Mary's delight, gallantly kissed her hand." Her ability to insert herself into powerful circles of political men came from early practice.

One oft-repeated tale involves a thirteen-year-old Mary riding a new pony to Ashland and finding Mr. Clay giving a dinner party. Despite the interruption, she was invited in, and Clay, apparently charmed by his familiar visitor, confided to the guests, "If I am ever President I shall expect Mary Todd to be one of my first guests." Mary then announced to the assembled gathering that she would enjoy living in the White House.

Her high regard for Henry Clay continued. When Andrew Jackson came to town on September 29, 1832, Mary rebuked one of her chums: "I wouldn't think of cheering General Jackson, for he is not our candidate." She added thoughtfully, "He is not as ugly as I heard he was." In this folkloric account, Mary Todd continues to confide that Henry Clay "is the handsomest man in town and has the best manners of anybody—except my father."

Mary divined that wit might be the best means for her to shine among adult males. Reflected glory was all she was expected to claim. The rules of the game stacked the deck against women. From an early age, Mary chafed against the restraints that fenced women in. This rebellious streak wreaked havoc on her adolescent personality, as a family member described her as having an emotional temperament "like an April day, sunning all over with laughter one moment, the next crying."

As Mary developed into a young woman, she became more, not less, unpredictable. If there ever was an age when she was sunny and tantrum-free, it would have been her earliest Lexington years. Her older sister Elizabeth recalled that she loved to wear flowers in her hair, and a childhood friend reminisced: "I always thought of tea roses in connection with Mary, they seemed so to suit her, to be a part of her." Blooms and fragrances brought her joy in youth, but youth would yield to maturity, and beneath these blossoms hid the thorns.

2.
Making Her Own Hoops to Jump Through

Ten-year-old Mary Todd wanted a swishing skirt of her very own and disdained the simple white muslin to which she was confined for Sunday dress. She had learned from an early age that asking permission often interfered with getting her own way. So she concocted a scheme secretly with her comrade and cousin, Betsy Humphreys. The girls made their own hoops (out of stripped willow branches) to place under their skirts so they might appear fashionable when they attended church. Before they escaped the house for Sunday services, the ruse was discovered. Betsy Humphreys Todd scolded them and forced them to abandon their home-made contraptions. Mary sobbed over this episode and was deeply wounded by her stepmother's displeasure. Robert's wife, Betsy, may have understood all too well what was involved as her strong-willed stepdaughter repeatedly attempted to flout the rules and call attention to herself.

Her cries for attention were generally disregarded by her father, and Robert Todd's preoccupation with other matters contributed to Mary's frustration. Much of this might have developed even if Eliza Parker Todd had lived to bear more children. We do not know how she would have weathered Mary's pranks and rebellions, nor how much of this was nature or nurture. But it was not Eliza who produced nine more heirs, who displaced the previous six Todd offspring within the household. Eliza's children became resentful of the attention and resources siphoned off by the brood produced by Robert's younger, second wife.

Mary eventually became close to one of her half-sisters, Emilie, whom she and her husband, Abraham, would call Little Sister. (Emilie became a family historian, and her daughter Katherine would publish the initial biography of her famous Aunt Mary, whom she first remembered from a visit to the White House during the Civil War.) One of Emilie's cousins—writing in the 1890s—recalled incidents from their youth that reflected "the other side of nostalgia."

Many postbellum white southern remembrances of times past included fond tales of black caretakers—and indeed, Katherine Helm's study was awash with such references. But privately exchanged confidences, found in an unpublished collection of letters between cousins, include dissenting perspectives.

Sally served as the children's nanny, appearing as a fond figure in family lore. But a Todd cousin, Elizabeth Norris, recalled the time when Emilie was just a toddler and disappeared one morning; the police were contacted, and all of Lexington turned out to search for her. With each passing hour, her mother agonized over Emilie's fate. Robert Todd was scouring the alleyways when he spied his daughter in a window, and retrieved her from a man who had found Emilie wandering the streets. This man had no children, and thought he would bring the little girl home to his wife so they might keep her. According to Elizabeth Norris, Emilie's "uncommon beauty overcame his sense of right."

How did Emilie end up in such a predicament? Elizabeth mentioned an explanation conspicuously absent from official versions: "I cannot think how it happened unless old Sally had you in the street and let you stray in one of her drunken spells, for she got drunk when she got the chance."

Slavery was a fact of everyday life for the Todds. Especially after the loss of her mother, Mary clung to those African Americans who represented both continuity and comfort for her during her youth. Her half-sister had suggested that Mary might "wonder" if her Mammy wanted to be free, but then "concluded she did not," because "How could we do without Mammy, and how could she exist without us?" This kind of cultural delusion was common for white women in the antebellum South.

Owners insisted that they maintained a paternalistic system, treating slaves as if they were "in the family"—a southern heritage of which Mary was taught to be proud. Yet there were obvious flaws to this system, flaws to which she was exposed. On the way to her Grandmother Parker's home, Mary might pass William Pullum's place, where the storekeeper kept slave pens and dealt in human chattel. She could hear if not see the auctioning of women and children at nearby Cheapside, where slavedealers came to show their "stock" and trade humans as commodities. Or she might witness the assembled townspeople at the public square when a slave might be affixed to a black-locust log "ten feet high and a foot thick," used as a whipping post, when the screams of lashed slaves punished for their offenses might gather a crowd.

If slavery was such a well-ordered system, why did it cause such disorder within southern society? From a very early age, Mary would have been tutored on the growing divide between those who favored the growth of slavery and championed it as a "necessity," and those who saw it as a growing problem and favored its decline. She might have hoped that her heroic neighbor, Henry Clay—the king of compromise—could broker peace, but events in even her own hometown disabused her.

Increasingly, feuds created conflict and divides within her community. Robert Wickliffe was a proslavery state senator from Fayette County, whose daughters Mary and Margaret were Mary's contemporaries and, although slightly older, intimate friends. Both Mary's distant cousin Cassius Clay and another leading Kentucky politician, Robert J. Breckenridge, favored emancipation and were involved in reform efforts to effect change.

Debates became more volatile, less civil, following the Missouri Compromise in 1820, when federal legislators drew a line across the lands acquired from the Louisiana Purchase and declared slavery legally protected in states organized for admission to the Union below this divide, while slavery would be outlawed in territories and future states located above it. Sectional disputes over the slave trade, the colonization movement (groups dedicated to repatriating descendants of slaves back to Africa), and the extension of slavery raged throughout the nation. Occasionally discourse could and did stray from the theoretical into the incendiary, and resulted in violence.

In March 1829, Charles, the brother of Mary's good friends Mary and Margaret Wickliffe, shot and killed the editor of a local paper, the *Gazette*, after he and the editor had traded vitriol in the pages of rival papers. According to Wickliffe, the *Gazette* editor had defamed his father, which prompted him to attack the editor at his press office. Wickliffe had shot the unarmed man in the back of the head with a pistol. He was acquitted by a jury of his Lexington peers.

Three months later, when the new editor of the *Gazette* offended him, he challenged him to a duel. This time, the editor was the scion of a wealthy Kentucky family, and the two editors had been childhood friends. George Trotter Jr. reluctantly had taken over the *Gazette* to honor his late father, General George Trotter, a military hero and a critic of slavery. George was the brother of another of Mary Todd's intimate friends, Catherine Cordelia Trotter. This ongoing battle over slave politics in the pages of the *Gazette* was riveting to all of Lexington. Trotter vigorously defended free speech and maintained a political challenge to slavery. He questioned the fairness of the Wickliffe verdict in print, and Wickliffe responded with a challenge for

"satisfaction," a well-known signal to those who subscribed to the *code duello.*

At nine in the morning on October 9, 1829, Trotter
and Wickliffe met at a field on the county line, and
fired at a distance of eight feet (not the customary
thirty). Trotter's bullet only grazed Wickliffe, but
Wickliffe missed his opponent entirely. This could
have been the end of it, but Wickliffe demanded a
second round. When they fired again, Wickliffe
once again missed entirely. But Trotter's bullet hit
its mark, and Wickliffe, carried to his home, died
shortly after—alive at dawn but dead by noon, killed
by his childhood friend in an affair of honor. Mary's
circle was traumatized by this kind of violence; young
Wickliffe's death brought home the personal costs
these battles might exact.

Mary would also have been aware of other consequences of slavery in antebellum Lexington. How could
she ignore the posses in search of enslaved runaways,
the scaffold for slaves convicted of crimes without
benefit of a jury of peers, the casual yet ghastly bru-
tality reserved for African Americans within the community? In 1834, Southerners were gripped by a story
emerging from New Orleans involving a respectable
couple, the Laluries, who had kept their slaves chained
up in the attic—and this chamber of horrors was just

one of the many atrocities to which these Louisiana slaves had been subjected. The details were splashed in headlines across the South. Just a few years later, in 1837, an equally horrifying scandal struck Mary's own hometown.

When Mary was nineteen, her father was on the jury during the trial of an abusive mistress, Caroline Turner, who, among other depravities, had thrown one of her slaves out a window. Her husband accused her of murdering half a dozen of their African-American charges—his tales were another catalogue of horrors, signs of slavery's brutal effects. The Lexington of Mary Lincoln's youth was cheek by jowl with slavery's most sordid aspects.

From the earliest years of settlement, the rising birthrate of mixed-race children induced authorities to attempt to resolve "miscegenations" with a stroke of the pen: In 1662, the Virginia assembly passed legislation declaring that slave offspring inherited the status of the mother (*partus sequitur ventrem*). This law provided white males with an *incentive* to prefer slave women as illicit sexual partners—as they could not be charged with bastardy. Racial differentiation had been a paramount concern in the colonial era, and when Kentucky began to tackle these issues, it followed its sister state, Virginia.

Other statutes and cases indicate that ~~
recognized the sexual dangers inherent in ~~
The color line was meant to be an effective mea~~
social control, and authorities did not want it tran~~
gressed. Elaborate systems of sexual and racial etiquette evolved to discourage liaisons between white
women and black men and to prevent white men from
legitimating unions with African-American women,
slave or free.

Regardless, liaisons persisted. Even if she might
not have known when she was younger, Mary became
aware of the fact that a distant cousin, John Todd
Russell, impregnated a slave who gave birth to his
only son. Upon her cousin's death, his grandmother
freed the mixed-race grandson and his mother, but the
affair embroiled the family in scandal, which dragged
through the courts for years.

These relationships were a consequence of life under
slavery, and did not need to be an impediment to status
or power within the slaveholder's world if rules were
obeyed rather than broken. Mary's father was close
friends with Richard M. Johnson, a wealthy planter
and former war hero who resided at his estate, Blue
Springs, near Lexington. Johnson would go on to serve
as Martin Van Buren's vice president from 1837 to 1841,
despite his penchant for ignoring local racial etiquette.

Johnson maintained a lengthy, intimate relationship with a woman, Julia Chinn, who was euphemistically referred to as his "mulatto housekeeper." The couple produced two mixed-race daughters, Imogene and Adaline. Problems arose when Johnson refused to deny this connection and chose to acknowledge his daughters in ways deemed inappropriate. He provided an excellent education for his light-skinned daughters, and the girls' tutor, Thomas Henderson, observed that "a stranger would not suspect them of being what they really are—the children of a colored woman." Indeed, when Lafayette paid his visit to Kentucky in 1825, he spent the night at Blue Springs, and a neighbor commented, "Evry thing that was necsary for the occasion was prepared in fine order. Johnsons Two Daughters they Played on the piano fine. They Ware Dressed as fine as money Could Dress them & to one that Did not no they ware as white as anny of the Laydes thare & thare ware a good many." This eccentricity might have been tolerated, until Johnson decided to marry off the two girls to white husbands.

His older daughter, Imogene, wed first. Then in November 1832, headlines sensationalized the union of her younger sister, Adaline: The Lexington *Observer & Reporter* exposed a "marriage extraordinary" between a "white man" and "a mulatto girl reputed and acknowledged daughter of the Honorable Richard M. Johnson." The Whig paper complained that "the people of Scott County" were shocked, and then outraged when Johnson deeded over estates and provided financial security for these daughters, which sought to confirm his biological connection. His private life was dragged into the headlines again when his name was put forward as a presidential running mate in 1835. Reportedly, Johnson was cohabiting with "a young Delilah of about the complection [sic] of Shakespeare's swarthy Othello who is said to be his third wife; his second . . . he sold for infidelity."

Johnson's liaisons became the subject of public scandal. In an 1835 editorial, George D. Prentice, editor of the *Louisville Journal*, explained the difference between Johnson and another well-known southern politician who had been involved in a similar scandal, Thomas Jefferson:

The author of the Declaration of Independence had his faults, but he was at least careful never to insult the feelings of the community with an ostentatious exhibition of them. He never lived in open intercourse with an "odoriferous wench"; He never bribed "his white fellow citizens" to "make such beasts of themselves" before the open eyes of the whole world as to stand up in the church, grasp the sable paws of negresses and pronounce the sacred vows of wedlock.

Johnson ignored social dictates about veiling his connections. Prentice raged: "If Col. Johnson had the decency and decorum to seek to hide his ignominy from the world, we would refrain from lifting the curtain. . . . His chief sin against society is the publicity and barefacedness of conduct, he scorns all secrecy, all concealment, all disguise." (Johnson's daughter Adaline reputedly died of "a broken heart" in February 1836, allegedly from the abuse heaped on her father because of his "colored" family.) Both these liaisons and their consequences created circumstances with which Mary would become increasingly familiar.

No matter how carefully slaveholders tried to shroud their relationships with slaves, they were common and visible throughout the South. Interracial liaisons represented moral corrosion and highlighted one of slavery's worst features—hypocrisy. Despite slaveholders' claims to the contrary, human bondage presented whites, especially white women, with a moral dilemma. In the decades following Mary's birth, these dilemmas would cause deep fissures within American politics and contribute to the escalating sectional divide. Some slaveholders believed the immoral aspects of slavery weighed too heavily and sought to lift the burden from descendants through private emancipation policies.

Mary received very clear signals about slavery's negative aspects from two women whom she most admired during her youth: her Grandmother Humphreys and her Grandmother Parker. Both of these women took the opportunity to emancipate several of their slaves. Mary Humphreys freed nearly a dozen slaves in her will in 1836, which reflected her family's and her own antislavery sensibilities. Humphreys had not only emancipated her slave John, but had educated him for the ministry and sent him to Liberia "to make Scotch Presbyterians of the heathen." Robert's mother-in-law even freed Jane and Judy, slaves serving in Robert's home, in her will. This clear and deliberate plan may well have influenced Mary's own developing attitudes toward the institution of chattel slavery.

Grandmother Parker not only freed slaves upon her death, but also provided an annuity for her servant Prudence, who was elderly and needed the financial assistance once liberated. These factors shaped Mary when she was at an impressionable age, and may have influenced her to migrate from Kentucky to live with her sisters in the "free state of Illinois," as the region styled itself. Illinois trumpeted that it was the first state to join the union where slavery had never been part of its heritage.

When she was only fourteen, in late June 1833, Mary's hometown of Lexington was devastated when cholera hit. Within ten days, fifteen hundred were struck down and nearly fifty people a day died. One observer noted, "The streets were silent and deserted by everything but horses and dead carts." The Todds were deeply affected by this ghastly outbreak: "Father had all the trunks and boxes taken out of the attic and hauled to 'Cheapside' to be given to the people who did not get coffins." Although there were no fatalities within the immediate family circle, the Todds were among the lucky ones. One account described a family in which seventeen out of nineteen died. Cousins and close friends suffered devastating losses. Summer festivities were suspended for the Fourth of July, as with five hundred dead, "the whole city was in mourning."

The trauma was evident in the evangelical zeal with which many ministers brought disbelievers into their flocks. Once the scourge was lifted, the pews remained full, but not quite as crowded. Mary Todd was a young girl afforded all the comforts religion might provide, but she was not steeped in the ascetic piety common among extremist southern evangelicals. She had been born into a family in which the ministry was as much a professional duty as it was a religious calling. Naturally, family heritage required church attendance and behavior consistent with social and religious dictates. She was confirmed in the Episcopal Church at the age of twelve. Yet neither this formal confirmation of her faith nor the plague that hit during her adolescence seems to have channeled her energies into religious activities.

The cholera calamity convinced the family to vacate their Lexington home every summer for reasons of health. Elizabeth Humphreys Todd was fond of these migrations to Crab Orchard Springs, Kentucky. One of her nieces claimed that she was delicate and in need of such respite.

Time at the springs was allegedly therapeutic. As one planter playfully wrote, "The water has the pleasant flavor of a half-boiled-half-spoiled egg and according to popular belief cures the following diseases: Yellow jaundice and white swelling, blue devils and black plague, scarlet fever, yellow fever, spotted fever and fever of every other kind and colour." Visits to the springs were not just for reasons of health, but for status as well.

While Mary enjoyed the waters at the springs during her early years, her family circumstances were in stark contrast to those of the man she would one day marry. Abraham Lincoln's youth resembled indentured servitude—as did the lives of many frontier offspring on small family farms during this period.

At the age of eight, Abraham Lincoln had relocated with his family to Little Pigeon Creek, Indiana, an outpost with few educational opportunities. His father put an ax in his hands and kept him busy with it as long as he remained on the family farm. The expectations Thomas Lincoln had for his son contributed to mutual dissatisfaction, as Abraham's rapid growth spurt between the ages of twelve and sixteen (when he reached six feet two inches in height and weighed about 160 pounds) meant he was expected to shoulder more and more chores for his father, abandon his book-learning, and endure relentless physical drudgery.

When Lincoln's cousin Dennis Hanks moved out of the Lincoln household following his marriage in 1826, fifteen-year-old Abraham was even more burdened by his father's demands. Dennis noted that Thomas Lincoln might "slash" his son for neglecting his work by reading. When he was seventeen, Abraham's beloved sister Sarah married a neighbor and moved a few miles away. His stepsister, Matilda (one of Sarah Bush Lincoln's daughters by a previous marriage), also moved out of the home. Abraham and his stepbrother John D. Johnston were in constant demand as their sibling circle was reduced. Also, when Abraham was weighed on the scales of his father's affections, he was found wanting.

Abraham had ambitions beyond Pigeon Creek, and participated in schemes to sell firewood to steamers passing on the Ohio River—which enabled him to earn enough to wear his "first white shirt." He hired himself out to local neighbors, working at hog slaughtering, plowing, and fence building, all to supplement the family income. In 1828, the lanky young man wandered farther afield when he hired himself out to James Gentry (who owned a store in a nearby village) and accompanied Gentry's son on a flatboat to New Orleans to sell some goods. The two young men encountered would-be robbers, lush plantations, and the allure of New Orleans. When he returned home, Abraham handed over his wages from the trip to his father, as required by both custom and law.

By this time, Abraham was clearly ready to break free of his family's hold—and even asked a neighbor to recommend him for a job on a riverboat. But until he came of age, Lincoln was legally beholden to his father—who was aging, blind in one eye, and indifferent to the hardship he imposed on his intellectually gifted son. Abraham was also perhaps unwilling to desert the household because of his warm regard for his stepmother, who had shown unusual kindness to him.

Sarah Bush Lincoln maintained an affectionate relationship with Abraham throughout his difficulties with his father, and even after her husband's death.

When his sister Sarah Lincoln Grigsby died in January 1828, shortly before his nineteenth birthday, Abraham Lincoln once again contemplated his bleak future in Indiana. Then in 1830, shortly after his twenty-first birthday, his father decided to pull up stakes once again, to emigrate to Decatur, Illinois, two hundred miles away. Perhaps recognizing that this would be his last chance to exploit his son's labor, Thomas Lincoln made the most of it. Once they laid claim to their land in Illinois, Abraham spent the spring and summer engaged in the hard work of homesteading: fencing in fifteen acres of land along the Sangamon River and using four oxen to "broke the prairie." By this time, Lincoln had reached his full height of six feet four inches.

Finally, Lincoln liberated himself from his father's farm and struck out on his own, moving to New Salem, Illinois, in 1831 at age twenty-two. He found the town full of engaging people who warmly welcomed him into the community. Local leaders applauded his talents and encouraged his ambition. Along the way, Lincoln found people eager to listen to his reasoning, with "argument so pithy and forcible that all were amazed."

Within two years of his arrival—"a friendless, un-educated, penniless boy" as he later characterized himself—twenty-three-year-old Abraham Lincoln put himself forward as a candidate for the state legislature in 1832. Even though Lincoln would lose this bid—albeit while obtaining 277 out of the 300 votes cast in his hometown of New Salem—the race signaled the ardor of his political ambitions and enthusiasm.

Though he would later play down his military ser-vice, Lincoln volunteered for the Black Hawk War in April 1832. This development was an important rite of passage as well, not only because of the opportunities for leadership it presented with his election as a militia captain, but because his stint with the army afforded him camaraderie with the rising young men of his state. His leadership role offered him the rare opportunity to meet as equals those who might be impressed by this rawboned young man on his own. His brief but well-timed military service provided Lincoln with invalu-able links to a larger political network.

In the summer of 1832, Lincoln served in a battalion with Major John Todd Stuart, one of the bright stars within party politics in Springfield. Stuart took note of the young captain from New Salem, who would prove a formidable rival when, as a Whig, he finally gained his seat in the state legislature in 1836. This acquaintance between the two Black Hawk War veterans blossomed into deep regard over the coming years, and the two young lawyers shared a bed when they both resided in Vandalia, when the state legislature was located there. Indeed, Lincoln moved to Springfield in 1837 to form a law partnership with Stuart. Thus 1832 marked Lincoln's first encounter with a member of the illustrious Todd family, a family that would play such a key role in his future.

Most accounts of Mary Todd suggest she first contemplated a move to Springfield after Abraham Lincoln arrived in 1837, which may well be true. But there is evidence that she visited the town even before Lincoln settled there. Sometime during the spring of 1835, both Frances and Mary Todd journeyed to Springfield to visit their married sister Elizabeth. Little is known of this 1835 visit. Perhaps Robert Todd took his daughters with him on one of his periodic visits to conduct business and visit his eldest daughter, Elizabeth (and her husband Ninian), and his nephew John Todd Stuart in Springfield. Ninian's father, Governor Edwards, had succumbed to cholera in 1833, and the young couple inherited an enviable social position, as well as property, which may have required Robert Todd's personal attention. Frances and Mary spent the summer of 1836 together, back in Kentucky. But shortly after,

Frances decided to take up the Edwardses' offer that she come to live with them in Illinois. Elizabeth was pregnant with her first child and welcomed her sister as a companion.

Frances left seventeen-year-old Mary behind to negotiate on her own in a household that now included sister Ann and brothers Levi and George, as well as stepbrothers Samuel, David, and Alexander, plus stepsisters Margaret and Martha. While Mary watched Betsy Todd turn out children, she also witnessed a parade of childhood friends march down the aisle to plight their troth. The fairs and circuses of yesteryear were being replaced by wedding suppers and christenings. Mary Mentelle, the daughter of her favorite teacher, married the son of Henry Clay, an idol of her youth, in October 1832. The next year, her older sister would marry, and one by one her circle of female intimates made the trip to the altar. Mary watched as her married sister, Elizabeth Todd Edwards, served as bridesmaid to Mary Jane Warfield as she walked down the aisle to marry Cassius Clay. Margaret Stuart, sister of her beloved cousin in Springfield, John Todd Stuart, became Mrs. Woodrow. Mary felt she was being left behind.

She had grown into an attractive brunette, with a pert nose, plump cheeks, and a vivacious manner that attracted favorable attention among young women and men alike. She was interested in fashion from an early age, and learned to show off her creamy complexion, which was often commented on by observers. Perhaps she employed some of the homemade concoctions women devised to keep themselves unblemished: Lemon, cucumber, and horseradish were popular ingredients. She might have rubbed on "Virgin's milk"—benzoin, glycerin, and water—to maintain her attractive skin.

Mary was bright, curious, lively—and clearly bored with the limited choices available for an educated woman. The only opportunity for her to make herself useful outside the Todd nursery was to become a teacher. She may have undertaken some assistance at Dr. Ward's school, having been a star alumna. But she could only postpone, not avoid, marriage. The role of oldest female sibling in her stepmother's burgeoning household weighed on Mary. She chafed at the restraints of female docility expected of young women of her class and status.

Mary's elders noted her ability to express herself vividly, and even to wound. A sympathetic relation observed that she could not restrain a witty, sarcastic speech that "cut deeper than she intended, for there was no malice in her heart." So she was growing both restless and perhaps embittered by her time on Short

Street. The birth of yet another Todd daughter in 1836 (who would, ironically, be the stepsister to whom she would become the closest) prompted Mary to reconsider her circumstances. She was often cast as her father's "favorite" among her mother's children, but how long could she hold on to this title in the face of such competition?

She spent an exploratory period in Springfield with her sisters in the summer of 1837, meeting her niece Julia, Elizabeth's daughter born in April, for the first time. She would have been charmed to resume her acquaintance with three young cousins, rising lawyers within Springfield: John T. Stuart, John J. Hardin, and Stephen T. Logan.

Mary was a cosseted houseguest at her sister and brother-in-law's home. Ninian Edwards was becoming a political force within the state who constantly hosted bright young men in his parlor. Along with Todd relatives, Mary might find Edward D. Baker and O. H. Browning, both leading lights in the Whig party, a new political faction formed in opposition to Jackson's Democratic organization, which had been on the rise since the 1820s. Edwards, an emerging Whig statesman, did not socialize exclusively with members of his own party, but hosted Democrats as frequent guests. His political gatherings attracted the flamboyant James Shields, the Irish-born state auditor, and the dynamic Stephen Douglas, the New England–born state legislator—nicknamed the Little Giant because of his prowess as an orator, which contrasted with his physical size.

Douglas came to prominence at the same time that Lincoln became known statewide for his role as a member of the Long Nine, a powerful group of Sangamon delegates to the Illinois statehouse—two senators and seven representatives. The Long Nine earned their nickname from the fact that they were all very tall, with a collective height of fifty-four feet. Lincoln's six foot four—nearly six foot ten in comparison to today's average—was viewed as gigantic.

Mary's summer in Springfield was a vibrant season of mixing with politicians, heightening her appreciation of current affairs. At this time, she may or may not have met Abraham Lincoln. All of Kentucky reeled from the scandals emanating from Washington in February 1838. A dispute between the honorable gentleman representing Kentucky in the House of Representatives, William Graves, and a member of the Maine delegation, Jonathan Cilley, resulted in their meeting for a duel near the Anacostia Bridge in the District of Columbia. They used rifles as weapons: pacing off eighty yards, firing and missing twice. But on the third try, Cilley missed while Graves killed Cilley. This might have reminded Mary of the Wickliffe-Trotter duel nearly a decade before.

Whigs defended this dueling episode, while Democrats vigorously attacked all involved, especially the two Kentucky senators. These men were blamed for their direct roles: Clay had written the formal challenge for Graves, and John Crittenden had furnished the weapons. The national outcry focused on the role of Kentucky in this terrible tragedy. Mary found it difficult to find any fault with her political hero, Henry Clay. Her father, Robert Todd, participated in a public meeting to defend Clay. These lively political disputes doubtless reached Mary Todd's circle.

Mary may have enjoyed her outings to the ballrooms of local entrepreneur Monsieur Giron and to the occasional wedding—but the sense of life passing her by must have accelerated following her sister Frances's marriage on May 21, 1839, to William S. Wallace, a Springfield doctor. Their sister Elizabeth took a great deal of interest and pride in her sisters' romances, and had married Frances off in a white satin dress.

The Todd household in Lexington included yet another baby, Alexander, born in February 1839. By this time, Mary had not landed a prospective husband nor even a likely suitor. Her social circuit in Lexington provided poor returns. She was only twenty, but with each year, her value in the marriage market diminished.

The median age of marriage for Southern women of Mary's race and class was twenty, but planter daughters often married as young as fourteen. By contrast, young women of similar backgrounds in the North most often married later, with the average Yankee bride nearly twenty-four. With both older sisters now comfortably ensconced in Illinois, it must have been welcome news when Elizabeth wrote to Mary inviting her to live with the Edwardses in Springfield. Frances would move out and Mary would move in.

Even if Mary was not concerned about casting her net more widely for suitors, she may have become more than her stepmother could or would want to handle. Years later, in the course of a family lawsuit, her younger brother would complain that he, like all of his father's "first children," would be forced "to abandon his house by the relentless persecution of a stepmother." Whatever Mary's reasons, whatever she felt was in store for her in Illinois, she headed to the Lexington depot in October 1839, to make a new life in Springfield.

3.

Athens of the West

S he could make a bishop forget his prayers," was how Mary Todd was remembered from her early days in Springfield. She swept in with charisma to spare, and soon the town was abuzz over the intensity of her blue eyes ringed by a dark fringe of thick lashes, over her creamy skin set off by the deep richness of chestnut hair, often decorated with flowers. She enjoyed being the center of attention and finding new admirers to fill out her dance card. One of the most prominent of her admirers was a lawyer and rising politician, Abraham Lincoln.

Lincoln could not capture such attention by physical attraction. He was odd looking, with his overlarge head and gangly limbs. His 160 pounds stretched on his six-four frame gave him an elongated appearance that cartoonists would later enjoy. Women found his looks more strange than alluring, with "eyebrows cropped out like a huge rock on the brow of a hill." His attraction was bound up with intellectual prowess and the intensity of his curiosity: "I love to dig up the questions by the roots and hold it up and dry it before the fire of the mind." But it took a perceptive woman to see beyond his eccentric looks to detect the fire within.

Lincoln had tied his fortunes to Springfield, the newly minted capital of Illinois. He had worked as a surveyor, a postmaster, a soldier, and even apprenticed himself to train as a lawyer after he failed to maintain a successful dry-goods business—all demonstrating his ambition and drive. Upon his election to the state legislature in 1834, the political neophyte had to borrow money from a friend in New Salem to buy himself a suit of clothes.

When Lincoln arrived to serve in December 1834, he was the second-youngest man in the statehouse. He quickly discerned that he was also the least experienced. But by 1836, hard work and flourishing skills helped him get reelected. Lincoln grew on the job and became the floor leader of the Sangamon delegation, spearheading efforts to relocate the state capital to Springfield from Vandalia.

Mary's brother-in-law Ninian Edwards was so pleased when the Sangamon gang pushed through a

Mary Todd Lincoln in the 1840s, earliest
known photograph

successful bill on February 28, 1837, that he treated the
legislators to champagne, cigars, and oysters. When the
Illinois legislature dismissed in April, the twenty-eight-
year-old Lincoln packed up his saddlebags and headed
for Springfield to try his luck lawyering.

Lincoln's faith in Springfield was much like the
young man himself—all elbow grease and promise.
An observer in 1835 reported that "Springfield is a
small village of about one hundred people and twenty

Abraham Lincoln in the 1840s, earliest
known photograph

or thirty shanties, a hotel—a hard-looking place."
Yet the town had significant boosters, and by 1840
the population exceeded sixteen hundred, including
state government officials who had moved in the year
before. Lincoln arrived in his new home considerably

bedraggled, with less than twenty dollars to his name. Yet the town opened up to him, and Lincoln had hopes that he might make something of his life, in spite of the gloom he demonstrated to intimates about his prospects. The contagion of optimism sprang from two key friends of his youth: his law partner, John Todd Stuart, and another enterprising Kentucky migrant, the young Joshua Speed, who offered Lincoln a roof over his head on the day they first met.

Springfield was a boom town plunked down on the prairie. The town center boasted a large central square of lawn fenced with railing. Besides the brick courthouse and market, the town held six churches (two Presbyterian, two Baptist, one Methodist, and one Episcopalian), an academy (Springfield High, with $200 tuition per year), four hotels, a large brick tavern, and the town jail. The commercial establishments demonstrated the new capital's potential: roughly twenty dry-goods stores, six grocers, four drugstores, two clothing stores, and (especially good for Lincoln's keen interest in reading) one bookstore. The two newspapers, the *Illinois Republican* and *Sangamo Journal*, afforded the town's rising professionals—nearly a dozen lawyers and eighteen physicians—suitable space for advertising. The papers showcased the city's blossoming cultural offerings, including the Young

Men's Lyceum, which by 1838 met regularly on Saturday evenings in the Baptist Church at the corner of Seventh and Adams streets. Following the arrival of the railroad, a newspaper editor trumpeted, "Our town will soon rival Lexington [Kentucky] in population, wealth and importance."

By the time Mary Todd relocated to Springfield, the village had been seriously transformed—in no small part through the migration of Kentuckians. Mary could feel right at home, as locals estimated that two thirds of the state's population were Kentucky-born. Springfield lawyer John J. Hardin wrote to a friend in Kentucky that any woman who arrived "would do well in Illinois," and wished for a bounty of females, like those boatloads once shipped to colonial Virginia from mother England. Mary had come out to escape the stifling rut of the premarital merry-go-round. She migrated to expand her horizons, not just to find a husband—since back home she "never at any time showed the least partiality" for suitors but "accepted their attention without enthusiasm," a descendant recorded.

Springfield gave her a much freer rein to entertain lots of possibilities, which, of course, eventually included members of the opposite sex with ambitions matching her own. Both of her married sisters, Frances Wallace and Elizabeth Edwards (with whom she lived), provided Mary with opportunities to shine. She was one of the brightest stars in the constellation of belles who kept Springfield men enthralled.

Like most of the unmarried women of her age and background, Mary Todd was absorbed with the trials and tribulations of courtship. The dance to the altar was often portrayed as a brief shining moment in what could otherwise be seen as a dull series of prescribed roles. By the early decades of the nineteenth century, eligible women could make their own selections of mates from available swains. Society increasingly regarded marriage as a symbol of love between two people, rather than an alliance engineered between two families. It was a woman's prerogative to find a fiancé from among the parade of eligible suitors. With a suitable choice, families would acquiesce to the young people's wishes. This was a transformative process, and one in which women presumably improved their chances of happiness in the marriage lottery.

Men, however, were the only ones allowed formally to propose, but women learned to wheedle, perhaps even to maneuver to their own advantage. And women, within this emerging pattern, had the right of refusal. But even if a woman did say no, then she could easily change her mind and say yes—or vice versa. At times, the proposals, demurrals, and acceptances might have appeared like some elaborate parlor game, but it was deadly serious to those involved. Betrothals reflected the soberness with which most women regarded the era's most sacred rituals. (And court cases revealed the most sobering consequences of this consideration, with payment for broken promises or defamed reputations.) Both Mary Todd and Abraham Lincoln were thus at a very critical crossroads on their personal journeys.

Lincoln's early days in Springfield, suffused as they are with folkloric elements as well as sober historical conjecture, provide a murky ground upon which to build an understanding of Mary and her suitor—or even of Lincoln and his tangled and perhaps somewhat tortured relationships with the opposite sex.

Though he was nearly thirty, Lincoln was uncomfortable and inexperienced in the realm of romance. He would rather have faced a hostile crowd on the political circuit than a woman his own age in an empty parlor. During informal conversations with men, he was easygoing, spinning yarns around a potbellied stove. He was even known for his eloquence on the stump or in the courtroom. But Lincoln found the niceties of polite conversation with ladies, let alone the banter necessary for sparking, discomforting. Even male intimates offered a characterization of him as a "shut-mouthed man." Lincoln could talk circles around his political opponents, but became tongue-tied in the face of doe-eyed females, especially any to whom he felt attracted.

Nevertheless, Lincoln was capable of deep affection for women, and he formed important attachments during his twenties. One was with Ann Rutledge, the comely, bright daughter of a tavern keeper in New Salem, the town where he first settled on his own as a young man. Scholars continue to debate the role of Ann Rutledge. There is no disputing the fact that they knew one another, and much to suggest that Lincoln's relationship with her was of an intense, romantic nature. When Lincoln showed up in New Salem in 1831 at the age of twenty-two, he could have been one of Ann's many admirers. She was the attractive daughter of one of the village's founders, James Rutledge. One young businessman, John McNeil (also known as John McNamar), apparently asked Ann Rutledge to pledge herself to him sometime in 1832, and there seems little doubt that she did, although the true nature of this bond is clouded, without any compelling evidence to suggest it was either a love match or a convenience. McNeil went East shortly after their engagement.

Some of the most fanciful of the stories about Lincoln's relationship with Rutledge (especially cinematic versions) assume that their attraction began when Lincoln was a boarder at the Rutledge tavern. More likely, intimacy developed nearly two years later, sometime in 1835, when Ann and her family were living on a farm outside the village. According to Ann's brother, Lincoln courted her after years of McNeil's absence, and his suit "resulted in an engagement to marry, conditional to an honorable release from the contract with McNamar." Another New Salem resident suggested that Lincoln visited Rutledge regularly (every two or three days), but the observer did not realize the understanding between them until Rutledge's death, when Lincoln "seemed to be so much affected and grieved so hardly that I then supposed there must have been something of the kind."

She was no myth but a flesh-and-blood female, getting stale on the frontier waiting for her man to return. McNeil's lengthy absence and the fact that he had stopped writing to Ann constituted a broken pledge, which might have convinced her to abandon hope. She became involved with Lincoln only after McNeil effectively had jilted her. Rutledge family members suggested that Lincoln rescued her from this unworthy fate and the two fell in love. But they postponed their own plans so that Lincoln could prove himself worthy of her faith in him: Ann would wait to marry Lincoln until after he completed his law studies, begun in 1834.

Allegedly they fell in love, planning a future together in the spring of 1835, and their wedding was only prevented by Ann's death in the summer of 1835. This deeply romantic story of Lincoln's first love (which was later repackaged by his biographer William Herndon as Lincoln's only true love) has caused controversy for more than a century.

Despite the ongoing disputes over Rutledge's importance, some factors remain undisputed. Heavy rains during the summer of 1835 contributed to a record mosquito crop, and outbreaks of malaria followed. When sickness crept into the country, at first families began to chatter, and exchange information about all the worries swirling around. Then a silence fell across the landscape, with the muffled cries of first one, and then another neighbor bowed with grief. Then fear's icy fingers began to grip the cabins scattered across the landscape.

Late that summer, Ann fell ill, and on August 25, 1835, she died. Certainly the death of a young, lovely woman whose company he cherished would have been a terrible setback for Lincoln. There are several indications that her passing had larger meanings.

After Rutledge's early, tragic death, Lincoln exhibited deep despair, and—as his depression deepened—many feared for his sanity if not his safety.

One acquaintance recalled, "I do know he was staying with us at the time of her death it was a great shock to him and I never seen a man mourn for a companion more than he did for her he made a remark one day when it was raining that he cold not bare the idea of its raining on her Grave." This sentiment was reported by more than one of Lincoln's New Salem circle, adding to its credibility. The death of a young woman whom Lincoln admired—whether or not they were engaged and whether or not she was the love of his life—could certainly have caused him to "run off the track," as one intimate described this bleak episode years later.

The death of his mother when he was just nine, the death of his beloved sister when he was eighteen, and then the sad circumstances surrounding the death of Ann Rutledge touched as sensitive a soul as Lincoln. It opened floodgates of grief. Each of the three women he had loved thus far had been taken from him.

Lincoln may have been temporarily unhinged by Rutledge's death, perhaps even suffering a nervous breakdown. But his collapse could have stemmed from several roots, especially his crisis over his career—his law studies, his looming debts, and his poor economic prospects—as much as from his sweetheart's death. His collapse could have been the result of a temporary inability to overcome his ongoing battle with melancholy as well as a particularly nasty bout of the "blues." Whatever he may or may not have felt toward Rutledge, her death coincided with his shedding any lingering ties with the New Salem community—releasing him out into the wider world, depressed and depleted, but also propelling him into the Springfield orbit.

The next familiar episode in this timeline of failed romances was Lincoln's inability to win the hand of Mary Owens, another eligible young woman with even more cultural advantages than Rutledge. Lincoln's romantic involvement with Mary Owens is confirmed by his letters—which afford fascinating insight into Lincoln's views on women and marriage—but also by Owens herself. She had been sought out by Lincoln's former law partner and biographer, William Herndon, and shared the sad story of Lincoln's badly botched courtship.

Biographer David Donald suggested that during Lincoln's New Salem days (and perhaps into his early Springfield period), Abraham Lincoln formed attachments "to older, married and hence unavailable, women." Many of these women made it their business to try to find Lincoln a wife, which is how he became entangled with Mary Owens, the sister of his New Salem neighbor and patron, Elizabeth Abell. Mrs. Abell was "a cultivated woman—very superior to the run of common women," who took a liking to Lincoln. When she failed to match him up with local Salem belles, Elizabeth Abell promoted him to her own sister during her visit to Illinois.

Mary Owens was a Kentucky-born, well-educated daughter of privilege and wealth, who came out to Illinois in 1836 to stay with her sister Elizabeth because she, like Mary Todd, did not get along with her father's new wife. The extent to which Mary Owens and Abraham Lincoln did or did not hit it off remains unclear. But Lincoln's interest in her as a marriage partner is revealed in correspondence with her back in Kentucky. Writing in December 1836, he worries about his most recent letter to her and fears her response: "The longer I can avoid the mortification of looking in the Post Office for your letter and not finding it, the better."

This exchange indicates that Lincoln had perhaps already made some overtures. But he confessed to Owens that his new surroundings in Springfield had raised a great deal of self-doubt. He described what her experience being Mrs. Lincoln might be like: "There is a great deal of flourishing about in carriages here, which it would be your doom to see without sharing in it. You would have to be poor without means of hiding your poverty." He asks with some impudence, "Do you believe you could bear that patiently?" His missive overflows with conflict and misgivings. Lincoln confessed that if he did all in his power to keep any woman happy who cast her lot with him, he needed for her to *be* happy—as he would be too miserable if he were to fail in his efforts.

Clearly, Lincoln was weighed down by thoughts of both *how* and *with whom* he might cast his lot. His series of letters offered awkward resolution by August 1837: "Our further acquaintance shall depend upon yourself. If such further acquaintance would contribute nothing to your happiness, I am sure it would not to mine. If you feel yourself in any degree bound to me, I am now willing to release you, provided you wish it; while on the other hand, I am willing, and even anxious to bind you faster, if I can be convinced that it will, in any considerable degree add to your happiness." This "understanding" seems full of doubt, rather than bad faith.

Mary Owens's second visit to Illinois had been engineered by her sister expressly to marry her off to Lincoln. Lincoln had agreed to this plan in conversation with Mrs. Abell because of Mary's intelligence and agreeability. Whether in earnest or half in jest, the plot was hatched. Lincoln had met Miss Owens before and said he "saw no good objection to plodding life through hand in hand with her." But Lincoln complained that when he saw Miss Owens again, he was taken aback by her appearance, for she was, in Lincoln's words, "a fair match for Falstaff" in size, as well as being weather-beaten and missing teeth. These physical attributes—which had either changed or been forgotten from before—apparently caused Lincoln a change of heart. Yet he would be forced to set aside any reservations and hold up his end of the bargain as a matter of honor. Because of Lincoln's idiosyncratic humor, interpreters are divided over the letter's intended meaning, as it is dated April Fool's Day, 1838. We are not sure exactly how much of this explanation of the breakup is tongue-in-cheek.

At the same time, Lincoln seems quite serious that he was thoroughly surprised by Mary Owens's rejection. She later confided that she rejected him because she felt he would be "deficient in those little links which make up the great chain of woman's happiness. . . ."

At the time of their courtship, Lincoln believed she was his for the taking, and thought that no one else would have her. But even more upsetting, Lincoln claimed that, at the end of the day, "I was really a little in love with her." This reinforces previous observations that Lincoln seemed attracted to the unavailable. Lincoln volunteered to friends that because of the

Owens debacle, he was giving up on marriage. He pre-dated the modern dictum by nearly a century with his explanation: "I can never be satisfied with any one who would have me."

Nevertheless, shortly after this failed courtship of Mary Owens, Abraham Lincoln met his wife-to-be, Mary Todd—most likely for the first time. Mary Todd would have heard about Abraham Lincoln before they were introduced—both as her cousin's law partner and, with her keen interest in politics, as a notable speaker for the Whigs. With his feature role in the lyceum and other local organizations, there is no doubt they would have been familiar with one another before meeting in person. In addition, her sister Frances Wallace offered accounts that she had known him well, before her marriage.

The famous first encounter between Mary and Abraham has not been recorded, only mythologized. Because Lincoln was listed as a manager of a cotillion held during December 1839, and Mary Todd would have attended such a glittering occasion, some have indicated this was their first meeting—even suggesting that this was the time and place where he confessed to her that he wanted to dance with her "in the worst way," about which Mary later recalled, with humor, that he did. But it is more likely that he first met her in making a pilgrimage to the Edwardses' home to meet this new Miss Todd from Kentucky, as he had done to greet Mary's sister Frances when she first arrived in town. Such a courtesy call was within his round of political and social outings.

Mary would have encountered Lincoln in her brother-in-law Ninian's front parlor during one of their Sunday gatherings. After all, Ninian Edwards's two-story brick home on what became known as Aristocracy Hill was a showplace for the Springfield elite. Edwards attracted the cream of Illinois society to his gatherings, including members of St. Paul's Episcopal Church, where he and Elizabeth worshipped. The group of young people who socialized together styled themselves the Coterie. They included Springfield couples like the Wallaces and Edwardses, but also the unmarried eligibles, visiting siblings and cousins who contributed to the lively matchmaking that preoccupied Springfield's young folk.

Mary made fast friends during her time in the Edwardses' home. Attorney Lawrason Levering, the son of Judge Aaron Levering, lived next door to the Edwardses. When his sister Mercy Levering came out west for a visit, she became a confidante of Mary Todd's, and the two were quite inseparable. They enjoyed pouring their heart out to one another and sharing adventures. Once, after the girls had been trapped by rain inside their houses for days, Mary convinced Mercy that they should make their way into town by scattering shingles to walk on—until they could reach the sidewalks lining Monroe Street. They escaped their perches on the hill, and frolicked into town. But after the visit, they had no way to get back home without sinking their shoes ankle-deep in mud. In desperation, Mary Todd hitched a ride back to her home with a local driver, Ellis Hart, jumping onto his donkey-pulled dray. But Mercy refused to go along, "failed to take advantage," fearing reprisals such impropriety might provoke. And Mary was spotted flouting convention, a bit reminiscent of her homemade hoops a few years earlier. The story of these antics were memorialized in verse composed by a local wag, Dr. E. H. Merryman, a poem handed down through the years.

So early in her Springfield career, Mary Todd was characterized as headstrong, pushing the boundaries. Not for this alone was she noticed by all the young men of the neighborhood, and one of them (who would one day come to despise Mary and become her severest critic), William Herndon, offered a lengthy description of his initial impressions. Herndon described Mary as "dashing, handsome—witty . . . cultured—graceful and dignified." Herndon also noted her sarcasm and haughtiness, but explained as well that "she was an excellent conversationalist . . . and soon became the belle of the town."

Mary Todd easily drew men young and old into her orbit. At 120 pounds on her five-foot frame, she could not be described as svelte. But her manner as much as her appearance made her attractive—not unlike a flirtatious, fictional descendant, the heroine of *Gone with the Wind*, whom Margaret Mitchell described in her opening line: "Scarlett O'Hara was not a beautiful woman."

Although other belles might exhibit more beauty, and many possessed a more substantial dowry, she was one of a handful in Springfield to boast such an enviable bloodline. But it was her wit and conversation that most people remembered. Elizabeth Edwards described Lincoln's enthrallment: "[He] would listen and gaze on her as if drawn by some superior power."

Perhaps Mary was able to draw out this bashful young suitor on some of his favorite topics. Mary would have been one of the few young women in town who would not have to sham her appreciation of Shakespeare. She displayed a lively interest in other literary figures, due to her superior education. Both Mary and Abraham adored poetry—and could quote by heart. One of Mary's childhood friends remembered that her passion for poetry caused amusement among her friends, as she could recite verse after verse and "liked nothing better."

Equally impressive, Mary discussed the finer points of politics and eagerly demonstrated her connections to the good and great. Here was a woman who told stories about Henry Clay from personal experience. Lincoln's god was Mary's idol as well, but she claimed him as a family friend. Why wouldn't Lincoln have been bedazzled by this new Miss Todd in town?

Mary also arrived on the cusp of a new era of change. In 1840, Springfield was afire with election news and spirited conflicts. Men and women were caught up in impending contests, especially the presidential election. In late February, Stephen Douglas berated Simeon Francis, editor of the *Sangamo Journal*, over something he printed in his paper, and a street fight ensued. In early April, Lincoln began making his rounds out on the Eighth Circuit. Shortly after he returned to Springfield, Mary headed out to Missouri to visit her father's brother, Judge David Todd. Judge Todd had left Kentucky and settled in Columbia in 1817, along with two other brothers, Sam and Roger. David's daughter Ann was close to Mary in age, and they enjoyed outings and house parties throughout the summer.

This may have been the period when Lincoln let his writing—more polished than his speech—speak for him, in an effort to impress the sophisticated Miss Todd. In between court cases and political meetings, Lincoln crafted epistles for this young lady whom he had known for less than a year. He posted a series of letters off to his absent friend. How did Mary Todd receive such attentions? Apparently, as she did everything else—with great aplomb.

Mary recognized Lincoln as striving, perhaps even a "diamond in the rough." What made him stand out from all the young men in town was not just his distinctive height, but his gigantic ambition. He was sending off legislative proposals to John C. Calhoun in Washington, making speeches across the region, and getting his name more widely circulated. He paid court through the post, and although we don't have his billets-doux, we have Mary's veiled response to them, in letters to others. On July 23, she wrote to a confidante, "When I mention *some letters* I have received since leaving S—, you will be somewhat surprised, as I *must confess* they were entirely *unlooked for*, this is *between ourselves*, my dear." (Mary's letters were always full of underlinings, odd use of commas, and general dramatic flair.)

Mary confided to Mercy that she was besieged by beaux, but none were attractive to her. Her uncle was pushing a suitor, a lawyer who was Patrick Henry's grandson—as bloodlines were of particular interest to Todds. Mary was uninterested in the suitable widower, Edwin Webb, because he was almost twenty years older and the father of "two sweet little objections," as well. Her younger sister Ann was shipped out to Illinois to live with her married sister Frances in May 1842, when she was just eighteen. Ann was expected to find a beau, even if Mary had not yet been able to land a fiancé.

Mary was approaching her twenty-second birthday—nearly an old maid by Southern standards. She maintained several adolescent fixations during her search for the perfect mate. One of her requirements was that her husband would allow her to go to the theater as much as she wanted. But she also cherished a tenet of her girlhood that many young women of her age and class were adopting: "My hand will never be given, where my heart is not." She wanted to marry for love.

Mary's concerns, shared in correspondence with other young women, barely match the intensity with which Abraham Lincoln and his most intimate friend, Joshua Speed, obsessed over the topic. Their fickle and sometimes ill-advised crushes were the talk of the Coterie, as Speed was well known as a romantic. In December 1840, Mary makes fun of him: "*Mr. Speed's* ever changing heart I suspect is about offering *its young* affections at her [Matilda Edwards's] shrine." Speed confided to his sister that he was happier in pursuit of an object of his affections than in the possession of what he sought in the chase. His rakish attitude was one that caused both heartache and consternation for his intimate circle, but it has not been uncommon among young men throughout the ages.

The Coterie exhibited plots and intrigue aplenty. Mary remained hopeful and deeply romantic throughout this period. She would try to tempt Mercy back to Springfield, promising sleigh-ride expeditions to Jacksonville—mentioning Lincoln as one of the party. Lincoln's attachment to Mary became common gossip by the holiday season in the winter of 1840. Most townsfolk who knew them suggested that they were secretly engaged.

The whole question of Lincoln's engagement and his relationship with Mary Todd during the two years leading up to their marriage is one of the hot-button issues and blind alleys within the Lincoln literature, and remains an unfathomable aspect of Mary's experience. Did Mary and Abraham declare their love for each other in the late fall of 1840? Did they decide not to expose themselves to the Edwardses' possible disapproval, and thus kept their matrimonial plans under wraps? Whether a wedding was planned for the 1840–1841 holiday season or not, clearly Abraham and Mary developed some form of mutual understanding—but perhaps neither anticipated just how bumpy their road to marriage would become and what would transpire before Mary would become Mrs. Lincoln.

4.

"Crimes of Matrimony"

The romanticized notion that a man and a woman meet, discover they are soul mates, and smoothly sail into marital bliss is bland even for fairy tales. Romance epics always include the drama of setbacks—love stacked against insurmountable obstacles—that fade by the requisite happy ending. Boy Meets Girl is nearly always followed by Boy Loses Girl. This story line fits the cliché that the course of true love never runs smooth.

How and why Abraham Lincoln and Mary Todd took two years after they first became engaged to become husband and wife remains a series of disputed ellipses.

The clashing explanations over what went wrong and what took so long were as obscure and conflicted to those who were directly involved as they have been to the generations of scholars who make it their business to weigh in. Missing pieces of the jigsaw puzzle can frustrate an investigator, but they do not preclude untangling the aspects of the pair's protracted courtship.

Sometime in the spring or perhaps early summer of 1840, the couple became more deeply involved in each other's lives. Mutual friends in Springfield took for granted that they were engaged or at least had some sort of "understanding" by the late autumn. Yet over the Christmas holiday season in 1840, Lincoln and Mary broke off their contact, and following this break, Lincoln fell into an awful slump.

Something happened between the two—but what? Some fanciful family folklore suggests that Mary Todd may have renewed her interest in rising Democratic star Stephen Douglas and purposely flirted with Douglas after Lincoln arrived late to a party. In this operatic scenario, Lincoln failed to appear on time to escort Mary to a social gathering to which she ventured alone. When Lincoln finally appeared at the party, he found Mary dancing with Douglas, his "arch rival," provoking him to jealousy and anger. Lincoln confronted Mary on New Year's Day: When he told her he would release her from their engagement, she accepted in a fury. Her niece described her response: " 'Go,' she cried with a stamp of her little foot, 'and never, never come back.' " The "little foot" stamp perhaps rings true, as we imagine Mary's disappointment. Lincoln made reference in a letter to "the fatal first," which several historians have suggested was Lincoln's nickname for the date when he and Mary suspended their relationship. However, reminiscences of this "fatal first"—collected at a much later date—raise as many questions as answers, like so many aspects of Lincoln folklore.

The question of romance between Mary Todd and Stephen Douglas remains for some a possibility or even a probability, while others claim this is unlikely. If Mary was the belle of Springfield that so many contemporaries described, then clearly Douglas would have been one of the many young men swarming round the Edwardses' home. Mary was attracted to intellectuals, and Douglas clearly fit this bill. Perhaps the passage of time, Lincoln's famous debates with Douglas, and then Lincoln's defeat of this star Democratic statesman during the 1860 presidential campaign stimulated exaggerated memories within family circles. There is no doubt that except for his New England origins, Douglas would have been deemed an eminently suitable match for Mary, and far more reliable than Lincoln.

But Mary was also someone who knew her own mind and acted on her instincts. Mary's cousin Stephen Logan teased her about the two men, and she contrasted them as a "Yankee" versus the "rough diamond" from Kentucky, volunteering that "Mr. Douglas differs from me too widely in politics." Logan argued that Mary saw Lincoln as a noble project: "to polish a stone like that would be the task of a lifetime, but what a joy to see the beauty and brilliance shine out more clearly each day!"

Both Ninian Edwards and his wife, Elizabeth, initially disapproved of Lincoln courting Mary, and their son remembered, "My mother did what she could to break up the match." A Todd descendant later commented, "Lincoln's people had not yet struggled up from dirt floors." Lincoln's frontier manners and lackadaisical demeanor were offensive to many of Springfield's aristocracy.

Still, Mary was independent enough to contract a liaison without obtaining the Edwardses' approval. And during the late autumn of 1840, perhaps she did. Her courtship blossomed via correspondence over the summer, as Lincoln found words on paper much easier than intimacy in person. The couple's correspondence may have caused them to become attached prematurely without a proper foundation, the necessary strong sense of one another.

Having reached the age of twenty-two, Mary was much preoccupied with thoughts of marriage. Two days after her birthday on December 13, Mary's letter to a friend was full of playful charm: "Speed's *grey suit* has gone the way of *all flesh*, in an interesting suit of *Harrison blues*, have replaced his *sober livery*, Lincoln's, *lincoln green* have gone to dust." But she touches on engagements and weddings: "Harriet Campbell appears to be enjoying all the sweets of married life . . . Miss Lamb, report says, is to be married next week . . . I am pleased she is about perpetrating the *crime of matrimony*."

Another year older, Mary Todd faced the 1840 Christmas festivities with high hopes—enjoying sleigh rides with Lincoln and others, and anticipating Mary Lamb's nuptials. Several Lincoln biographers have concluded that, once Lincoln and Mary faced each other in person during the white heat of the social season in Springfield that winter, their romance cooled. What triggered the cooling off? Was it Mary Lincoln's weight gain during her Missouri trip? The appearance of a younger, more attractive belle on the scene? Or was it cold feet, plain and simple?

Lincoln's cold feet could have stemmed from a variety of worries, most especially financial. He was a constant worrier about money, and was perhaps reluctant to take on the responsibilities of marriage when he had no fixed income beyond his legislative pay, even though he was also bringing in roughly $1,000 a year practicing law. Other disruptions disturbed him as well: Joshua Speed was selling his share in the store on the Springfield town square and moving back to Kentucky. This displaced Lincoln from the loft he had shared for years—and he had to move on. Yet the emotional and economic considerations of taking a wife weighed heavily on him.

That Lincoln became smitten with Matilda Edwards is certainly possible. Apparently Matilda (staying with the Edwardses and sharing a bedroom with Mary) had a noticeable impact on the town's bachelors. She sent Joshua Speed into a tailspin, and allegedly bewitched nearly two dozen men into proposals. The idea that Lincoln would break off with Mary in hopes of winning a prize like Matilda seems far-fetched. Yet his attraction to another woman would undermine his commitment to Mary, causing a crisis of conscience. However, Matilda Edwards, even unintentionally, did not lure Lincoln away from Mary, although her presence may have triggered a fickle heart, derailing plans to wed Mary.

Both Ninian Edwards and Joshua Speed later volunteered identical stories about the crisis with Mary

Todd: Each claimed that Matilda Edwards was the root cause of the breakup. Edwards even claimed that Matilda turned down Lincoln's proposal. Implausibility intervenes after examining a letter from Matilda reporting that Mary enthusiastically invited her back to Springfield. If Mary had any serious rivalry with this younger woman, or if Lincoln had been so beguiled, she would hardly have suggested Matilda's return. Matilda Edwards Strong later recounted that during her stay in Springfield, Lincoln barely managed to pay her any compliment—which hardly supports the marriage proposal claim. Lincoln might have harbored a passionate crush that he dared not express, but this interpretation seems tortuous at best. Matilda was most likely the vessel into which he poured his doubts—and did not figure significantly otherwise.

Whatever the cause, townsfolk believed that Lincoln and Todd had been a couple in 1840, but sometime shortly before or after New Year's Day, 1841, they had broken up. This was the version circulating among friends, acquaintances, and casual observers. Subsequently, Lincoln exhibited signs of suffering a severe breakdown—and his deterioration was the subject of gossip.

Lincoln described his state of mind during this period in the bleakest terms: "I am now the most miserable man living. If what I feel were equally distributed to the whole human family, there would not be one cheerful face on earth. Whether I shall ever be better I cannot tell; I awfully forbode I shall not. To remain as I am is impossible; I must die or be better, it appears to me." This was the second glaring episode of Lincoln's mental problems, when he supposedly had a kind of collapse, if not a breakdown.

Was this yet another episode of clinical depression, which would come to haunt him his entire life? Or was it an example of Lincoln's acute emotional empathy, which periodically sank him into a state of melancholia? What was the root cause of this January 1841 episode, which was so troubling that medical experts interceded?

The loss of a fiancée would have been traumatizing, as would have been the impending departure to Kentucky of his best friend, Speed. The money due on a debt, the extreme cold weather, and several other factors might have exacerbated Lincoln's mental instability. Whatever unmoored him, his friends held him in fearful regard, with yet another suicide watch. Naturally, in a town the size of Springfield, his condition was much discussed. All of this was considerably "unmanning," to use a phrase to which Speed resorted.

James C. Conkling confided to his fiancée, Mercy Levering: "Poor Lincoln! How are the mighty fallen. . . . Surely 'tis the worst of pain to love and not be loved again.' " So Conkling, among others, hinted in correspondence of Lincoln's sad fate. But these correspondents do not provide explanation—and the object of all these thwarted affections still remains unspecified. Lincoln had characterized his own suffering as a form of "hypochondriasm," or "hypos." The hypos were something with which men of Lincoln's world were extremely familiar. Melville sketched out the hypos in his classic *Moby-Dick*: In the very first paragraph, the narrator describes them as "a damp, drizzly November in my soul."

Lincoln consulted daily during this period with Dr. Anson Henry, who became "necessary to my existence." Lincoln was absent from the legislative roll call; he was confined for more than a week, and when he finally appeared in public, he was "reduced and emaciated." Henry could have used the commonplace treatments of his profession: bleeding, purging, and other near-tortures designed to revive in such extreme cases. Jane Bell, a casual acquaintance, reported that Lincoln was up and around on January 27 after spending nearly a month "dying with love they say." She went on to claim, "He came within an inch of being a perfect lunatic for life. He was perfectly crazy for some time, not able to attend to his business at all."

Lincoln's episode was so severe that Dr. Henry proposed radical change—seeking a consular position in Bogotá, Colombia. Congressman John T. Stuart (Mary's kinsman) attempted to help him obtain such an escape, as Lincoln confessed, "I fear I shall be unable to attend to any business here, and a change of scene might help me." Whatever caused his melancholy, the first weeks of 1841 were dispiriting. Yet by February 3, he was able to joke about his condition, confessing in a letter to Stuart that he was "neither dead, nor quite crazy yet."

Throughout the spring, Lincoln remained, as one friend described him, moody and silent. Sadly, he was the likeness of his best friend, Joshua Speed, who was downcast by his own failed courtship of Matilda Edwards. Speed confessed to his sister, "There is more pleasure in pursuit of any object, than there is in its possession. This general rule I wish now most particularly applied to women. I have been most anxiously in pursuit of one . . . but the possession I can hardly ever hope to realize." Speed slunk away to Kentucky in defeat, and was felled by the hypos on the boat home.

Intriguing as well has been the suggestion that Lincoln suffered yet another romantic or matrimonial setback due to involvement with another Springfield belle, Sarah Rickard. Rickard was the sixteen-year-old sister of Lincoln's landlady, Mrs. William Butler. Claims have swirled around her name and solidified into speculation, which has been transformed by select scholars into fact: "At some point in the winter of 1840–1841, Lincoln proposed to her, but she turned him down, saying she was too young to marry." Whether this was an offer made in jest (replete with biblical references to Sarah and Abraham) or Lincoln's serious effort to secure a mate, Sarah's role in this tumultuous period remains a mystery.

Mary Todd seems considerably lost in this revolving blur of romantic possibilities, misunderstood and often maligned by Lincoln biographers. Her brokenhearted state was clear to family and friends, something her family had a stake in promoting: "Mary had tears in her eyes" when she told her sister about her broken engagement. Whether she felt she was losing the love of her life or simply agonizing over public humiliation, Mary became extremely vulnerable following this episode. Her wounded pride and apparent despondency were underlined by the limbo into which she tumbled.

In Speed's rendition, Lincoln wanted to call off his understanding with Mary by writing to her, but

Speed convinced him to face her in person, calling on his sense of honor. In this version, when Lincoln confessed to Mary that he could not go through with the engagement—using we don't know what excuse—Mary took the news badly, breaking into tears. In comforting her, Lincoln took her in his arms, sat her on his knees, and kissed her. Clearly, this was not what Speed would have recommended. Lincoln's physical intimacy considerably muddled matters—and perhaps rekindled dashed hope. Mary may have officially released Lincoln from his promise but denied that her own feelings had changed.

Because of Lincoln's anguish over breaking Mary's heart, this parting caused considerable grief to both parties. His honor seemed at stake, and Mary was allegedly willing to drive this sliver of guilt deep into his heart. The public nature of Lincoln's distress in the weeks and months to follow might have given Mary every reason to believe her former beau regretted his actions.

One scholar suggested that Mary wanted Lincoln to suffer for humiliating her, and purposely kept him tied to her by invoking honor. But this ignores another possibility—that she had fallen in love and wanted him to love her back. This simpler explanation remains plausible.

Immediately following the rupture, there is every reason to believe that Mary Todd hoped her ability to feign indifference could improve the situation. Indeed, she could wait him out. Lincoln wrote Speed in June 1841, making reference to "Hart, the little drayman that hauled Molly home once." "Molly" was Lincoln's pet name for Mary; apparently, six months after the breakup, he could not banish memories of her from his thoughts.

In public, Lincoln recovered his equilibrium enough to fulfill his duties and continue on the circuit. Speed had written to friends in July, "I'm glad to hear Lincoln is on the mend." In August, Speed invited his friend to convalesce at his Kentucky home, where Lincoln spent five weeks in the summer of 1841.

Lincoln was comfortably ensconced at the Speed plantation—presented with his own slave attendant for the duration and surrounded by all the luxuries slavery allowed. He was also given a Bible by the Speed family matriarch, Lucy Gilmer Speed. This was not bestowed with evangelical zeal, as she was not particularly religious nor intent upon chalking up conversions. The Bible was donated because Lucy Speed herself tended to melancholy, and she offered her son's friend the Oxford edition as a kind of primer, urging him to seek relief in its pages. That he might find consolation in this way "made a deep impression on him," Joshua Speed confided, adding that "I often heard him allude to it—even after he was President."

Lincoln wrote to Speed's sister in September 1841 that he intended to read the Bible her mother gave him "regularly when I return home. I doubt not that it is really, as she says, the best cure for the 'Blues'" but added with his trademark skepticism, "could one but take it according to the truth."

Other cures presented themselves when Joshua Speed courted Fanny Henning, a deeply religious young woman from Louisville, whom Lincoln described as having "heavenly black eyes." Certainly they were eyes Speed got lost in—and thus he found himself engaged to Fanny by the fall of 1841. This was surely a wonder to Lincoln, who had agonized with his friend over flirtations and affections galore. Even though he expected his friend to plan a more settled life once he landed back in his hometown of Louisville, it must have come as quite a shock that Speed would actually pin himself down to one woman, and welcome the altar looming on the horizon.

Lincoln got to know the pious young woman during his visit with Speed and shared with Joshua's sister Mary his single reservation, that "there is but one thing about her, so far as I could perceive that I would have otherwise than as it is. That is something of a tendency to melancholy. This, let it be observed, is a misfortune, not a fault." Lincoln might well have been thinking of his own misfortune when he penned this—having just gone through a lengthy period of bleakness, leaning heavily on Speed throughout.

Lincoln's pleasure at his closest friend's return to Springfield was dampened when Speed slipped into a deep funk, bedeviled by doubts over his impending marriage. Joshua Speed confessed, "Something of the same feeling which I regarded as so foolish in him [Lincoln]—took possession of me." The two began to trade confidences and swap remedies, their correspondence reminiscent of swooning schoolgirls.

As Speed rode back to Louisville in early 1842, intent upon marching up the aisle, he carried with him a long missive from Lincoln. This letter offered bracing advice, as Lincoln suggested, "You will feel very badly some time between this [his departure from Springfield] and the final consummation of your purpose." Lincoln's thoughtful, measured prose was intended as bravado to bolster Speed's failing nerve, so he wouldn't give in to his worst fears. He wanted Speed to savor his reminders as the wedding approached. He predicted that "in two or three months, [you] will be the happiest of men."

Lincoln eagerly awaited Speed's first message as a married man—and tore into the letter, literally and figuratively. Lincoln felt "intense anxiety and trepidation" as he eagerly consumed the contents. When he wrote back to his friend ten hours later, he confessed, "I have hardly yet . . . become calm." Lincoln was buoyed by his friend's improved spirits, but worried: "I feel somewhat jealous of you both now; you will be so exclusively concerned for one another, that I shall be forgotten entirely."

By the spring of 1842, Lincoln confided to Speed, "There is *one* still unhappy that I have contributed to make so," adding mournfully, "That still kills my soul, I can not but reproach myself, for even wishing to be happy while she is otherwise." Reflecting on Mary Todd, Abraham Lincoln told Speed in July 1842: "I believe now that had you understood my case at the time as well as I understood yours afterwards, by the aid you would have given me, I should have sailed through clear." Even with this new insight, Lincoln couldn't act on his feelings, as he felt the need to regain confidence in his own abilities. Lincoln's blues continued when his close friend Bowling Green died; he was so choked up at the funeral in early September 1842 that he was unable to deliver a eulogy.

During this same period, a series of satirical letters appeared in print in the *Sangamo Journal.* They ridiculed Irish-born politician James Shields, who took deep offense at "articles of the most personal nature and calculated to degrade me." The first of these "Lost Township letters" appeared on August 19 and was signed "Rebecca." Lincoln wrote an attack on Shields and sent it to the newspaper on August 27. This unsigned piece appeared on September 2. Finally, Shields sent an accusatory letter to Lincoln on September 17, insisting upon a formal retraction. Lincoln and Shields traded exchanges, and within two days, Shields issued a challenge to a duel.

Remarkably, Lincoln did not duck this confrontation. He usually wriggled free of such tangles with wit and resourcefulness. But in this case, perhaps because he was acutely concerned about issues of honor, Lincoln did not back down. However, he may not have felt that he "rose" to the occasion but perhaps "stooped." Lincoln never made public his feelings about this incident, and he refused to speak about it to those who brought it up in later years.

Lincoln instructed his second, Dr. E. H. Merryman, to negotiate a settlement with Shields's second. But in case all else failed, Lincoln selected cavalry broadswords as his weapon of choice—using his wits, as the sword would give someone with his height and long limbs an advantage if any contest actually took place.

On September 21, still at a standoff, the adversaries headed off to Alton, Illinois, where a nearby island, technically in Missouri, provided a dueling field. Before the two assumed gladiatorial positions on the field of honor, a satisfactory solution allowed the duel to be called off.

Mary Todd became involved in this particular imbroglio because she and her friend Julia Jayne composed the poetry and satirical letters lampooning Shields, and signed their pieces as "Rebecca." By refusing to reveal the author of some of these letters, Lincoln theoretically might have been protecting Mary's honor. At a minimum, he chivalrously deflected Shields—while the whole town gossiped about the affair. Lincoln described the event in an October 4 letter:

You have heard of my duel with Shields, and I have now to inform you that the duelling business still rages in this city. Day-before yesterday Shields challenged Butler. . . . Yesterday Whiteside chose to consider himself insulted by Dr. Merryman, and so sent him a kind of quasi challenge.

However, Lincoln had something much more important on his mind: settling a matter that had been weighing on him for months, soon to be years.

A week earlier, on September 27, both Mary and Abraham attended the wedding of Martinette Hardin to Alexander McKee, which took place in Jacksonville. Sarah Rickard recalled that rather than "bring constraint" upon the company, the couple spoke to each other, and "that was the beginning of the reconciliation." Others have suggested that the wife of the editor of the *Sangamo Journal*, Mrs. Francis, deliberately maneuvered Mary and Abraham into renewed contact.

Whatever the exact circumstances, reconciliation was clearly on Lincoln's mind when he pointedly wrote to Speed, married eight months: " 'Are you now in *feeling* as well as *judgment*, glad you are married as you are?' From any body but me, this would be an impudent question not to be tolerated; but I know you will pardon it in me. Please answer quickly as I feel impatient to know." In any case, Lincoln's impatience stemmed from his renewed intimacy with Mary Todd. Her feelings for him had not changed—despite the eighteen-month hiatus in their relationship, and he was stirred into emotional agitation.

After their renewed contact in Jacksonville the last week of September 1842, the two continued to meet back in Springfield—clandestinely and privately. They spent time together at the home of the editor of the

Sangamo Journal, Simeon Francis. The couple rekindled their feelings for each other. Later when Elizabeth Edwards questioned her sister about why she had kept their renewed relationship secret, Mary replied that after all that had occurred, it was best to keep the news from "all eyes and ears."

Mary Lincoln intended for their courtship to be thus remembered: *his* crisis and *her* fidelity. Mary clung to the notion that she was the only one to whom her husband had ever truly been attached, that she was destined to be his. Mary Todd had been courted by many attractive prospects over the years, spurning advances by Patrick Henry's grandson, by politician Edwin Webb, and unnamed others. Her station in life afforded her contact with a wide range of eligible suitors. Even though Lincoln had cut himself off from her, she had ample opportunity to travel—back to Kentucky, to visit other family in Missouri or in Illinois. Mary clearly had access to any number of attractive locales available to find another mate, someone to rescue her from this state of spinsterhood, especially after she and Lincoln split apart.

In the public arena, rumors had circulated that Mary had jilted Lincoln. She managed to keep up appearances throughout the spring of 1841: "Miss Todd is flourishing largely. She has a great many beaux."

Although she maintained gaiety in public, Mary privately shared her feelings with Mercy Levering, complaining in June 1841 that Lincoln "deems me unworthy of notice, as I have not met *him* in the gay world for months . . . I would that the case were different, that he would once more resume his station in Society." After weeks stretched into months and another year passed, neither seemed able to move on.

What must have gone through Mary's mind as she watched friends making their way on a steady parade to the altar?—first James Conkling and Mercy Levering in September 1841, then Joshua Speed and Fanny Henning in February 1842. Mary endured these ordeals until Lincoln, at long last, renewed his courtship and asked her to marry him, sometime in October 1842.

Who asked whom, and what prompted this reconciliation continues to stimulate debate. The wedding was, by all accounts, a hasty affair. Lincoln and his bride did not give Mary's family proper notice, partly to avoid the elaborate wedding preparations that had bedecked the Edwards mansion in May 1839 for Frances and William Wallace's nuptials. But also surely they both wanted to forestall any further objections about Lincoln's unsuitability. There have been elaborate attempts to show deceit or betrayal at the heart of this haste, but perhaps—once Lincoln decided to take Mary for his wife, whatever the reasons, and once she decided to have him as her husband, whatever her motivation—they simply did not want to waste any more time.

The real motives behind their haste remain unknown, and perhaps were obscure even to the couple themselves. Evidence indicates their wedding was hatched with less than a day's notice. On November 3, 1842—a month after Lincoln sought reassurance from Speed about the benefits of marriage—he visited Charles Dresser, the Episcopalian minister, to ask him to perform a wedding service for the couple at the reverend's home that evening. Lincoln ran into Mary's brother-in-law Ninian on the street and revealed their plans; whereupon Edwards insisted that the exchange of vows take place in his home. When Elizabeth Edwards was informed of Mary's plan to marry, she also insisted that the ceremony take place in their home, and begged that it be postponed until the next day, Friday, November 4.

Elizabeth rushed around making arrangements to accommodate. Invitations for nearly three dozen were extended, including bridesmaids Julia Jayne, Elizabeth Todd, and Anna Caesaria Rodney. James H. Matheny, a local attorney and fellow circuit rider, served as Lincoln's best man.

The Edwards home in Springfield, where Mary Todd and Abraham Lincoln were married on November 4, 1842

Folklore has been handed down giving several different versions of events that wedding day, with considerable variations in interpretations. Several stories say Lincoln had an attitude of resignation rather than jubilation, and his best man claimed Lincoln appeared as if he was being led to "slaughter." One version has Lincoln putting on his best clothes and being asked where he was going, only to have him reply, "to hell." "Tying the knot" has been interpreted by many Lincoln scholars as slipping a noose around Abraham's neck. At the same time, Mary's sister Frances stated that during the wedding festivities, Lincoln was as "cheerful as he had ever been." The truth perhaps lay somewhere in between.

It was not a storybook wedding, as Elizabeth Edwards had only hours rather than days to prepare. She had hosted an Episcopalian sewing circle the night before and was vexed at having to make such hasty preparations for her own sister's nuptials. Friction between the two sisters was evident when Elizabeth complained that she might only have time to send out for gingerbread. Mary tartly replied, "Ginger bread [is] good enough for plebeians," alluding to the Edwardses' objections to Lincoln's low origins.

Mary's white dress skirt, a muslin she already owned, was selected for the ceremony with care; she wore neither a veil nor any flowers in her hair— perhaps because it would be an unnecessary expense for a November bride. Elizabeth Edwards baked a wedding cake for the guests. A descendant reported that the cake turned out badly, because "it was a Friday, which was an unlucky day." The occasion was also marred by torrents of rain, which beat ferociously against the windows during the whole of the exchange. Yet when Mary Todd repeated her vows from the *Book of Common Prayer*, pledging to spend the rest of her life with Abraham Lincoln, when he finally slipped the ring on her finger, the gold circle engraved "Love is Eternal," Mary must have felt that despite warm cake and wet weather, Friday, November 4, 1842, was her lucky day.

5.

"Profound Wonders"

The small room—only eight by fourteen feet—was no place like home for either of the newlyweds. And yet a little upstairs bedroom at the Globe Tavern became the Lincolns' first marital home. Mary had probably never lived in such cramped austerity since her boarding-school days back in Lexington. But Lincoln would have found the room he shared with his wife a relative luxury—especially after years at the loft over Speed's store, which he had sometimes shared with three others.

A week after they were wed, the *Sangamo Journal* announced:

MARRIED—in this city on the 4th instant. At the residence of N.W. Edwards, Esq., by Rev. C.

Dresser ABRAHAM LINCOLN, Esq., to Miss
MARY TODD, daughter of Robert Todd, Esq.

That same day, November 11, Lincoln wrote to a
friend and concluded his epistle, "Nothing new here,
except my marrying, which to me, is a matter of profound wonder."

For Mary, life was more challenging, as most mornings, her husband headed off to his office, escaping four
walls and fellow boarders, while Mary was confined
and left behind—even more so when Lincoln went
out on the circuit. Mary had not just been reduced to
a hausfrau, she was a wife without a house. She would
have echoed the words of a child who confessed when
asked where her parents lived that "They don't *live*,
they BOARD."

Rooming at the Globe was not squalid but quite respectable. Yet it was a trial to have to pinch pennies
and conserve candles, seeking a way out of this relative scarcity. Lying in the darkness, hearing the sounds
of the other boarders, Mary plotted to live once more
among the comforts and privilege to which she was
born.

At the same time, Mary recognized that their private bedroom, with a shared dining table and parlor,
was certainly better than the accommodations of many

struggling young couples. And indeed, she was in the same place where her sister Frances had lived for years with her husband, William Wallace, before they were able to move into their house. Newlyweds were often forced to cram themselves into family households cheek by jowl with in-laws and other relatives. The Lincolns, by contrast, were on their own.

Mary Lincoln had distanced herself from the Todds more than geographically. She had not married any of the Kentucky scions or well-connected suitors proffered. She was extremely *thrane*—"stubborn," in the Scots vernacular. True to her Scots-Irish heritage, Mary made her own choices. She would live with them, and with the poor lawyer she had finally landed. She had perhaps been a huntress, like Diana, aiming straight for the heart, and not for the faint of heart, as she was clearly ignoring warnings from the gods. After she cheated her critics out of spinsterhood, she would have to settle for an elevation of her status in theory rather than practice.

Married life presented a new series of challenges to the former Miss Todd. First and foremost, she wished to prove her doubters wrong. She would push herself—and her husband as well—from the trenches up onto the barricades, where she might regain some perspective on the battles ahead. Her tall, rawboned husband, she was sure, would one day rescue her from the ordinary and lift them both to the heights of which she imagined him capable. Lincoln was a self-made man whose enormous reserves of empathy, ambition, and emotional intelligence blended with abundant talents and abilities. His poetic soul and dark psyche were balanced by his easygoing wit and earthy sense of humor. He was a political genius, as much new scholarship attests. Yet at the time he married Mary Todd, this was just a potential, not a given.

Mary's greatness was to be among the *first* to see beyond the rough patches and ragged edges of the Springfield lawyer with whom she fell in love. When she looked into his future, Mary became enthralled. Despite the initial obstacles to their union, she fought for a shared future, and they eventually made a life together.

Betrothal was a very serious consideration for a nineteenth-century woman; one Southern bride-to-be called her trip to the altar "the day to fix my fate." The man she married would become the defining element of her life. Mary's troubled path to her wedding was idiosyncratic, to say the least. But her sense that her groom would determine her own future was not.

The earliest months of her marriage were a challenge for the new Mrs. Lincoln. She might have been alarmed to discover what roommates had long endured—that Lincoln would often awaken her from sleep "talking the wildest and most incoherent nonsense all to himself." Lincoln's roommates on the circuit suffered his black moods on the road. David Davis, later a distinguished jurist (and executor of Lincoln's estate), treated Lincoln gingerly in the mornings, when he was "sombre and gloomy." So one of the first obstacles would be that her new husband was not just the man she knew him to be but also the cranky, tossing-and-turning sufferer of the hypos. Considering their delayed and blighted path to matrimony, it could not have come as a complete shock. But surely Mary hoped she could banish his blues with the sheer force of her love. At the same time, Mary was a bandage, not a cure. She could ease but not erase the melancholy that periodically entrapped her husband.

In 1842, Lincoln had no hope of any congressional seat. He withdrew from the race for Whig party candidate—deferring to his good friend Edward Baker. Baker, in turn, lost out to a cousin of Mary's. Lincoln had hopes he might be able to rotate into the congressional slot in a future year. But in the interim, as a lawyer, and without his pay from the state legislature, he would have to redouble efforts to earn all he could on the circuit. Lincoln knew he needed to milk cash from the countryside, while tending to the people's business in the state capital to enhance his professional reputation.

Money was at the root of many of their troubles. Lincoln not only had a wife to provide for, but was still followed around by a cloud—obligations from his time in New Salem. He referred to this burden as the National Debt. Financial responsibilities colored the couple's early prospects. Lincoln was so poor when they married that he and his bride could not make the visit to Kentucky they both keenly desired.

Springfield offered little relief. While her sisters, Mrs. Edwards and Mrs. Wallace, enjoyed spacious homes, Mrs. Lincoln's marriage to a humble lawyer with no family money turned her into a poor relation. She deeply resented this status, and likely blamed her sisters rather than her husband for the situation.

Mary Lincoln's pregnancy added to her discomfort in the first few months of 1843. There she was, tucked away in this rather bustling boardinghouse, with the bell clanging at all hours to announce the stagecoaches arriving. While increasingly indisposed, Mary was forced to deal with strangers and guests at mealtimes, again, yearning to escape.

A home in a tavern was not what she had dreamed of when just a young girl, or even when she settled in Springfield. Lincoln wrote to his friend Joshua Speed in January 1843, "Mary is very well and continues her old sentiments of friendship for you. How the marriage life goes with us I will tell you when I see you here, which I hope will be soon."

Speed surely raised the issue of an heir—which was much on his mind, as his wife Fanny had not yet become pregnant after over a year of marriage (and, as it turned out, the Speeds never had children). Lincoln wrote to his dear friend in March, confessing, "About the prospects of your having a namesake at our house, can't say exactly yet," which leaves matters vague. Perhaps he was not willing to mention Mary's pregnancy—or perhaps he was unwilling to commit to the possibility of naming a baby after Speed. He was most likely sensitive to his friend's feelings at this delicate juncture, as he realized Speed would be most eager for his own little namesake, and perhaps chagrined at learning Lincoln's news.

During the early years of their marriage, Mrs. Lincoln believed in her husband's legal talents. She became a devoted audience early in his career, and attended court when Lincoln appeared in Springfield. She was perched on the bench (next to the judge) in a crowded courtroom in December 1842 when the arrest (by order of the Illinois governor) of Joseph Smith, the Mormon prophet, precipitated a hearing. This caused quite a stir within the community. However, the life of a prairie lawyer's wife was not always so eventful.

Indeed, it could be not only humdrum but downright unbearable—especially with her husband away on the circuit for weeks on end. Lincoln was one of the few to attend nearly every circuit, no matter how distant, for at least three months in spring and three months in fall, making him an absentee husband nearly half the year. It must have been particularly difficult for the first year of her marriage. Even during the honeymoon months, Mary had to endure long separations, made more poignant by the anticipation of their first child.

Mary Lincoln was eager to welcome a baby into her life, but she, like most women of her age and era, dreaded the precariousness of pregnancy, and most especially the ordeal of childbirth. Complications of childbirth were the most common cause of death for Southern women of childbearing age, and were a fear of all women, even the rich and wellborn. Statistics show that even with excellent medical care available, giving birth was a risk. Pregnant women prepared for the birth of a child and faced the possibility of their own deaths, while attempting to "trust in the Almighty," as family and friends counseled.

As Mary's time grew near, Abraham Lincoln may have been thinking of the tragic death of his sister

Sarah in childbirth, while Mary had the vivid example of her own mother's death following childbirth nearly twenty years before. Naturally, both were apprehensive, like the Southern woman who complained "that the happiness of the conjugal relation was obliged to be bought at so dear a price." Both Mary and Abraham were anxious by midsummer, as Lincoln wrote to his friend Speed, on July 26, "We are but two, as yet."

While Lincoln felt they were two, his wife Mary, like most pregnant women, experienced two-ness all by herself. As a young bride, she barely had time to contemplate the sense of a shared life before new life began stirring in her own womb. And with this strange new sensation, she might have been visited by conflicting emotions: pride and eagerness, but also loneliness, as her time approached. Almost exactly nine months after their wedding, a son was born: On August 1, 1843, Robert Todd Lincoln arrived.

With parenthood, both Lincolns felt new priorities, and Mary carved out for herself a new identity, to a greater extent than Abraham. After their child was born, Lincoln rarely called his wife by his pet name for her, Molly—although he still might affectionately refer to her as Puss, joking about her as a feline in front of intimate friends. After she bore him the first of four sons, Lincoln reverently addressed his wife as Mother.

Mary had always, in the Victorian style of the day, referred to her husband as Mr. Lincoln, but following their son's arrival, she most often addressed him as Father. Years later, she fondly referred to her memory of him bending over to peer at their newborn, staring with awe and affection.

Naming their first son after Mary's father was a sensible decision, and one that brought reward. Whether Lincoln had been serious or not about naming a male offspring after his friend Speed, the Lincolns' first son was prudently named after his Grandfather Todd. There had been two previous Robert Todd namesakes (one by each of his wives), who had not survived. None of Mrs. Lincoln's siblings had paid their father this tribute. Mary sensed an opportunity and seized it. She had always curried her father's favor, and now was at a particularly important juncture.

Boardinghouse life with a baby was hard on any mother, but especially a new one, even if the house provided experienced older women like the owner, Sarah Beck, a widow. Luckily for Mary, Alfred Taylor Bledsoe, in legal partnership with Edward Baker, also lived at the Globe Tavern with his wife and family. Mrs. Bledsoe "washed and dressed the baby, and made the mother comfortable and the room tidy for several weeks till Mrs. Lincoln was able to do these things for herself." Furthermore, Bledsoe's daughter recalled, "I was very fond of babies and took on myself the post of amateur nurse . . . I remember well how I used to lug this rather large baby about to my great delight. . . . I have often since that time wondered how Mrs. Lincoln could have trusted a particularly small six-year-old with this charge; but he was never hurt or allowed to fall when in my care." It was common for younger children to be taken care of by older children in a family—and thus the boardinghouse provided a kind of extended family setting, especially for new parents.

However, the presence of these substitute sisters and cousins underscores the lack of support from blood relations. As biographer Ruth Painter Randall tartly observed, "One misses a record of Mary's sisters helping her at this trying time."

There is a record of her father coming to the rescue. Robert Todd paid a visit to Springfield before the new year was out. Pleased with the baby boy, he presented Mary with $25 in gold. Robert Todd's visit to see his namesake was fortunate, because it not only cemented closer relations between a father and favorite daughter, but it gave Lincoln a chance to impress his father-in-law. Robert Todd apparently became reconciled to Mary's choice. He entrusted his new son-in-law (and not his nephew or other sons-in-law who had taken care of such matters before) with the recovery of a debt. As a bonus, Robert promised that Lincoln could keep the $50 owed. These gifts, plus an annuity for Mary of roughly $120, gave the Lincolns the means they needed to leave the boardinghouse behind.

The couple temporarily removed to a small, three-room cottage of their own on South Fourth Street, where the rent was $100 per year. This cottage was just temporary, until they could purchase a permanent residence. But these moves demonstrate the increased costs associated with marriage, as Lincoln's boarding prior to his wedding was $1 per week, which rose to $4 per week when he and Mary were at the Globe.

They had their eye on a house on the corner of Jackson and Eighth streets, which Lincoln had visited to ask Rev. Charles Dresser to perform the marriage. The Greek Revival cottage, built in 1839, was a modest five-room, oak-framed house, which the minister's brother, Henry Dresser, a noted local architect, may have designed. An eighty-acre plot of land southwest of Springfield, a gift from her father, had given Mary Lincoln more financial security. They paid $1,200 in cash, and signed over a lot on Adams Street (with a value of $300) to buy a new home. The house was a far cry from the more lavish homes her Springfield sisters occupied, but it was a measurable improvement. The cottage boasted walnut doors and interior trim, with wooden pegs and hand-wrought nails for construction.

Most typical wood-framed homes were divided into three separate sections: public rooms, private rooms, and workrooms, and the Lincolns' new home was no exception. Mary was especially appreciative to finally have her very own parlor in which to *parler*—the French from which the name is derived. She longed to boost her husband into a more prominent role in Whig political circles.

Although they might have had identical aspirations, the couple often were at odds over how to achieve their goals. They battled on the home front, disputes that spilled over into earshot of their circle of friends. Clearly Mary was happiest trying to create a salon, with lively political gatherings reminiscent of her father's Lexington dinner table. Her husband treated many aspects of her social agenda with rough indifference. His carelessness in appearance and demeanor were legendary—answering his own front door in shirt-sleeves or wrestling on the front-room floor with the children. This had more to do with his easygoing nature than deliberate disregard for Mary's insistence upon fine manners.

As a couple, they did seem to work out a rough division of spheres. Lincoln encouraged his wife to be in charge of home life, including the setup of their house. Mary was relieved when Lincoln finally freed himself from the debts he had accumulated before moving to Springfield. As a husband, he gave her a fairly free hand with household expenses. This was not a common practice during the antebellum era, when men both held the purse strings and generally ruled over domestic matters. Lincoln's friend and ally, Lyman Trumbull, bought furnishings for his new house while out on the circuit. He wrote to his new wife, hoping she would like what he purchased.

Mary Lincoln must have felt considerably privileged to be able to select on her own for her new abode. She may even have had sympathy for her bridesmaid Julia Trumbull, even if Julia had much more disposable wealth available. Mary's husband allowed her to hold sway over "the little things," as he called them. She set to work once they planned to move, and even prepared the yard for May, when they would take possession. The yard would be Mrs. Lincoln's domain. As the daughter of a nearby neighbor later remembered, "My mother does not recall ever seeing a hoe or a tool in his hand." Lincoln had served his time as a manual laborer and had left all but splitting the firewood behind.

The ledgers of shopkeepers for the next two decades mark Mrs. Lincoln's purchases (although her husband was just as likely to do the shopping), all of which not only kept the household running, but eventually allowed Mary the privilege of a home where she could entertain, a luxury denied her in her boardinghouse days. She became adept at economical hosting during these early years of marriage. She also used her guest list to her own advantage. Mrs. Lincoln was the one wholly in charge of social invitations, and her company was only of her own choosing. For example, William Herndon, whom she disliked intensely, was never a dinner guest in the Lincoln home, despite being her husband's law partner. She often filled the house with Todd connections and others deemed suitable. From early in the marriage, she wielded invitations like a military strategist. Mary once gave a party and left out a cousin because the woman had had the audacity to suggest that although her son Robert was a "sweet child," he was not good-looking.

One sure sign that she was coming up in the world: Mrs. Lincoln acquired servants. Her hired girl was collecting goods by October 30, 1844. At the time, servants were becoming a more common feature and less a conspicuous sign of wealth, especially in the small towns sprouting across America.

The progress of equality put its stamp on the continent, especially in the domain of dress. American clothing, even fashion for women, became less a reflection of class and more a matter of taste. By midcentury, paper patterns, ladies' fashion magazines, and other developments created some standardization in women's wardrobes, so much so that *Harper's Bazaar* noted that "the mistress and the maid wear clothes of the same material and cut." This development caused Mary Lincoln discomfort. As remedy, she took to the needle and whipped up flattering and distinctive confections for herself. At first, of course, this was done for reasons of economy as much as originality. But she did pride herself on her appearance—fastidiously creating her own wardrobe and that of her children, as well.

Mary Lincoln seems to have been quite parsimonious in the early years of her marriage. She was painfully aware of the financial limitations that her marital choice had imposed. As a Todd, she might be able to draw on her father's largesse, while Lincoln was more likely to be leaned on by relatives seeking loans. This disparity in family resources caused Mary to pinch pennies, often in ways that local shopkeepers and peddlers resented.

Mrs. Lincoln was charged with all sorts of faults by folks in Springfield (especially during interviews conducted during the period after she had turned her back on the town following Lincoln's death). Most commonly, Lincoln's wife was branded as someone unable to hold on to her servants, unable to manage them effectively. Throughout her life, she was plagued by gossip that she was whimsical or abusive with those she employed.

The wife of Benjamin Edwards (Ninian's brother) moved to Springfield in 1840, and reported: "It was almost impossible to get servants. I had brought a woman from St. Louis but found her so intemperate that in less than a year I was obliged to discharge her and my troubles in housekeeping began." The letters of nineteenth-century American housewives persistently reflect this refrain, a veritable chorus of complaints about the "servant problem." Mary Lincoln, too, was consistently challenged by domestic matters.

She was thrust onto the treadmill of the nineteenth-century household routine in 1844, just a few months past her twenty-fifth birthday, with a baby and a nomad of a husband. Once they moved into their new home, Lincoln escaped to the office and left Mary to "the little things," which a new matron and mother might have found looming large rather than little. The world of the nineteenth-century housekeeper was a steady stream of daily labors, weekly tasks, and seasonal challenges.

Even a woman with servants had a long list of chores to perform. Someone was expected to check the mattresses for fleas and bedbugs. While servants might make beds, empty chamber pots, dust and polish, swab and sweep, most housekeepers prepared their own family meals.

Monday was one of the most challenging of weekdays, as it was traditionally laundry day. The household laundry was an onerous task, and women were expected to not only use starch and bleach, but to hang clothes to dry and press most of the washing with hot irons, as well. Thus the "Monday blues" had an additional meaning for most housewives, alluding to the compounds used to counteract the yellowing of white fabrics.

Monday also was the day to shop, to complete any extra cooking, and to carry on with everyday tasks, which included mending, knitting, and other ongoing and productive activities. The house was not just the "refuge" of the family, but a woman's workshop.

Food preparation was a constant daily and seasonal burden. Yes, the woman in town, unlike her rural sister, did not have to slaughter or pluck, plunge her arms up to the elbow in brine, or grind her own flour. But many kept chickens to collect the eggs, and cows to ensure unadulterated milk. The Lincolns kept both a cow and a horse (for Lincoln's travel) housed in one of the outbuildings on the property. Their property included a carriage house (eighteen by twenty feet) and a combination wood house and privy (thirteen by fifty feet).

The female head of household had a full plate of daily and weekly domestic duties, from baking bread to churning butter. Health issues and pride kept women from "store-bought" baked goods well into the twentieth century. Many nineteenth-century women shuddered in horror at the urban penchant for commercial cakes and pies. Mary Lincoln's cake recipe appears in a late-twentieth-century White House cookbook, attesting to her preference for homemade baked goods and her talents in the kitchen.

The Mason jar was introduced in the 1850s, aiding women in their arduous summertime chore of scalding fruit and vegetables in preparation for canning. Many kitchen activities were aided by new gadgets, billed as time-saving. Most homemakers were spectacularly happy to purchase the metal eggbeater, an unqualified success in the nineteenth-century kitchen. Yet as a matron's pantry became filled with apple parers and coffee grinders and other modern conveniences, these objects introduced new responsibilities.

Almost all the fetching and hauling of water fell onto women's shoulders. The weekly bathing of the children was quite a chore. Back-kitchens were the site of

Saturday-night baths in most respectable nineteenth-century households.

Some Americans considered unclothing and submerging for baths a dangerous practice, one that might lead to chills and premature deaths. In Pennsylvania in the 1840s, some legislators attempted to prevent these deaths by banning bathing between November and March. Presumably the good people of Illinois, the Lincolns included, did not brook such nonsense. Disrobing and washing up in a sitz bath or small washtub was a common practice. It is almost comic to imagine Abraham Lincoln contorting his limbs to fit himself into the bathing tub.

Simple daily household tasks like cleaning and cooking became doubly burdensome with water drawn from a well or carried up from the basement. One woman complained that her dishes stared down at her from the shelf, as she wearily contemplated her repetitive ritual of rinse, scald, scour and wash, rinse, and dry. Kitchen fireplaces were filled with trivets, pots, and skillets requiring heavy scouring. Most women welcomed the cast-iron stove, which allowed cooks to conserve fuel and reduce hazards. At the same time, tending a stove was time-consuming, requiring at least an hour a day.

Mary Lincoln did not have to make her own soap and candles like her foremothers, but was still likely to beat her own carpets, air her own linens, and restuff her own pillows for freshness. Her husband might have split their firewood for cooking and heating, but it was the wife's job to keep the fires banked. This was a vital matter of economy.

Once, when the fire went out, Mr. Lincoln went to a neighbor to borrow some coals. The neighbor apologized for not having any burning, and offered him matches. Lincoln responded in a way that indicated that he considered matches a luxury item. Consider Mary's challenge if her husband considered a household item such as matches something they could not afford. And perhaps they could not, with another baby on the way— which was the case in the fall of 1845, when Robert was just two. Edward Baker Lincoln, named after his father's political ally, was born on March 10, 1846.

To introduce some relief, Lincoln engineered a scheme to relieve domestic pressures. He brought Harriet Hanks, his cousin Dennis's daughter, from Charleston, Illinois, to live with them. The plan was for her attend school in Springfield, which Lincoln felt would be a big help. He imagined this young girl might also provide child care in exchange for room and board. He hoped extra hands and welcome companionship would offset any expense. As is often the case with best-laid plans, the scheme backfired. Hanks reported that not only her arrival, but the treatment she endured, more "like a servant, a slave," caused a "fuss" between the couple. This would not have been the first time married folks became embroiled in squabbles when relatives came to live with them. Hanks was followed by a parade of female servants—black and white, but predominantly Irish immigrants.

Despite her difficulties with visiting relatives and the servants she would later refer to as the wild Irish, Mary's ability to work alongside her African-American servants earned their loyalty, as later reminiscences would attest. Ruth Stanton, who was a young girl in the early 1850s, working in the Lincoln household, recalled that although she would scrub floors and clean dishes, Mrs. Lincoln would do all the "up stairs work" and cook the meals. She also remembered that Mary sewed her own clothes and those of the children. Her recollections were of someone who was "very plain in her ways," who did not wear the silks and satins of the other churchgoers, but donned a calico dress and sunbonnet for worship. Furthermore, Mrs. Lincoln held a sewing circle of Episcopal ladies on Thursdays, stitching up clothes for the poor.

This portrait of a modest matron in an economical home is fairly consistent with other Springfield reminiscences of the Lincolns during the 1850s, especially the limited number of servants. But prior to this, the Lincolns made do with a series of day servants. Without permanent help, they hired on an ad hoc basis. When Lincoln became a presidential candidate in 1860, a fourteen-year-old boy, Phillip Dinkell, was hired to work alongside a serving girl who lived in as well.

Any well-run Victorian household included hot meals on the table for the breadwinner three times a day. This was an incredible challenge for Mary. Her husband frustrated household routine with his fits of forgetfulness. He rarely arrived back home for his noon meal on time. His habit of erratic homecomings has prompted some scholars to ascribe darker causes—that Lincoln avoided coming home to evade his wife's company. Jumping to such conclusions is both unwise and unwarranted.

Whether interpreters have been harsh or sympathetic toward Mary Lincoln, nearly all agree that Lincoln's informal attire and manner gave his new bride enormous challenges and remained a headache for her—figuratively if not literally—for most of their married life. The long list of grievances included lying prone on the ground for reading—which may have been a holdover from his log-cabin days, when furniture was scarce. But no matter how many suitable and comfortable pieces of furniture were purchased—and multiple scoldings administered—a visitor to the White House was just as likely to see Lincoln stretched out on the floor as was a caller at their new Springfield parlor.

Mary rightly believed that no matter how tasteful her menus and wall coverings, her husband needed to spruce up both his appearance and his manners to correct the impression that he was fresh off the flatboat. Even Lincoln's admirers saw this dilemma. As Gamaliel Bradford suggested, "She had the strange, incalculable, most undomestic and unparlorable figure of Lincoln to carry with her, which would have been a terrible handicap to any woman."

So while Mary had to learn to tame many unruly elements in her life, keeping Abraham Lincoln polished and burnished was the real challenge of her domestic routine. Two sons under the age of three by the summer of 1846 and her husband's eagerness to serve in Congress gave her even more to juggle.

Lincoln had been struggling with his political ambitions, trying not to let doubts douse the intensity of his desires. Even while reading by candlelight in his childhood cabin, then in his loft on the Springfield square, he dreamed of fulfilling a greater destiny. His partnership with Stephen Logan, begun in 1841, was on its last legs by the fall of 1844, when Logan decided to go into partnership with his own son. Lincoln and Logan was formally dissolved by March 1845. There were prominent and eminent attorneys in Springfield who might have welcomed Lincoln into their firms, but he selected William Herndon, a younger man who looked up to him. Clearly, Mr. Lincoln was tired of being the junior partner in his professional collaborations. Just as clearly, he never consulted Mary about his choice of partner.

In local political circles, Lincoln had been derided as someone who had gone "silk stocking" with his marriage into a clan not only associated with the Kentucky mafia, which controlled many Illinois factions, but also with the Edwards elite who held sway in statewide politics as well. Lincoln wanted to maintain his political identity as someone of the people, and William "Billy" Herndon was just the person to forge such a link. To Mary, he was "common" and also too commonly drunk for her taste. She was not pleased by this choice, and did not warm to her husband's law partner over the years: The Herndons were never dinner guests at the Lincolns' home, for example.

But when they first solidified their partnership, Lincoln felt confident in his choice—especially in terms of his political reward. He knew what the state needed, and knew that Billy Herndon was a key to this populist constituency on the rise. Lincoln had spent years crisscrossing the eleven thousand square miles that comprised the Eighth Circuit of Illinois, canvassing in small towns with less than one thousand inhabitants. Robert Lincoln's principal memory from early childhood was of his father packing his saddlebags.

After his fourth term in the Illinois legislature, Abraham Lincoln ended his official service in the statehouse in Springfield. As a Whig, he knew Democrats' tight hold on the state would prove challenging—and providential. Lincoln needed to seek higher office. He had lost out in his bid for Congress in 1843, but the Whigs decided to hold conventions, something that favored him. He became a party stalwart, an expert campaigner, and his friend David Davis called him the "best stump speaker in the state." Lincoln's voice— that is, the *sound* of his words—was not his strongest feature: Many commented on his high-pitched tone and his Indiana-inflected speech patterns. But he impressed his audiences with the power of his words and the polish of his delivery as he found his true voice— the communication with an audience that was his gift.

Indeed, it was during his speaking tour on the circuit on behalf of the Whig candidates in 1844 that he returned to Rockport, Indiana, his former home, where both his mother and sister were buried. This nostalgic journey included encounters with old friends with whom he was happy to catch up and, in some sense, face his past. The experience prompted an outburst of verse, as Lincoln was an occasional poet during this period. Love of poetry was something he and Mary shared. They often read poetry aloud to each other, and both penned verses. Lincoln's compositions occasionally appeared (unsigned) in the local paper.

Lincoln's sentimental visit to his childhood home, as he described it, "aroused feelings in me which were certainly poetry." Indeed, he penned a poem in response, which included the following verse:

The friends I left that parting day,
How changed, as time has sped!
Young childhood grown, strong manhood gray,
And half of all are dead.

Marriage and the arrival of his own children in many ways opened up a new vein of emotional depth.

Both Lincolns doted on their children. Neither was much of a disciplinarian, but apparently Mary was even worse than her husband at handling her boys. She was extremely indulgent where her children were concerned. If her husband corrected their firstborn and Robert became unhappy, Mary kicked up a terrible fuss. At the same time, Mary was almost paranoid about the children disappearing, and her husband and neighbors reported her frequent alarms. When Robert was "lost" at age five, Lincoln claimed in a letter that most likely his mother had "found him, and had whipped him," adding playfully, "by now, very likely, he is run away again." So Mary's parental mode was mercurial. Lincoln was more unperturbed, if not lax, about child rearing. As a father, he confessed, "It is my pleasure that my children are free, happy and unrestrained by parental tyranny. Love is the chain hereby to bind a child to its parents."

Young Robert was notoriously precocious, and Lincoln once confided that he feared this young son was a "little rare-ripe" sort who might peak at the age of five. As a boy, Robert proved articulate, feisty, and mischievous—but seemed to become more somber as he matured. His intellectual prowess pleased both parents, who both might claim he took after them in his studious proclivities. In personality and appearance, the family agreed that Robert was clearly a Todd and it was apt for him to have this as his middle name.

The birth of Mary's second son, Edward Baker Lincoln, brought even more responsibilities. Having two young boys presented challenges for the mother of a growing family, especially as Lincoln was making a run for Congress later that fall.

Mary had just about gotten used to the patterns of her life, the ups and downs of the Illinois capital, and having her husband gone so much of the time, when along came the possibility of a grand adventure, a great escape. When Mr. Lincoln won his seat in Congress and headed to Washington, Mrs. Lincoln intended to accompany him.

6.

Playing for Keeps

The shift from Springfield to Washington was a momentous move for both the Lincolns. For Mary, it meant that her husband was now on the road to the political career she had envisioned for him. His rise to a congressional seat would demonstrate that her faith had not been misplaced, both to Springfield friends and to her Kentucky relations. Leaping into the national arena after waiting so patiently in the wings, Lincoln believed, provided an opportunity to make his mark. Both were eager to head to Washington. Mary's accompanying her husband was unusual, as most congressmen did not take wives, and especially not their entire families, along to the District. It was a sign of Mary's sense of entitlement and Lincoln's recognition of her partnership that she came along with him, determined to have a front-row seat.

The Lincolns did not take their leave of Springfield until October 25, 1847, and stopped off in St. Louis, where they were greeted by Joshua Speed, for a brief reunion. Next they made their way to Lexington, where Mary could show off her husband and sons to family and friends. The particulars of their arrival in early November provided for mirth. A Humphreys relative (one of her stepmother's nephews) had been on the train with the family of four, and had bolted for the Todd household with the stories of his trials on the journey to Lexington: He had been bothered by two awful boys and their indulgent parents. Imagine his chagrin when these people showed up on the doorstep as guests at the Todd home.

After her self-imposed separation from her father's house, it was a poignant homecoming for Mary—reconciled with her stepmother, returned to her father, and surrounded by delighted siblings. She was happy to greet her brother Sam, who came back from his studies at Centre College in Danville to romp with his nephews. Abraham and Mary were especially charmed by eleven-year-old Emilie, who was nicknamed by them Little Sister during this visit.

Mary was a few weeks shy of her twenty-ninth birthday, back in her hometown for the first time since going into exile over eight years before. During this absence, she had fallen in love with a man who was at long last her husband and borne him two children, and they were on their way to the nation's capital. Stopping off in Lexington as a proud matron and mother, as the wife of a congressman-elect, Mary Lincoln happily faced Todd relations and enjoyed rare repose during this interlude.

The Lincolns spent nearly a month ensconced in Lexington, making day trips and basking in affection and attention—especially from a retinue of slaves. This might have been reminiscent to Lincoln of his long stay with the Speeds nearly a decade before, although now his rest was full of purpose, to ready him for the exciting next stage of his career. Lincoln spent a lot of time sprawled on the parlor floor, reading the *Niles Register* and borrowing books from the Todd library. He could not help but be impressed by his wife's familiarity with legendary political figures. He was inspired by Henry Clay's public speech on the U.S.-Mexican War on November 13, when Clay denounced American aggression. (Clay's son would serve and die in the conflict.)

On November 25, the Lincolns left Lexington, finally arriving in Washington on December 2, 1847. The freshman Congressional delegate looked forward to rubbing shoulders with some of the great politicians

Lincoln as a rising politician

of the day—senators like Thomas Hart Benton and
John C. Calhoun. Lincoln had long admired Daniel
Webster's verbal pyrotechnics, and had met him when
the great man came to Springfield in 1837. They

would renew their acquaintance when Lincoln arrived in Washington. Webster joined Lincoln in attacking President Polk and his military adventures in Mexico. Funding for the war had been the centerpiece of Polk's address to Congress in December 1847.

The Democrats were belligerent in support of their warrior leader, and Stephen Douglas quoted Frederick the Great: "Take possession first and negotiate afterward." By the time Lincoln got to Congress, the government had spent over $27 million on the war, and more than 13,000 soldiers had died. Staunch Democrats like Jefferson Davis, a West Point graduate who had returned from the Mexican battlefield to fill a Senate vacancy in Mississippi, advocated keeping a strong military presence in the conquered territory. The national government, and especially Southern slaveholders, feared what the British might do and wanted to hold on to what had been so dearly won by bloodshed.

On January 3, 1848, Lincoln joined eighty-four other Whig representatives to condemn the actions of President James K. Polk. While Lincoln had been a young man struggling with debt in New Salem, Polk had been buying vast tracts of land in Mississippi, determined to make more money. He planned to stock his new estate from his Tennessee holdings, and confided to a friend, "The negroes have no idea that they are going to be sent to the South and I do not wish them to know it." While Polk was in the White House, he counted on plantation income for his retirement.

Political differences more than class issues stimulated Lincoln's attack on the president. The Whigs needed to attack "Polk's war" without offending their constituencies. On December 22, 1847, Lincoln vigorously questioned the rationale for the war, requesting that the president identify the "spot of soil on which the blood of our citizens was so shed." Lincoln continued his campaign with a January speech suggesting that "the blood of this war, like the blood of Abel, is crying to Heaven against him [Polk]." He hoped these attacks would not tarnish the sacrifices of U.S. soldiers. His friend Edward Baker—after whom Lincoln had named his second son—had become a war hero. But Lincoln's strategy misfired.

Herndon, his law partner back in Illinois, regretted that this "Spot" speech elicited ridicule, saddling Lincoln with the nickname Spotty. Understandably, Illinois friends of his old rival John J. Hardin (who died in the war) howled in outrage, calling Lincoln the "Benedict Arnold of our district." Herndon reported that most Illinois Whigs opposed Lincoln's position; thus these attacks on Polk eroded his political support.

All of this culminated in Lincoln's eventually losing his seat in Congress.

While opposition to his dissent erupted back in Springfield, Lincoln faced some small mutinies on other fronts. Mary had gladly abandoned boardinghouse life in 1843. She found the transition back into rented rooms difficult. Upon arrival in D.C., the couple took up residence at a boardinghouse run by a widow, Ann Spriggs. The Spriggs home stood in sight of the unfinished capitol dome—where the present-day Library of Congress stands. The dinner table was presided over by Mrs. Spriggs. Whigs filled the establishment, including five delegates from Pennsylvania and the abolitionist firebrand Joshua Giddings of Ohio. Other residents included a young doctor who fondly remembered "scenes of merriment" when Lincoln would lay down his knife and fork, place his elbows on the table, and launch into a story during communal dinners.

Lincoln would daily march into the halls of government. On his way home, he might stop off with friends at the local bowling alley or visit congressional buddies, such as Alexander Stephens of Georgia, who boarded nearby. His daily routine was full of variety, social contacts, and keen challenges. Mary, wife of a low man on the totem pole, had little other than mealtime banter to entertain her.

Mary felt herself being left behind once again, trapped in her rooms with dull routine as occupation. Here she was, abrim with anticipation, thrilled at her husband's swearing in but unable to throw any of the parties or entertainments that would have allowed her to shine. Out on promenades, she might worry that her wardrobe, hand-sewn and quite fashionable in Springfield, was a bit dowdy for Washington. She might make an occasional escape with an invitation to the famed salon of Dolley Madison, a former First Lady who had once been married to a kinsman. But even though Mrs. Lincoln could accept such invitations, her circumstances would not allow her to receive callers. Being such a small fish in quite a large pond proved frustrating. She was not free to pursue the social agenda she desired. While her husband sallied forth, she felt weighed down.

Despite the lack of housekeeping duties, Mary had heavy responsibilities. She was saddled with a pair of rambunctious boys to corral—Robert, at five and a half, and Eddie, less than two. Keeping her sons out of trouble, especially in the close quarters of a board-inghouse in wintertime, was a full-time occupation, and not one she managed easily.

The glittering social scene to which she aspired was beyond the couple's limited financial means. Lincoln's congressional salary and the meager rent of $90 per year paid by Cornelius Ludlum (their tenant back in Springfield) did little to advance the Lincolns in the capital. Mary's allowance did not stretch as far as she had hoped. Her husband was being sucked into committee work and the political merry-go-round of campaigning. Lincoln was extremely heady when he confessed to Herndon on January 8 that although he had said he would not be a candidate again, if he *were* drafted, he would consider serving another term. Only a month after being sworn in, he dreamed of reelection.

Mrs. Lincoln was decidedly less enthusiastic about the situation—and eventually headed back to Kentucky. It is not known if she was still in Washington on February 23, when former President John Quincy Adams died following his collapse two days prior on the floor of the House. (Adams was a member of the Massachusetts delegation, the first and only former president to serve in Congress.) Lincoln did have to postpone his party's Washington's Birthday ball, an occasion for which he played an organizational role.

It is also not known if Mary was at the Spriggs house when two slavecatchers arrived and kidnapped an African-American house servant. The enslaved man had been saving to buy his freedom when he was bound and gagged, then carried off to a district jail before being shipped to the auction block in New Orleans. This violent reminder of slavery, if she was a witness, would have been upsetting to Mary. Perhaps by the time this incident occurred (Joshua Giddings demanded a congressional hearing about it), Mrs. Lincoln had already taken her leave. Not everyone missed her, as Lincoln slyly indicated in his comments on fellow boarders: "All the house—or rather all with whom you were on decided good terms—send their love to you. The rest say nothing."

Whether her departure for Kentucky was a mutual decision or Mrs. Lincoln left in a fit—of anger? in despair? with resignation?—remains a mystery. The couple clearly had misgivings about their separation after only a short while. It was certainly a good thing that Mary had renewed ties with her Todd relatives, and she enjoyed a warm greeting back in Kentucky once again. The memories of good times in her hometown may have drawn her back as much as her unpleasant situation in D.C. pushed her out.

By mid-April, the congressman wrote to his wife in Lexington, "When you were here, I thought you hindered me some in attending to business; but now, having nothing but business—no variety—it has grown exceedingly tasteless to me . . . I hate to stay in this old room by myself." And so it was, as with many other couples, absence made for apologies, and the heart grew fonder. Lincoln wished his wife could be back with him rather than in Lexington.

Yet Lincoln remained absorbed by business, attending the Whig national convention in Philadelphia, canvassing widely for General Taylor (Whig contender for president), and voting on critical resolutions in Congress. The few letters extant between the Lincolns during this period are full of jest and sentiment, and demonstrate they were besotted by their children. When Mary hinted that the baby might have forgotten his father, Lincoln replied with worries, which she then soothed. Details about mundane matters and dreams dotted the pages. They reveal just how much the couple cared for one another.

Mary conveyed with some relief that she was not suffering from her familiar complaint of migraines. Lincoln wrote back: "You are entirely free from headache? That is good—good—considering it is the first spring you have been free from it since we were acquainted." He continued with some joviality: "I am afraid you will get so well and fat and young as to be wanting to marry again." This kind of banter suggests an easy and comfortable relationship, built upon a solid foundation—as in other correspondence Mrs. Lincoln might joke about her "next husband" or wanting to be rich enough to travel, which might not have been mentioned if they were sore points. Lincoln even added playfully: "Get weighed and write how much you weigh." This confident intimacy shows the depths of the couple's bond.

By late July 1848, the Lincolns were all together again, with Bobby and Eddie back under their father's watchful eye and tangled up in other boarders' feet. The family planned to return to Springfield, but only after Lincoln went on a New England speaking tour. This would allow him to broaden and enhance his reputation as a stump speaker and would give Mary a chance to see more of the urban seaboard. The Boston *Advertiser* praised Lincoln's skills on September 12, lauding his "able arguments and brilliant illustrations." His experience on the political circuit had a significant impact. Lincoln was impressed by William Seward's fiery speeches and believed that the slavery question deserved more attention in future political discussions.

His roundabout trip allowed a family holiday, seeing Niagara Falls with his wife and boys. They sailed from Buffalo on the steamer packet the *Globe*, landed in Chicago, and finally arrived back in Springfield by mid-October. No sooner had he gotten home than Lincoln was back out on the road, campaigning in

Illinois, offering ten speeches in less than two weeks to promote Whig candidates. With Taylor's resounding victory early in November, Lincoln had less than a month to skedaddle back to Washington, this time *sans famille*. He was involved in the organization of the inaugural ball, having been one of the stalwarts contributing to Taylor's victory.

After Lincoln's hard work on the presidential election, he anticipated a reward but was disappointed to be passed over. Lincoln was not supported for federal appointments by either Webster or Clay—who instead put forward their own candidates, including one who had not even supported Taylor. Lincoln believed that instead of these party traitors and slackers, he should have been the one to be offered a land-grant post. He was offended that none of the positions went to him nor any of his nominees. Indeed, the Whigs failed to use patronage to cement loyalty in the western states. Lincoln expressed regret: "It will now mortify me deeply if Gen. Taylors administration shall trample all my wishes in the dust." He found himself besieged by requests he had no hope of fulfilling. The letdown and frustration grated on him.

Finally, Lincoln was offered the governorship of Oregon, as a sop. He declined, citing Mary's objections. Indeed, she would not have gone out to the

Pacific coast to be settled down among Democrats—too far to go to be so wide of the mark. But it did not suit Lincoln, either, and later he reflected it was a blessing that he did not accept this honorific post. Back in Springfield in 1849, he reinvigorated his law practice. Any worries Mary might have had about her husband's foundering career were completely overshadowed by personal losses that followed.

First, her father collapsed suddenly during a cholera epidemic in July 1849. He fell ill after a political speech on July 7, and his doctors could do little for him. As he lay dying, Robert Todd made a will and passed away quietly on July 16.

This was a harsh, cruel blow to Mary Lincoln. She had been separated from her Kentucky home for over a decade and had felt especially deprived concerning her father's affections for so very long. Then, just after she renewed her ties, reassured of her place as a favorite daughter, death robbed Mary of her remaining and beloved parent. She would feel the depths of despair over being orphaned. Her father's death would stir up fears of abandonment, tapping into her feelings about her mother's death many years before.

Devastated by the loss, Mary was thrown into even more disarray when her brother George challenged her father's will in probate. George had moved out of the Todd home in 1846 while attending medical school, a sign of his early conflicts with his father. George believed that Robert's marriage to Elizabeth Humphreys had displaced his children by Eliza and that the younger brood of Todds were receiving more affection and favor. When his father's will was found to have only one signature instead of the requisite two, he challenged its validity in court, which caused his stepmother to fear for her inheritance. Lincoln was selected to go to Kentucky to protect the interests of the four sisters settled in Springfield—Elizabeth, Frances, Mary, and Ann (Ann being the youngest of the quartet, who had married merchant Clark Moulton Smith in October 1846).

Her father's untimely death allowed Mary to return to Lexington in October yet again—the third time within the past twelve months. The four Lincolns made an extended visit, but this was a difficult period. Divides widened among the older Todd siblings, as well as between them and the children of Mary's stepmother. A lawsuit concerning Robert Todd's estate would drag on for years, with one of Mary's brothers claiming that the money she and Abraham had been given by her father needed to be repaid to the estate. The couple returned to Springfield disheartened over family affairs, as well as the health of their younger son. In December, Eddie came down with what was believed to be a bad case of diphtheria. Near the end of January 1849, Mrs. Lincoln received another letter with terrible news: Her Grandmother Parker had died. This grandmother's death—after a long life— was sad but not unexpected. However, it did exacerbate Mary's grief and sense of loss.

Shortly after, Mary was felled by another trauma: Eddie grew worse, and after fifty-two days of suffering, his body gave up the struggle. The boy died a few weeks shy of his fourth birthday. His death certificate blamed chronic consumption. Whatever the cause, their son's death on February 1 was a terrible blow. Yes, many of Mary's women friends had lost a child, but the reality come true from this statistical likelihood proved ghastly. Lincoln suffered stoically, writing to his brother three weeks after his toddler's funeral with poignant simplicity: "We miss him very much."

While Lincoln mourned with reticent dignity, his wife found herself consumed by her grief—and suffered severe spells of weeping and a lack of appetite. Alarmed by his wife's condition, Lincoln sought out Dr. James Smith, a cleric, not a physician. Smith had been a rather wild boy growing up in Edinburgh, but became not only a beloved pastor of the First Presbyterian Church in Springfield, but a respected author, having recently published *The Christian Defense*. Smith's work was designed to counter the arguments of a skeptic—a category that clearly included Abraham Lincoln. Smith conducted Eddie's funeral service and offered both the Lincolns pastoral care. The reverend proved an effective counselor and got Mrs. Lincoln through a difficult period. She penned verse to commemorate her boy's death: "Bright is the home to him now given / for of such is the Kingdom of Heaven." Although she might not have felt Christian resignation at the time, she trusted that she might grow into piety under Smith's tutelage. Mary officially joined Smith's church in 1852. Lincoln's faith and the meaning of his church attendance remain a subject of lively engagement. Whatever his spiritual proclivities, Abraham Lincoln seems to have had a genuine appreciation of Rev. Smith's pastoral care and enjoyed theological discussions with him.

Burying Eddie might have been one of the hardest things the Lincolns had ever done. For each of them, the death of a loved one represented the return of a dark shadow. Lincoln's penchant for melancholy might have allowed him to sink into gloom, but apparently he reined in his emotions and forged ahead with legal work. His professional zeal was a crucial necessity to rebuild his law practice. President Zachary

Taylor had died suddenly in July 1850, and Lincoln's political star was waning, unless he could regain lost ground.

But for Mary there was no other avenue to throw herself into. She was trapped within her domestic world, and every little item of Eddie's seems to have refreshed her endless wellspring of grief. Within weeks of the burial, Mary passed along his clothes to a neighbor boy, Henry Remann. It might have pained her to see the toddler next door in Eddie's plaid shirt, but it was preferable to letting these garments remain within her own household as a daily reminder of her loss.

The situation was particularly difficult as Robert was an only child again, a most puzzling and painful time for the precocious six-year-old. His mother tended to dote on him again at this time, trying to find comfort in her living child. It was not long until Mary was pregnant again, and seeking comfort at Rev. Smith's church. She employed a live-in Irish serving girl, Catherine Gordon, to help her with housework. She tried to look forward to this new child growing within her.

The Lincolns could not forget this most treasured boy, but clearly wanted to climb out of the deep trench of sorrow into which his death lowered them. The family joyfully greeted the birth of William Wallace Lincoln four days before Christmas in 1850. His safe arrival, so close to the holiday celebration, proved a godsend for the entire family, and lifted the gloom. Mary's brother-in-law William Wallace looked after his fragile sister-in-law, and both Lincolns were overly cautious about the health of this new baby.

The Lincolns had weathered several crises on the road to matrimony, and their union was severely tested in 1850. Even though infant mortality was common— one in ten children died before the age of one at midcentury—the death of a child often caused permanent rupture. Mary's mother and stepmother had each lost babies, and her sister Elizabeth had lost her firstborn in 1836, so this was a fact of life among her most intimate circle. Such a loss might bring a couple closer together, or might provoke them to drift apart, each enveloped by diverse or even conflicting coping mechanisms. The birth of Willie Lincoln within ten months of Eddie's death seemed to signal a renewed commitment to their marriage.

Both were incredibly ambitious, Lincoln for political advancement and Mary for the social cachet that would accompany her husband's political rise. A stint in Washington had not slaked either's thirst but rather presented them with more vividly frustrated desires.

Lincoln realized that he was vulnerable, especially after his crisscrossing the state for the party had not paid off handsomely. He recognized that there were other elements to the political game, and perhaps learned that now, more than ever, he needed Mary as his helpmeet.

Lincoln's strategy to build his base in Illinois continued with his canvassing the state, adding to his core network, regaining lost momentum as an attorney. Mary also regained her footing as a materfamilias, first with Willie, and then with her fourth and final child, Thomas Lincoln, born in April 1853. This son was nicknamed Tad—presumably because his very large head at birth had reminded his father of a tadpole. Tad's birth was fraught with medical complications, which caused Mary future gynecological ills and may have contributed to his being the couple's last child. Mary also remained melancholy about her dead little Eddie, and even after the birth of two more babies, she confided to a friend in July 1853 that she did "not feel sufficiently submissive to our loss." Mary found it difficult to wrestle with her faith on these matters. Her strong temperament resisted any passive acceptance required by Christian doctrine. She struggled to find a moral balance for her living children, whose needs multiplied during these years.

Giving birth to two sons in a little over two years required renewed vigor, and devotion to raising her children. Observers often drew different conclusions about the child rearing techniques of the couple—but with the younger boys, Abraham Lincoln was viewed as "tender" and Mary as the disciplinarian of the two. Having apparently mischievous boys could turn her into the martinet that some portray, since what mother would not have been forced to induce more order in a household with three rambunctious boys? Her older son, although often depicted as studious, was also challenging. As one neighbor recalled, "Bob Lincoln was an awful tease. He never made trouble at school but at home he delighted in tormenting his mother. I've seen her fly into a rage at his pranks."

Bob was also a boy who needed more care than most boys during this period, as he suffered from teasing at school over his looks. Robert was apparently born with a lazy eye, which made him the object of ridicule. Mary did not want her son to suffer cross-eyes or teasing. Perhaps an eye-patch treatment was tried at first or perhaps not, but eventually Robert had eye surgery to correct the problem. If the condition was left uncorrected, not only would its cosmetic aspects plague him, but his depth perception would be permanently distorted. That was a very unpleasant prospect to endure, but the surgery was very nasty and difficult to undertake as well: it involved using a scalpel to alter his eye muscles and force realignment. Mary was the one who accompanied him to the procedure and stayed with him throughout in the doctor's office, during the awful ordeal and his terrible wailing. But, again, she was fully committed to her role and responsibilities, as Lincoln always addressed her as Mother, her highest calling during these years.

It was no small irony that the final Lincoln child was named after Abraham's father, Thomas. Abraham Lincoln, to use modern parlance, had many unresolved issues with his father. When Abraham was back in Illinois after his time in Congress, his half brother had sent an urgent message to Springfield in May 1849 that Thomas Lincoln was dying and Abraham must come at once—which he did. When Lincoln arrived at his father's house, Thomas Lincoln was recovering from his illness and in no danger. Months later, when he fell ill once again, Abraham was sent for in the winter of 1850–1851. At this time, Lincoln might have reasoned that his brother had cried wolf before, and thus did not respond. When he heard from another source that his father was indeed seriously unwell, he sent a letter explaining he could not visit because Mary was bedridden with "baby-sickness." He felt he could not leave her. The first few weeks of a baby's life were critical and with the loss of Eddie less than twelve months before, Lincoln was rightly apprehensive about his family.

When Thomas Lincoln subsequently died in 1851, perhaps both Lincolns shared the guilt that Thomas's son had not undertaken the three-hour buggy ride to see his dying father. But Abraham did not attend the funeral, either. Lincoln's feelings about his father's death, and even about his father generally, remain a topic rich with speculation and diverging interpretations. But he always maintained warm regards toward his stepmother, Sarah Bush Johnston Lincoln. And in 1853, he named his last child after his father.

Lincoln's thriving law practice and growing family led to a period of relative prosperity. Some later accounts include episodes of family disharmony during these years—Mary yelling at her husband, chasing him down the street, lashing out physically, angered by his indifference or neglect. However, one of Mary's Kentucky sisters visited their home for six months in 1854–1855 and portrayed a warm, intellectually companionate relationship: "I heard him say he had no need to read a book after Mary gave him a synopsis. He had great respect for her judgment and never took an important step without consulting her."

It is quite impossible to determine the truth of any marriage, and there will always be more than two sides to the story, especially in a small town where everyone was exposed to everyone else's private affairs. But it is safe to say that the couple were devoted to one another, and remained faithful to one another. If either had strayed, gossips would have seized upon this matter, and it would have wormed its way into interviews about the great man after Lincoln's death.

Although she was clearly the less popular of the two and he was viewed as one of the most caring of men, several Springfield neighbors reported Mary's kindness. A Mrs. Dallman, who was quite ill following the birth of her child, remembered Mrs. Lincoln serving as a wet nurse, breast-feeding the Dallman child—an image that does not mesh neatly with the harsh, narcissistic image most scholarship projects for her.

A garrulous man named James Gourley, who lived next door to the Lincolns for nineteen years and was a friend of Abraham's since 1834, was probably as good a witness to the couple as any. Gourley offered a variety of insights, including a decidedly backhanded compliment: "She is no prostitute—a good woman."

Gourley provided William Herndon, Lincoln's law partner, who conducted many interviews for his

biography of Lincoln, with some unflattering anecdotes about Mrs. Lincoln. He observed, "Lincoln &
his wife got along tolerably well unless Mrs. L. got
the devil in her." This devil might have had to do
with her own feelings of neglect, as Gourley reported,
"She always said that if her husband had Staid home as
he ought to that she could love him better."

Gourley was not an unbiased observer. He said that
Mrs. Lincoln had once or twice dared him to kiss her,
and he rather ungallantly added that he had refused.
Did he see Mrs. Lincoln as the needy housewife next
door? Or as the shrew who splurged when her husband left town? Or as a woman who suffered from
wild mood swings and crippling fears of storms, fires,
and burglars, to name but three of her phobias?

In any case, *Rashomon*-like accounts of Mrs.
Lincoln's personality and character can cloud the
issues. Whatever her faults—and there were many—
Lincoln indulged his wife, always finding excuses for
her. Even the critical James Gourley added the ambiguous comment that "Lincoln yielded to his wife—in
fact, almost any other man, had he known the woman
as I did, would have done the same thing." So despite
the ongoing battles over Mrs. Lincoln's personality,
she was her husband's choice. Even if she was the
bargain that he might have hugged harder as a "bad"

one, as some have suggested, she was also a bargain in other ways. He had struck a good one on his climb up the ladder, for she was his mate and his match, a woman to guide him on his struggle for advancement.

Mary Lincoln was a partner who struggled *with* her husband and *for* him. At times she showed poor judgment or bad temper, at other times generosity and shrewdness. But she made considerable sacrifices to marry him. Her life reflected a wide range of satisfactions and setbacks. She endured much as the wife of such an ambitious lawyer, a gifted politician cooling his heels on the Illinois prairie during most of their married life. She also endured much because her husband's rough manners were difficult to tame; he spent more time joshing his wife about her refined airs than adopting them for his own purposes.

His ambitions often took the form of wanderlust, something to which his wife never adjusted, but accepted as a necessity of life with Lincoln. For a woman who feared thunder and lightning, who ran screaming from the house during paranoid episodes— as when she thought an umbrella mender who came to the door to seek work was going to murder her— she was remarkably ill suited to a husband out on the circuit half the year. Clearly these were important absences, such as those involving cases in Chicago and farther afield, all to enhance and expand his professional prospects.

Abraham Lincoln relished his forays on the hustings, learning on the job as a railway lawyer, especially during trips to Chicago. His ties in the business and legal community would provide him with an important leg up in political contests to come.

Then when he was home in Springfield, he managed to make the rounds of male hangouts, as Lincoln enjoyed his time at William Florville's Springfield barbershop. Indeed, this Haitian refugee, "Billy the Barber," had been an acquaintance from New Salem days. The white lawyer and black barber renewed their friendship in Springfield early in both their careers. Lincoln became Florville's legal adviser, though he seldom sat in the barber chair: His unruly hair was the subject of amusement for decades around Springfield. Nevertheless, Florville's shop was an informal clubhouse where prominent men might swap stories, and Lincoln's career as a raconteur flourished at Florville's, as well as at other spots around the capital and the state.

When her husband was in residence, working at his Springfield office, Mary often sent the younger boys to town to fetch him, for he was notoriously forgetful about his meals. (Joshua Speed said Lincoln was "regularly irregular.") Tad often perched on his father's shoulders for the walk home. The tall attorney was a favorite among the neighborhood children, who frequently provided an escort for him to and from work. Immigrants from England, Ireland, and Germany, as well as African Americans, single boarders, widows, and growing families like the Lincolns, all settled chockablock on streets between Lincoln's home and his office on the square near the statehouse.

Despite the abundance of other young matrons in Springfield, Mary Lincoln seems to have found only a few with whom she could share confidences. She might perhaps on occasion enjoy visits with her sisters, but friction with both Frances and Ann caused a distinct coolness. She seems to have gotten along better with her brothers-in-law, Dr. Wallace and Mr. Smith, than with their wives. Ninian and Elizabeth Edwards remained Mary's surrogate family, and her older sister had a way with Mary, yet personality clashes also kept *them* at a distance. Mary was, for a long time, close to her former bridesmaid, Julia Jayne Trumbull. But in 1855, they, too, had a falling out. Intimacies were dearly bought in small towns, and expensive to maintain.

Mary was a cordial neighbor to Julia Sprigg, a widow who lived in the house north of the Lincolns;

Mary used Julia's daughter as a babysitter. She was fairly good friends with widow Mary Black Remann, the sister of a local merchant living on her street. Mrs. Remann had several young children near the age of Mary's.

Mary became particularly attached to only one of her neighbors, a widow named Hannah Rathbun, ten years her junior. Perhaps it was Hannah's vivacious style or her wit that distinguished her from others. When Mrs. Rathbun moved to Springfield to live with her sister, she lived in the Miner home, diagonally across from the Lincolns. Mary and Hannah became intimates. Hannah's sons, Edward and James, were attended by a new doctor, John Shearer, who had moved to Illinois from Philadelphia. Shearer soon courted the widow, under Mary's watchful eye, and the couple were married in 1858. Hannah Shearer was, like her friend Mary Lincoln, attractive, temperamental, and ambitious for her husband. The Shearers lived in Springfield until the doctor discovered he might have tuberculosis, whereupon the Shearer family relocated to Pennsylvania in 1859—by which time Hannah Shearer had become a close confidant of Mrs. Lincoln's.

Hannah had come into Mrs. Lincoln's life in the mid–1850s, just when there was a distinct letdown for her on two fronts. First, in 1855, Mary had lost the company of a favorite houseguest, her sister Emilie Todd, and second, that same year, she broke off her friendship with Julia Trumbull, with whom she had been close for over a decade.

Mary Lincoln had followed the family tradition of bringing Kentucky belles out to Springfield to find a husband. Emilie, her Little Sister, eagerly accepted the Lincolns' invitation to visit—perhaps to find a match. She arrived in Springfield late in 1854 or early

Young Emilie Todd

in 1855. This Kentucky houseguest both provided an excuse for festivities and added to the joys of the household. Deprived of a daughter, Mary found this sister, eighteen years her junior, just the right companion. Emilie became a favorite of Mary's husband as well. Mrs. Lincoln had plenty of young girls within her community—nieces, for example—who might have filled this substitute-daughter role, but Elizabeth Edwards's older daughter, Julia, did not elicit any preferential treatment, nor did Frances's daughter Mary, who might have been favored because of her name. Actually, both were given less than cordial regard by their Aunt Mary over the years, while she was loving and giving with other younger women, such as Emilie Todd and Hannah Shearer.

During this visit to Springfield, Emilie became the petted and cosseted young acolyte, invited to a whirlwind of parties, balls, and entertainments. She adored her sister Mary and looked up to her as everything from fashion icon to clairvoyant. Emilie remembered being at a neighbor's party when Mary insisted that she wanted to go home, so they left the gathering, went back—and found the house on fire. This apparently saved the children, whose caretaker had been sleeping. Emilie reported that Mr. Lincoln "said he was glad he had a wife who could 'sniff a fire a quarter of a mile away.' "

Emilie also reported that the family loved to tell the tale of Mr. Lincoln taking his son out in a baby wagon, when he became so absorbed in his thoughts that Mary caught him pulling an empty wagon—with the baby tumbled out onto the street, "kicking and squalling in the gutter." Lincoln fled the scene, not waiting to hear the piece of Mary's mind that he knew he would receive.

Much of her scolding involved this legendary neglect, as Lincoln would not respond to calls in the home, regularly ignored the clock, would walk home in the rain without an umbrella, and would perform other feats of absentmindedness. Dr. William Wallace, her sister Frances's husband, warned Mary to look after her husband's precarious health, as Lincoln's frailties had been the subject of local concern for years. Emilie witnessed her sister's deep pride in her husband, their affection for one another, and the couple's contentment in Springfield.

Mary seemed to have come into her own on a certain level. She was more comfortable with her status once she returned from Washington. Although she was impatient for her husband's political career to take off, she was absorbed with her daily round of social activities and her rising status in the community. Meanwhile, her oldest son, Robert, was moved from Mr. Estabrook's academy in Springfield—where he had spent three years—into the newly opened "Illinois State University," a preparatory school with four ministers teaching a crop of eighty boys. When Robert entered the school in 1853, he became acquainted with another studious lad, John Hay (who left in 1855 to attend Brown University). Hay became a lifelong friend, and later worked in Lincoln's White House. With Robert on a steady course, and her two younger boys thriving, Mary felt happy to host visiting family (as long as they were her relations rather than any of Lincoln's less-than-appealing kin). She deeply appreciated Emilie's company and tried to keep her interested in Springfield.

Despite her best efforts, however, Mary did not land a beau for this favorite sister, who went home midyear in 1855. Emilie welcomed Mary's long letters, full of newsy gossip, urging her to return soon. And she might have, had a young suitor named Ben Hardin Helm, the son of a Kentucky governor, not caught her attention. Mary suggested that perhaps Emilie's younger sisters, Elodie (Dedee) and Katherine (Kitty), would follow the Todd example and spend a winter season socializing with in-laws in Illinois.

However, it must have been a mixed blessing for Mary to have Emilie as her devoted companion during the winter of 1854–1855, when Lincoln suffered one of the bitterest defeats of his political career. He had begun 1854 content to focus exclusively on his burgeoning legal practice. Lincoln had deferred his political dreams, but developments drew him out of retirement—most spectacularly, the Kansas-Nebraska Act in 1854. This reversed the policy of the 1820 Missouri Compromise, which had restricted slavery's expansion to territories south of Missouri's southern border, and thus upended years of established accord. For decades, the nation was divided into North and South by a geographical line. Yet the country was also united by compromise on the slavery question, most recently with the Compromise of 1850, which included provisions that allowed the vast tracts of land obtained during the U.S.-Mexican War of 1846–1848 to be organized into territories and to vote on their own whether they would be free or slave states. This newly acquired territory thus got special treatment, differing from the treatment of land obtained through the Louisiana Purchase.

Agitation in the West pressured Congress to reopen deliberations and let all territories determine their own futures. Thus the Kansas-Nebraska Act exploded onto the scene in 1854. This act allowed any states organizing out of territorial land to vote on slavery, putting the decision into the hands of the people and creating what was called the doctrine of popular sovereignty. The fact that his old rival, Stephen Douglas, was the champion of popular sovereignty doubtless contributed to Lincoln's aggressive response to this development. His hibernation came to an end when national debates erupted.

Perhaps Lincoln first spied the opportunity to reemerge when Mary's cousin Cassius Clay came to Illinois to campaign against the Kansas-Nebraska Act in July 1854. Both Mary and Abraham attended his Springfield speech, where he lectured that Free Soilers, Whigs, and Democratic opponents of Douglas must unite to keep slavery sectional. Meanwhile, Frederick Douglass pleaded on behalf of enslaved African Americans, lamenting to his audience in Chicago that his people had no Stephen Douglas to fight for them, only their humanity. Lincoln could sense something stirring across the country, and felt ready to jump on the bandwagon.

Nerved by the righteousness of the cause, Lincoln found in this renewed antislavery campaign a fresh, more urgent voice. He would raise his tinny, high-pitched cadences in protest. His voice would roll over rural audiences and give him a reignited political agenda. In subsequent speeches that year, Lincoln assailed slavery as both a "monstrous injustice" and a "cancer," clearly stepping up either his own estimation of its dangers or the rhetoric he hoped would stir supporters. When Lincoln decided to relaunch his political ambitions, to strike out on another speaking campaign, Mary became even more valuable to him. She was not just a wife knitting by the hearth and murmuring the occasional polite "Yes, dear." She read to him from the paper, and then they discussed the headlines, exchanging ideas on issues and strategies. She also clipped articles out of the paper and kept him apprised of local gossip when he was out on the road. With the Whig party dissolving into disarray, abolitionists, Free-Soilers, and members of the newly formed Republican party began their march into political prominence, and her eyes and ears were essential to him during his renewed efforts.

Lincoln was a writer particularly attuned to oral expression, who wanted to read prepared speeches aloud to see how they sounded before delivering them. Thus we know he sometimes auditioned before an audience of one. Mary served as a key sounding board during this formative stage of his career. She had been his champion, in the courtroom while he addressed the bench, in the gallery at the Illinois statehouse, and in the House of Representatives. Her rapt attention and sharp critiques were an invaluable asset as he reentered the political fray.

Their joint efforts in 1854 were aimed at increasing his regional support and national visibility, by mobilizing attacks on the expansion of the slave power. However, if he was gearing up to capture the junior senator's seat from Illinois, then he would not have allowed his supporters to put his name forward as candidate for state representative. The new Illinois constitution forbade anyone serving in the legislature from being elected to the office of senator.

Lincoln had not even seriously considered another Senate race until he saw the returns from the November elections—and the strength of the "anti-Nebraska" sentiment. Politicians who opposed the Kansas-Nebraska Act made dramatic inroads, and Lincoln had easily won his seat from Sangamon County. When he realized that he had an opportunity to try for a place in the Senate, with Mary's encouragement, he resigned from his seat in the statehouse in order to politick for the Senate nomination.

Lincoln's resignation led to a special election. When the Democrats scooped up the seat, the "anti-Nebraskans" carped that Lincoln was placing his own ambition ahead of party interests. Many of his former allies criticized him for abandoning their cause.

Yet Mary sensed the time was right and backed her husband in his bid for higher office.

Lincoln worked the hallways of the General Assembly, garnering support. He was up against Democrat incumbent James Shields, with whom he had nearly fought a duel twelve years earlier over a matter concerning the honor of Mary Lincoln, while she was still Miss Todd. The Democratic governor, Joel Matteson, indicated he was willing to be drafted, and he sat in the catbird seat, having taken no public position on the Kansas-Nebraska Act.

The vote for a new Senator, scheduled for January 31, was postponed due to the ferocity of a snowstorm. When the joint session met at the statehouse on February 8 (only four days before Lincoln's forty-sixth birthday), Mary had hopes, even if her husband didn't, that he might emerge from the chamber as Senator Abraham Lincoln. He went into the contest several votes short of the majority required. On the first ballot, Lincoln captured forty-five votes, Shields garnered forty-one, Lyman Trumbull received five votes from anti-Nebraska Democrats, and eight votes were scattered among other candidates, including Governor Matteson. Mary Lincoln, sitting in the gallery, was buoyed by the fact that her husband needed only six votes more to win the requisite majority of fifty-one, and simultaneously annoyed that Trumbull took any support away from her Abraham.

However, imagine her distress when on the next tally, Lincoln lost four (but gained two), then on the next go-round, slipped another two notches. The neck-and-neck race was a nail-biter for observers during these first hours. On the fourth ballot, Lincoln's count dropped to thirty-eight, while Trumbull now had eleven supporters.

The fluctuation of the tallies, the setting of the sun, and the tedium of this horse-trading of votes disheartened participants and the audience. Finally, on the seventh ballot, a seismic shift took place: All of Shields's votes migrated to the governor, making his total forty-six—more than Lincoln had ever mustered. By the eighth ballot, Lyman Trumbull had collected eighteen votes. This dramatic sea change posed a dilemma for the anti-Nebraska agitators, including Lincoln.

Lincoln had been savaged as the man who put personal gain above principle. But he knew that his chances to prevail were fading. He also knew that what had begun as a contest between him and Shields was now clearly a battle between Trumbull and Matteson—and without Lincoln's votes, Matteson would win. On the next ballot, the governor held his

ground at forty-seven votes, and Trumbull continued to gain ground with thirty-eight. Matteson might
chip away and capture the necessary additional votes
to win, and rather than allow this to happen, Lincoln
took his fifteen remaining votes and threw his support behind Trumbull. This was a major show of party
loyalty and a personal sacrifice for Lincoln.

Unhappy about the outcome, he later characterized
that long day as "Matteson's double game." Lincoln
took pleasure in the governor's defeat. The women
in the gallery might have known something was up
when they saw Governor Matteson's wife and his
daughter enter the gallery earlier in the day—as indeed
women may have played roles behind the scenes, and
been as aware as the men of what was in play.

Back in the privacy of her own home, Mrs. Lincoln
learned of her husband's defeat—not at the hands of
the Democratic contender, but by former supporters
in coalition with anti-Nebraska Democrats. Lyman
Trumbull would be senator, and her husband would
not. Although her sister Emilie—still a Lincoln
houseguest—said everyone worried about what a
blow it would be to Mary, she reported that her sister
kept her disappointment strictly to herself.

Of course, revenge is a dish best served cold. Mary
Lincoln held a special grudge against Norman Judd,

one of the men who went for Trumbull on the first ballot. She believed he robbed her husband of the position he deserved. Mary Lincoln never welcomed Judd into the fold—and kept him from her husband's inner circle years later. It was a private vendetta, and she was able to use her contacts with Lincoln's closest adviser, Justice David Davis, to block Judd's petitions for advancement. In public, however, she maintained the requisite cordiality and even wrote solicitous letters to Judd's wife.

Mary Lincoln's rupture with Julia Jayne Trumbull became a permanent ringing down of the curtain. Julia had been a girlhood friend, someone whose youthful antics were a part of Mary's most beloved memories of early days in Springfield. The two girls were belles together; then, as young matrons and mothers, they had much in common. Julia had been in Mary's wedding, and their husbands were colleagues on the circuit. Both women had politically connected fathers, and had married ambitious lawyers. Julia's husband's status allowed them to lead a more economically comfortable life, which her friend Mary may have envied. But their small Springfield coterie remained a tight-knit group. Mary and Julia joined Rev. Smith's Presbyterian church at the same time, and hosted alternating sewing circles for the poor. Several scholars suggest that it was petty for Mary to break off relations with Trumbull's wife just because he won the vote. But this perhaps ignores what Mary might have suspected about how events unfolded.

She was both a rival and a friend of Julia Trumbull's. She likely knew exactly what Julia's role was in this Senate campaign. Mary had been her husband's close adviser, and knew his support was strong and growing stronger. He had been on the front lines of the anti-Nebraska campaign and taken all the weight of the crusade onto his shoulders. He had worked hard to line up his votes, with little lists he kept in his notebooks that they both studied in the evening. Mary recognized the way in which each vote was based on a relationship borne of time and work. Mary must have been bitter to think of all her husband's efforts in vain, while Lyman Trumbull, being in the right place at a fortunate time, snatched up the victory.

Mary and Julia had once aggravated Shields when they anonymously attacked him in their published "Rebecca letters." Julia was not a passive observer in the world of politics. Perhaps because of her intimate knowledge of Lincoln's moral and ethical principles, Julia could have predicted his capitulation when it seemed the Democrats might prevail. She could have advised her own husband to hang back, let Lincoln's support waver, then move in to rescue the party hopes. If the candidate to beat became Matteson, then her husband might usurp Lincoln's front-runner status. Mary Lincoln was well aware that Julia Trumbull's insights and talents made her as effective a political wife as she herself. Julia might have recognized that to land the position, gaining a toehold would be sufficient—and certainly Trumbull's five votes on the first ballot did not constitute much else.

Mary's anger might well have been based on her appreciation or imagination of Julia's role in the drama as it unfolded. Their friendship might not have been collateral damage, but a direct response to events. Mary, who had desperately wanted her husband to prevail, knew Julia would be the one to be heading to Washington, not her. Perhaps other wives might have just accepted that it was not Mr. Lincoln's time to serve in the Senate, but Mary had been impatient from an early age: She wanted things when she wanted them. Julia would be given the prize, and Mrs. Lincoln could not forgive her.

7.

Enlarging Our Borders

With the loss of his Senate race, Lincoln was forced to throw himself back into professional matters, as he himself described, "to pick up my lost crumbs of last year." It was a personal setback, but Lincoln soon recovered—and indeed, was amicable toward Lyman Trumbull, who had won the seat. He also could take both refuge and pride in his legal practice. His earnings averaged at least $2,000 per year by the late 1850s. He also gained a reputation on the circuit as a hard worker and a fair and honest attorney. In 1856, when a client sent Lincoln $25 for some work, he returned $10, maintaining, "Fifteen dollars is enough for the job." When he felt that he had not done enough to earn his fee, as in the McCormick reaper patent case, he returned the check. (He later pocketed his pride and cashed the check when it was sent a second time.)

However, after nearly twenty years in Springfield, the accomplished lawyer in Lincoln allowed him to demand what he felt was due, as when the Illinois Central Railroad failed to pay its bill of $5,000 in 1854. He had gone up against his former partner, Stephen Logan, and won an important case for the railroad— which saved the company thousands in local taxes— but they wanted to skimp on the fee. He finally sued for payment, winning a verdict in his favor in January 1856. Eventually the company not only paid Lincoln his $5,000 (a sum he split with his partner, William Herndon), but hired him for other cases, signaling his formidability and esteem.

Settling back into the routine of a Springfield haus-frau was satisfying to Mary at this stage only because her husband had been improving his prospects and earnings. By 1856, Mrs. Lincoln decided to undertake some expansions of her own.

Her home was wholly inadequate for a family of five, occasional guests, and a live-in servant. She wanted to add a full second story to the house. In 1854, Mary had gotten $1,200 for some farmland her father had given her early in the marriage, and the promise of Lincoln's large railroad fee might have prompted this spate of home improvement. In April 1856, builders invaded the house to begin renovations.

Adding a second story was a major expansion. Previously a person of average height could only stand erect at the peak of the gable, but after renovations, the upstairs had eleven-foot ceilings. Mr. Lincoln had his own bedchamber, adjoining his wife's, a symbol of their elevated status, not of any diminution in marital affection. The boys' bedroom stood across the hall from their parents' rooms. In addition, the two front bedrooms had false fireplaces installed as a decorative background for Franklin stoves. At the back of this new second story, the house boasted both a guest bedroom and a maid's room, accessible also by a back stairway.

The downstairs interior was considerably reconfigured as well, although visitors were still greeted with a wide stairhall upon entering from the front door. A door on the left led to a formal parlor, and on the right was a family sitting room for informal gatherings. Mary provided her husband with his own separate library and study, through double sliding doors from the parlor. Across the hall from her husband's study, at long last she would have a dining room that, although small, would be suitable for entertaining.

Mary took the opportunity to decorate her home in a lavish Victorian manner, with floral carpeting in the formal rooms, elaborate wallpaper throughout, plus ceiling-to-floor drapes and heavy swags. The effect was that her home looked completely "done," tastefully appointed with the latest home furnishings, like a spread from a ladies' magazine.

Besides the refined interior decor, which has been preserved in the current National Park Service home museum, visitors might be struck by the set of false windows on the exterior walls of the formal parlor.

Abraham Lincoln home in the 1850s, Springfield, Illinois (Lincoln and his son Willie behind the fence)

These windows, and indeed the entire style of the home are vaguely familiar. They were definitely familiar to anyone who had visited the Todds' home in Lexington, and it must not have escaped Lincoln's attention that his wife was trumpeting their good fortune by renovating their Springfield residence with architectural homages to her girlhood home.

She explained to her former houseguest Emilie that "we have enlarged *our borders*, since you were here." She wanted to convince her favorite half-sister, recently married to Ben Hardin Helm, to pay another visit to the more commodious Lincoln house. Emilie did not get to see Mary's remodeled home in Springfield, but she sent her husband, Ben, who paid a visit on his own while pursuing a legal case in the Illinois capital in 1857. Mary welcomed this new relation with open arms and grand hospitality. During this period, the Lincolns' exalted affections for Emilie expanded to include Ben Helm, and Lincoln took a brotherly interest in this young man.

The size and splendor of Mary's expanded home were well noted by the Springfield community. Mrs. John Todd Stuart, the wife of Lincoln's former partner (who was a kinsman of Mary's), commented, "I think they will have room enough before they are done, particularly as Mary seldom ever uses what she has." By now, Mrs. Lincoln's airs were a part of local lore. She had been penny-pinching as a young bride, not surprisingly, for her husband had brought a large debt to their marriage. Her parsimony persisted, even after Lincoln had paid off this debt. She was unwilling to offer top wages to servants, she bargained with tradesmen, and she exhibited other measures of stingy domestic economy. For years only her husband had his clothes made outside the home, as she would sew for herself and the children. Perhaps economizing was a defensive mechanism. Her husband had tried to loosen the purse strings on more than one occasion, but Mary had a manic streak about money, and a deep fear of impoverishment. After Lincoln's prospects had improved so dramatically, Mary's fears about their poor circumstances were exaggerated and perhaps unwarranted. But this was a difficult habit for her to break. She instead began to seesaw between giddy, excessive acquisition and guilty penance for indebtedness. This would be a pattern that would develop into disastrous dimensions over time.

Once Lincoln's legal practice began to flourish, Mary's fears about poverty would wane. Renovating her house was the most visible manifestation. She was no longer in the poorhouse, and could even afford a show house. Once Mary got her house in order, she naturally threw a party—with five hundred invitations no less. This lavish display was doubtless intended to inspire envy by allowing crowds to tramp through her home. Bad weather and a wedding in Jacksonville kept many away on the February evening in 1857 when she staged her viewing. Yet half of those invited did come to gawk.

With the inauguration of James Buchanan and the Dred Scott decision in March 1857, forty-eight-year-old Abraham Lincoln became even more committed to his involvement in Republican Party politics. His core base in Illinois was strong, and he responded to their hearty solicitations for his leadership. But he was looking beyond his state for an audience, for political recognition. The fierce fires of antislavery fervor were being stoked in the East, with small wildfires spreading out to the prairie. Free-soil agitators and Republican zealots reconfigured political landscapes during the volatile period following the Kansas-Nebraska Act. With all these changes in the air, surely Lincoln would have a chance to shine again.

Both Lincolns profited when Abraham decided to pay a call in person at the New York headquarters of the railway company that owed him money. This jaunt to the East turned into a family holiday in the summer of 1857, and Mary's second visit to

America's cultural capital was a great success. She had whisked through as a congressman's wife, accompanying Lincoln on tour, but this time she got to stay in Manhattan and see the sights. She was enamored of this glittering island and the wonders it afforded. Her vacation gave her a fresh new perspective, but she caught a terrible case of envy.

At the end of summer, she wrote her sister Emilie a long letter, which included the anecdote that when she saw the European-bound steamers in New York harbor she sighed "that poverty was [her] portion." She laughingly reported that she told "Mr. L" [the way she often identified her husband in letters] "that I am determined my next husband *shall be rich.*" Next she gossiped about the new "palace" being built by Governor Matteson ("worth a million") and a paragraph later remarked that her niece Julia Edwards Baker "has nothing but her dear Husband & silk quilts, to occupy her time." She added a bit wistfully: "How different the daily routine some of our lives are." The letter closed on a sad note, begging to be remembered to Emilie's husband—while her own, she forlornly confessed, had been in Chicago for nearly a month.

Lincoln's long absences would only increase in the weeks to come, as he embarked on a renewed quest—

to win a seat in the U.S. Senate. It is hard to imagine how the Lincoln of 1841, navigating the shoals of indecision and frail health, dismissed as a lunatic by some and a lovesick Cyrano by others, could become the robust Senate hopeful of 1858. Even more remarkable, Lincoln rose to the challenge of running against an incumbent, the respected statesman Stephen Douglas.

Lincoln was convinced that he might unseat Douglas, his old rival, in the 1858 election, and at least earn an enhanced reputation trying. Mary was in a unique position to support her husband against this incumbent. Nearly twenty years before, when Douglas, Lincoln, and Mary were all young and unattached in Springfield, most assumed that Douglas would become the superior statesman, and thought Lincoln considerably outclassed. Lincoln himself was struck by Douglas's knack for success, contrasting it with his own cursed fate.

This was not Mary Todd's opinion: She frequently declared Mr. Lincoln's superiority—something about which she was so sure that she staked her future happiness on it. And the very determined Mrs. Lincoln intended to prove she was right. Two decades later, she encouraged her husband's run for Douglas's seat, feeling that once people recognized what she already knew, her husband would prevail. Mary poked fun at Douglas's nickname, saying he was a "very little, little giant by the side of my tall Kentuckian, and intellectually my husband towers above Douglas just as he does physically." Lincoln was the undisputed favorite of old-line Whigs and newfangled Republicans, while Douglas maintained his core constituency, Democrats. Douglas knew he was vulnerable from the backlash to the Kansas-Nebraska bill, but depended upon his considerable charm and his incumbent status to secure his reelection. He knew he was an influential national figure and during the campaign would remind the Illinois voters of his influence and superior credentials.

Lincoln accepted the Republican Party's nomination (an unprecedented ironclad endorsement) in June 1858, offering a stirring speech: "I believe the government cannot endure permanently half slave and half free." This "house divided" speech—a beautifully crafted and well-rehearsed piece of political theater— signaled that Lincoln had found a powerful voice. This campaign fired him up as nothing before had done. He pursued his goal with gusto, traveling thousands of miles and giving scores of speeches, not just the famous ones with Douglas in attendance.

The two candidates went on the stump together for formal events in the months leading up to the

November 1858 election—the famous Lincoln-Douglas Debates. These marathon sessions were each three hours long, beginning on August 21 and concluding with the seventh and final event on October 15. Stenographers faithfully reproduced the two men's speeches, and many appeared verbatim in Illinois journals, then were reprinted in eastern papers. Such a public and protracted discussion focused on the expansion of slavery attracted national attention. The *New York Times* declared Illinois "the most interesting political battle-ground in the Union." (Of course, the *Times* was unabashedly partisan, a Republican paper.)

For the many thousands of citizens who witnessed these debates, the spectacle was often thrilling. The disparity between the two politicians was stark. The tall, lanky Lincoln in plain clothes was a sharp contrast with the short, puffed-up Douglas in his dandified wardrobe. Douglas arrived in town in a private railway car, while Lincoln glad-handed his way through passenger cars, comfortable with plain folk. Douglas was accompanied at each debate by his trophy wife, Adele Cutts Douglas, a beauty nearly twenty years younger whose haute bearing signaled her aristocratic pedigree. Lincoln kept his own wife under wraps, until she made a final appearance at the last engagement, at Alton, Illinois. Historian David Donald has suggested that emphasis of these contrasts, even Mary's absences, was part of Lincoln's strategy. Perhaps Mary insisted upon coming to this final encounter because she had to see for herself what everyone was talking about, and because she needed some boosting of her own spirits. She spent their entire marriage worrying over Lincoln's health, and here he was engaged in a debilitating marathon that she felt he had little hope of winning.

The debates were a roller-coaster ride for Lincoln and his opponent, as the election became a gladiatorial war of words. Even though more Illinois voters turned out for the 1858 midterm elections than had appeared at the polls for the 1856 presidential contest, and even though Lincoln's supporters won the popular vote, the Democrats maintained their hold on the state legislature—Lincoln had lost. Republicans complained about Democrats gerrymandering districts and bitterly protested immigrants shipped in on railcars to cast ballots illegally. Whether it was fair and square or not, Douglas was returned to the Senate—yet another political defeat for Lincoln. This took a lot out of him, although he knew it had been an important battle, regardless of the outcome. Lincoln knew that despite his setback, the "fight must go on," yet he expressed his fears that he would "sink out of view" as a sacrifice to the anti-Nebraska campaign.

It seems even more incredible to realize that Lincoln did not sink out of view, but rose to national prominence and won the nomination of his party over rivals with impeccable credentials—and then was elected president less than twenty-four months later. Many developments contributed to this miracle of 1860, most notably Lincoln's Cooper Union speech in February 1860, showcasing his skills from previous campaigns and his ability to rise to the occasion.

Many have outlined the remarkable features of this period—Richard Current, David Donald, Richard Carwardine, Douglas Wilson, and, most recently, Doris Kearns Goodwin. Goodwin's *Team of Rivals* traced the intertwined fates of William Seward, Salmon Chase, and Edward Bates in their bids for the Republican nomination, and the contributions of these men, as well as Edwin Stanton, to Lincoln's presidency. But many modern readers are surprised to learn that Lincoln did not even attend the Republican convention where he was nominated for president, and once Lincoln earned the nomination on May 19, 1860, his speechmaking days on the campaign were over. As Mary's niece reported, "From

Friday, May 18, 1860 . . . until the day of the election, November 6, he remained quietly at home."

Nineteenth-century presidential candidates were conspicuously absent from the campaign trail. At the same time, other speakers went on the stump: Charles Sumner and Cassius Clay, as well as Lincoln's former rivals for the nomination, Chase, Blair, and Seward. All these leaders were working to secure Republican victory. Visitors of every description came to Springfield to visit—many bearing gifts, which was a campaign tradition as yet unassociated with any taint of corruption. From the old woman from New Salem who brought a pair of socks she had knit to New York machine bosses promising to deliver votes, the doors were wide open. Correspondents flocked to Lincoln's Springfield home as profiles and biographies were published. Yet he was denied his most powerful tool, the podium, as others were delegated to plead and state his case.

Lincoln had recognized that he entered the Republican convention as a dark horse, letting his rivals pick one another off. Furthermore, his prime opponent for the prize, Seward, had lost the nomination as much if not more than Lincoln had won it. Seward was defeated as a known quantity, while Lincoln's qualities as an unknown served him well in the Chicago con-

Brady's "Cooper Union" portrait of Lincoln in 1860

vention. The hullabaloo concocted by his Illinois supporters went a long way toward impressing the thirty thousand gathered in Chicago to select Lincoln as their leader. But he also knew that, although running as an unknown had helped him to win the nomination, it might not be as useful in the presidential election.

While Lincoln's supporters were jubilant, they anxiously watched developments in the other parties. Democrats had not been able to settle on a candidate at their convention in April 1860, and they met again in June in Baltimore, where Stephen Douglas, after a walkout by Southerners, emerged with the endorsement. Southern Democrats had bolted because Douglas had refused their demand to include a plank in the platform guaranteeing slavery in the territories. They reconvened in Richmond and put forward the current vice president, John Cabell Breckinridge of Kentucky, as their nominee for president. This brought the tally of candidates up to four, as another breakaway group, the Constitutional Union Party (a coalition of disaffected Whigs and Know-Nothings), had met earlier in May to propose John Bell of Tennessee for president. The realignment of political parties, as well as these third-and fourth-party breakaway groups, suggested an electoral free-for-all in November. Lincoln would have to put his best case forward to win key states like Indiana, Pennsylvania, and Ohio. Perhaps keeping a low profile back in Springfield was a better strategy than he knew.

Veteran political kingmaker Thurlow Weed made the trek out to Springfield, at the invitation of Lincoln's Chicago advisers. He wanted to meet this man who had snatched the nomination away from Weed's designated candidate and dared hope to succeed. Once Weed discovered Lincoln was more astute than he had been led to believe, he came aboard to offer campaign savvy and support. In Weed's wake, troops of Eastern correspondents and other politicos came out to inter-view this "Honest Abe," "the Railsplitter," a mythic candidate in the making.

And so began the race to define Lincoln to the lar-ger public, who knew little about this prairie politi-cian. As scores made their way to the Lincoln home in Springfield, it would seem that Mary was ready for them. If some of her family are to be believed, she had been waiting for this moment more than half her life: As a little girl growing up in Kentucky, she had dreamed of living in the White House. Now here she was with her husband running for president, against John Breckinridge and Stephen Douglas, two politi-cians with whom she was well acquainted, and an upstart named Bell, whom she was sure her husband could handily outpoll.

Journalists and politicians began their parade of visits to the Lincoln home in the summer of 1860. Mary Lincoln convinced the census enumerator to record her age as thirty-five, although she was forty at the time, in what was the beginning of lifelong dissem-bling about her correct age.

Mary recognized that her husband had become a national figure, and she must host and charm those who made their way to inspect him and the Lincoln household. She was particularly attentive to members of the Eastern establishment. Her husband was put on display, and several portraits were made of him that summer, few of which pleased his critical wife. Mrs. Lincoln knew that she and her boys were also on display, and tried to manage accordingly.

The president of the Republican convention, George Ashmun, who headed the delegation that came to formally offer Lincoln the nomination, wrote favorably about his visit to the candidate's home, complimenting its "solid substance" rather than "showy display." Another visitor observed "a colored man" setting out a banquet table, as well as a sideboard loaded with decanters of brandy and a basket of champagne—which he worried might offend prohibitionists. But Mary intended to wine and dine members of the press, politicians, and important visitors. One night in June, she hosted nearly two dozen party leaders at her dining table. She projected the image of a gracious hostess who disdained ostentation.

As a result, a recurring theme ran through the published findings of many visitors that summer, including John Scripps, editor of the *Chicago Tribune*, who would write Lincoln's campaign biography: The Lincolns were not the country hicks the Eastern establishment might expect. The *New York Herald* reporter suggested the tasteful décor of the house's interior was courtesy of Lincoln's wife, "who is really an amiable and accomplished lady." A later report in the *Herald* emphasized the resemblance of the Lincoln house to Longfellow's residence in Cambridge, Massachusetts. And a Pennsylvania writer found that the family dwelling wore a "Quaker tint of light brown." These reports were meant to reassure voters along the sophisticated urban seaboard that they would not be electing a wild Westerner. Lincoln's move from the capital of Illinois to the capital of the nation, authors argued, did not represent as wide a gulf as detractors might paint. Further, as *Frank Leslie's Illustrated Newspaper* reported, the main house had a roof like a Swiss cottage, and was surrounded by a fence and several outbuildings. The message was clear: Lincoln was the master of an appropriate domain, and might be a fit leader for the nation.

Mary's grand illusions, Springfield would have to recognize, were paying off, with so many great men coming to pay court during the summer of the campaign. Even though Norman Judd had once dashed her husband's hopes for higher office, Mary wrote solicitous letters to Mrs. Judd that summer to maintain cordial relations with the man helping her husband get elected. (Once her husband did win, however, she would work against him behind the scenes.)

She kept the children clean and orderly, and tried to manage her public rooms, which sometimes resembled the waiting room at a train depot. She wrote lively letters and struggled to keep up her husband's spirits. On June 12, the Lincolns were both saddened when their Springfield in-laws, Ann and Clark Smith, lost their son to typhoid fever. This ten-year-old boy was named Lincoln, just as the Lincolns' eight-year-old Willie was named after one of Mary's other Springfield brothers-in-law. A family funeral in the middle of jubilant politicking was a sobering reminder that kept Mary both prayerful and grateful for her own family's good fortune.

She longed especially to see her son Robert, who had been sent away to school in the East and had been gone for nearly a year. She confessed to being "*wild* to see him." It had been a great disappointment when Robert went to take his qualifying exams for Harvard in August 1859 but had failed miserably. When a friend's son, George Latham, also failed his exams, Lincoln wrote sympathetically to the boy: "In your temporary failure there is no evidence that you may not yet be a better scholar, and a more successful man in the great struggle of life, than many others, who have entered college more easily." He was endorsing the school of hard knocks as well as suggesting that he try, try again. This subtext and message conveyed much, and was perhaps a comfort to Robert as well. Lincoln's oldest son was not sent back home, but was shipped off to Phillips Academy in Exeter, New Hampshire. He hoped he might be able to head for Harvard the following year if he could pass his exams the next spring.

His father had visited him in New Hampshire during his trip east to offer what came to be known as his Cooper Union address. Lincoln was pleased to see Bob doing so well, and conveyed this on the home front. Later that spring, Robert let his family know that Harvard would admit him unconditionally, as his hard work at Exeter had paid off.

Mary longed to see her older son and congratulate him as he realized not just his own dream, but his family's highest hopes. Although his father's nomination by the Republican Party gave Robert some celebrity when he entered Harvard, many students took no notice, in that the college had a long tradition of admitting the sons of famous men. Robert himself made fun of his status as the son of a presidential candidate when questions about his illustrious father became part of his freshman hazing. But in the weeks before the election, all but Robert were frazzled by the limbo into which this campaign had spun them.

Mary especially was weary, confessing to her close friend Hannah Shearer in the fall of 1860 that she didn't know how she could bear defeat, presumably sharing her doubts with a confidante and trying not to express such anxieties to her husband. She tried to behave coolly about the upcoming contest, remarking that she hoped his election would give her the opportunity to visit her old friend in Philadelphia. She refused to let energies flag in pursuit of this goal, and less than a week before the election, she wrote to a minister who was head of an anti-Masonic group, passionately avowing that her husband was definitely *not*, and *had never been* a Mason. She knew election night would be a long ordeal, but tried to convey to her husband and his supporters that she felt the best man must win—and that the best man was decidedly her husband.

8.

Hope That All Will Yet Be Well

My friends—No one, not in my situation, can appreciate my feeling of sadness at this parting. To this place, and the kindness of these people, I owe everything. Here I have lived a quarter of a century, and have passed from a youth to an old man. Here my children have been born, and one is buried. I now leave, not knowing when, or whether ever I may return, with a task before me greater than that which rested upon Washington. Without the assistance of that Divine Being, who ever attended him, I cannot succeed. With that assistance I cannot fail . . . let us confidently hope that all will yet be well. To His care commending you, as I hope in your prayers you will commend me, I bid you an affectionate farewell,

—ABRAHAM LINCOLN, *February 11, 1861*

As the train pulled out from the Springfield station on February 11, 1861, one day short of his fifty-second birthday, President-elect Abraham Lincoln was finally headed for the White House. Mary Lincoln had to stay behind for an extra day to finish up details, but she would join him for this triumphant journey east within twenty-four hours. She found the realization of her dreams overwhelming, moving from junior partner in her husband's political career to organizing her move to the District of Columbia. Yet his election, a Lincoln friend observed, was the most gratifying moment of Mary's life.

She could feel exalted by this victory, having struggled against daunting odds to attain such a pinnacle. All her hopes and dreams were pinned on her husband, and she intended to ride his coattails into glory, and to keep her sons on the towline as well. Despite this complex victory at the polls, the battle was far from over. With the rough-and-tumble of national political intrigue, there would be more challenges ahead, of that Mary was sure. To her discredit, at this moment of supreme triumph after such a long siege, Mary was unable to rise to the occasion—to gracefully relinquish her place at the table. She desperately clung to her role as consigliere. Mrs. Lincoln could not rest easy with others who failed to recognize her primary and vital role as her husband's sounding board. Her reluctance to allow others more intimate and constant access, to feel herself losing more and more control with each passing week, if not each passing day—this problem became exacerbated during her time in Washington.

Mary, Willie, and Tad in 1860

When Lincoln, the candidate, found himself president-elect, the terms of his political allegiances shifted as well. The first Republican president, the first man from the prairie to capture this high office, he encountered enormous resistance to his rise. To the Eastern establishment, Lincoln was an interloper who provided a nasty surprise. Mary Lincoln was wounded when she and her family members were seated in the dining room of the Metropolitan Hotel in Manhattan in January 1861 and overheard New Yorkers discussing her husband in disparaging terms. The men loudly questioned: "Could he, with any honor, fill the Presidential Chair? Would his western gaucherie disgrace the Nation?" These insults rattled Mary and hinted at discord ahead.

Lincoln's election was greeted with skepticism, even by his own party leaders, as his political savvy was not widely acknowledged—the "Honest Abe" image prevailed. Additionally, party stalwarts felt his antislavery appeared lukewarm. However, allowing himself to be underestimated was a familiar weapon within the Lincoln arsenal, and one that he would skillfully wield to his advantage during his presidency.

Lincoln's electoral victory took place during a period of seismic shifts, with renewed threats of secession by the South. Congress had labored intensively to hold the center with the Compromise of 1850 and the Kansas-Nebraska Act (1854). Yet slaveholders' aggressive politics and the rising Free-Soil movement continued on a collision course. South Carolina's declaration of secession in December 1860 and the formation of the Confederacy in the early months of 1861 appear inevitable only in retrospect. Lincoln himself had every reason to believe that as the duly elected president, endorsed by the Electoral College (several of whose delegates he would meet during his train ride east), he would take office and rule over the entire nation, not just those states that supported his victory. But, alas, here Lincoln was wrong.

Mary had already been on one long journey to the eastern seaboard the previous month. She traveled from Springfield on January 8, accompanied by her brother-in-law Clark Smith. Ostensibly, she went east to make purchases in anticipation of her needs as wife of the president. She headed for Manhattan and then to Cambridge to collect her son Robert from Harvard, enjoying the notice she garnered along the way.

Her first trip to Manhattan after her husband was elected president elicited a barrage of negative publicity. Mary was chided for her politicking along the way, divulging details about her husband's choices for his cabinet. One Republican complained, "The idea of the President's wife kiting about the country and holding levees at which she indulges in a multitude of silly speeches is looked upon as very shocking." This hinted first at Mary's significant role in Lincoln's political circle and the attacks such significance attracted. It also signaled that her unconventional attachment to politics was viewed as a threat, not an asset, by many of Lincoln's supporters and, indeed, the wider world.

It might have been with some relief to Lincoln's closest advisers that his wife went away, however briefly, during those few weeks between her husband's election and his inauguration. Clearly Mrs. Lincoln had strong ideas about whom her husband should appoint, and voiced her opinion in ways that were not condoned by many contemporaries, nor by later generations of critics.

Her absence from Illinois, however, did not prevent Mary from making her wishes known. On January 17, she wrote a strongly worded letter (marking it CONFIDENTIAL) to David Davis, one of Lincoln's closest advisers, reporting that she had overheard some gossip about Norman Judd, a Republican party leader from Illinois. She did not want Judd in her husband's cabinet and joined forces with Davis to block his appointment. In her letter, she suggested that Judd's rise to the cabinet might harm her husband's sterling

reputation "when honesty in high places is so important." She flatters Davis to use his influence to prevent such an appointment, and concludes with the
obligatory apology that she would not intrude, but
"for the good of the country." In reality, Judd was
Mrs. Lincoln's enemy from the days of Lincoln's
failed Senate bid in 1855. Although Lincoln forgave
Judd, who became a strong Republican ally in the
years to follow, Mary Lincoln did not. In her letter
to Davis, she supplied him with ammunition for their
mutual goal of knocking Judd out of the running.

She lingered longer than she planned on this
eastern sojourn, meandering back through Buffalo,
Cleveland, and other stops along the way. Mary basked
in attention and gifts as she made her way home to
Springfield. She was much pleased by a "Sewing
Machine, mounted in a solid rosewood full case, and
altogether a bijou of an affair, destined as a present
to the wife of the President elect, and to find a location in one of the apartments of the White House. It is
richly silver plated and ornamented with inlaid pearl
and enamel. It is worthy the possession of a duchess, and indeed the very companions of this superb
sewing machine have actually been finished and
sent to the English Duchess of Sutherland, and the
Russian Duchess of Constantine." The bestowing of

gifts, and other gestures, allowed Mary to swaddle herself with the faux trappings of royalty, cushions she learned to enjoy, came to expect, and, at times, demanded. Both the Lincolns were showered with attention and gifts, with the president-elect attracting notable hats.

His political confidants may not have missed her, but Lincoln clearly did: He went to meet incoming trains two days in a row, in vain, before his wife finally arrived back on January 25. He was eager to see his son Robert as well, home triumphant from Harvard.

Mary was only gone a short while, but during those few weeks between her husband's election and his inauguration, Mrs. Lincoln's worldview sharpened. She should have been basking in reflected glory, enjoying the bounty of her husband's new financial security, ready to arrive with great pomp and dignity at her Washington destination. But calm was in short supply.

Mary struggled to change her expectations to meet the challenges ahead. When she returned from her trip east, she found the men overcrowding her parlor still bickering over her husband's administration. Mrs. Lincoln was daunted by the hundred tasks she was required to undertake and by the mixed messages her Springfield community conveyed concerning her ascension. She was burdened by the dawning realization that things on the national political scene were more disturbing, even more dangerous, with each passing week. All was not well at all, she was forced to admit.

The veil was most cruelly torn away from Mary's eyes by the hate-filled mail and packages that began to appear—first seeping into Springfield. How could Mary Lincoln maintain equilibrium when, according to Henry Villard, a few weeks after the election "a scandalous painting on canvas was received by Mrs. Lincoln, expressed from South Carolina. It represented Mr. Lincoln with a rope around his neck, his feet chained and his body adorned with tar and feathers." She, along with others, became increasingly concerned about the safety of her husband, as dire threats accompanied the secession of South Carolina in December 1860, and continued as states from the Deep South followed suit. These seceding states promised to form a new nation and to elect Jefferson Davis, former senator from Mississippi, as president of their Confederacy.

During early 1861, Mrs. Lincoln's mood ranged from giddy to frantic. While political harmony unraveled, she plunged into domestic tasks, believing that social obligations must not be abandoned even as political tensions intensified. As states peeled away one by one and joined the Confederacy, Mrs. Lincoln's mood must have darkened. She knew the consequences this would have on families across the nation, including her own. As families divided, Mrs. Lincoln adopted the stance of loyal wives throughout the nation: "My husband is my country."

Mrs. Lincoln had a timetable as well as an agenda. She knew her family would have to move out of their home, store what wasn't being shipped, and make their way to Washington well in advance of the inauguration festivities in the first week of March. She also knew it would take some of their father's smooth talking to convince the boys to leave behind their dog, Fido, which they left with the Roll family, who had two boys near the ages of Willie and Tad. The first week of February, Mary and her sisters organized a grand celebration for political supporters in Springfield, a levee for seven hundred, to bid farewell.

The family finished packing up for the move a few days after the levee. Sadly for future generations, part of the preparation involved consigning private family documents to the flames of an infamous "burn pile." Many items were designated for storage, entrusted to friends. The Lincolns decamped to the Henry House in downtown Springfield on February 9. Lincoln tied up his own trunks and unassumingly lettered "A.

Lincoln/White House/Washington, D.C." on hotel cards that he pasted on the outside of his luggage.

His advisers had initially objected to the transportation arrangements. Some believed that Mrs. Lincoln, with the three boys, should ride separately from her husband for security reasons. Mary refused and was vindicated when General Winfield Scott wired from Washington that despite rumors of plots to harm him, Lincoln should, for safety's sake, "be surrounded by his family." We do know that the couple did not depart on the same train. Lincoln made his touching "Farewell Springfield" address on February 11, and boarded his own luxurious train car. The railway company décor included an interior of "light colored tapestry carpet" and "richly dark furniture" with sidewalls covered in "crimson plush, while between the windows hung heavy blue silk studded with thirty-four silver stars." On the outside, the presidential car was shiny, with varnished panels of orange, bedecked with flourishes and streamers galore. Flags decorated the engine, and red, white and blue festoons abounded.

The president-elect steamed out of town on his own, but Mary and the boys joined Lincoln in Indianapolis on February 12, his fifty-second birthday. At the time of his election, he was the second-youngest president to take office, older only than Franklin Pierce.

The Lincolns made their way slowly eastward, stopping in Columbus, Ohio, for a party at the home of Governor William Dennison, where the whole family spent the evening being entertained. Some have suggested this Ohio stopover may have prompted the famous feud between Mary Lincoln and Kate Chase, the charismatic daughter of Lincoln's Republican rival, Salmon P. Chase. Rumors would circulate that Mrs. Lincoln was angered by her husband's dancing with the beguiling twenty-year-old beauty at a military ball given for him that night. This kind of gossip would become commonplace. Tales of Mary's unreasonable jealousy abounded during her husband's presidency.

In this case, the story was patently false. Kate Chase was not even in Ohio. Instead, she had already left for Washington: "Mrs. Lincoln was piqued that I did not remain at Columbus to see her," she later explained, "and I have always felt that this was the chief reason why she did not like me in Washington." She might have been disingenuous about her explanation, but at least it shows that the gossip about Mary's pique over her husband's attentions to this younger woman in Ohio was both fictitious and malicious.

While in Cleveland, at a dinner arranged by the president of the railroad, a company of Zouaves (soldiers who imitated certain French military units who wore distinctive uniforms, which might include a fez, sash, or Turkish pantaloons) ceremoniously stood guard. When a gun salute shattered the windows, many diners were startled. Although splinters of glass were sprinkled on her as a result of the accident, Mrs. Lincoln remained uncharacteristically composed.

The next day, the group departed for Buffalo. Pushing out of Ohio, the presidential train continued to be greeted by enthusiastic crowds. At Ashtabula, Ohio, the assembled crowd called for Mrs. Lincoln but were disappointed when she did not appear. Lincoln remarked that "he had always found it very difficult to make her do what she did not want to." A group of nearly five thousand mobbed the Lincolns in Pittsburgh. In Buffalo, former President Millard Fillmore escorted Lincoln to a Unitarian Church, then invited the couple to his home for Sunday dinner. At their next stop, in Albany, the Lincolns dined at the governor's mansion before greeting one thousand at a levee held in the president-elect's honor at Delavan House.

By now, fed and feted amid demanding throngs, Abraham Lincoln recognized the great weight of responsibilities he shouldered by becoming "public property." He had been enthusiastically embraced by crowds along the rail journey—so wildly in Buffalo that one of his guards, Major David Hunter, suffered a sprained arm. When they arrived in New York City on February 19, the jostling continued. A crowd estimated at 250,000 watched the parade of eleven carriages ferrying the Lincoln family and entourage to the Astor House. Dinner with Vice-President-elect Hannibal Hamlin (whom Mrs. Lincoln had not met before), meetings with dignitaries, receptions, serenades, and other festivities kept Lincoln and his family in a whirlwind. Daily bulletins of Lincoln's schedule appeared in the local papers, so everyone knew about his meeting with members of the Electoral College, his encounter with the eminent Brooklyn minister Henry Ward Beecher, and other significant comings and goings.

It is not known how and when Lincoln slipped away to shop, but he apparently did make a visit to Tiffany's. In an extravagant gesture designed to reward his wife for her faith in him, he presented Mary with a magnificent six-piece seed-pearl parure (necklace, earrings, bracelets, and brooch) that cost $530.

As they headed down the coast, the mayor of Trenton met the train, and a procession of one hundred mounted men escorted Lincoln to the New Jersey State House. Mrs. Lincoln was so fatigued from all these exertions that she did not receive callers during their stopover in Philadelphia. The scrutiny was wearying; one reporter wrote that "the entire female population are in ecstasies of curiosity to know who she was, what she is, what she looks like, what her manner is, and if she has a *presence* of the sort necessary in the exalted station to which she will soon be introduced." In addition to the sheer physical stamina required, Mary Lincoln received word of more threats to the president-elect, increasing concern for her husband's safety.

The Lincolns were separated for the final segment of travel to Washington when General Winfield Scott reported that intelligence concerning assassination plots required a change of plan before approaching Baltimore. The military did not want to take any chances and demanded that Lincoln travel alone and in disguise. Moving under cover of night for the last few miles to his destination was a precaution he later deeply regretted.

Mary and her sons arrived in Washington the next day and were escorted to Willard's Hotel, where Lincoln was ensconced. While the president-elect worked to organize his government, Mary Lincoln launched her own campaigns—hosting family and friends, greeting diplomats and statesmen. It was, of course, difficult, coming out of a completely different milieu. The coldness and snobbery of Easterners were as legendary as the friendliness and familiarity of Westerners.

Even before she moved into her new home, Mrs. Lincoln was exposed to one of the most peculiar and idiosyncratic of American institutions—Washington society. Those at the heart of the city's beau monde—the toughened corps of social arbiters—were known as "cave dwellers." Their tenure and tenacity gave them sway over the parade of newcomers straggling into the city at regulated intervals. This inner circle of D.C. society was surrounded by the "moneybags," those whose rung on the ladder was bought, and by the "highbrows," whose station was secured by talent regardless of wealth—although it was considered felicitous when the two went hand in hand.

Three outer rings of the socially powerful applied steady pressure, jockeying for improved positions: the diplomatics, the army-and-navy crowd, and the politicals. But clearly it was the cave dwellers, particularly women like Mary Anne Clemmer and Laura Holloway, who influenced the pecking order.

While the city had some legitimate claims of social gentility, the physical attributes of the nation's capital did not recommend it to visitors. Noah Brooks described the streets as "canals of liquid mud." John Hay concurred: "It would be difficult to conceive of a meaner street in architectural adornments than Pennsylvania Avenue."

There were, of course, areas of the city that boasted palatial homes like the mansion built by Senator William Gwin from California (who spent $75,000 to furnish his new home) and the fine estate of Senator Stephen Douglas near the corner of I Street and New Jersey Avenue.

Retiring President James Buchanan supplemented the White House entertainment budget with his own personal funds, as he needed more than his salary to keep up with the demands of his position. The Buchanan White House had been run with great efficiency, with ten servants to take care of household needs: The butler was Belgian, but all the other staff were from Ireland, England, and Wales because Buchanan felt that these British-trained servants would best suit his needs. Harriet Lane, James Buchanan's niece and White House hostess, had wangled $23,000 for new furnishings for the mansion, which included a circular divan (which remained in the White House into the twentieth century) and new chandeliers for the state dining room.

In 1857, when the extension of the Treasury building forced the removal of a greenhouse from the grounds, Buchanan authorized a conservatory, with a passage connected to the White House. This expensive addition was completed at a cost of $16,000 (more than the original appropriation of $12,000). The Buchanan White House staff left the Lincolns a detailed list of protocol on the running of the Executive Mansion. Harriet Lane met with Mrs. Lincoln in advance and arranged a meal for the Lincolns following the inauguration. But she was not impressed and wrote cattily that Lincoln resembled the Irish doorkeeper, Thomas Burns, and reported, "Mrs. Lincoln is awfully *western*, loud & unrefined."

People were open in their contempt toward the Lincolns, and especially toward this uncouth frontier usurper who had come east to take his seat at the table. Class and status anxieties abounded. British journalist William Russell observed on a train ride that one man made "very coarse jokes about Abe Lincoln and Negro wenches, which nothing but extreme party passion and bad taste could tolerate."

Arriving in a town with rigid social hierarchy and intense snobbery, Mrs. Lincoln decided that a top priority for her would be finding a dressmaker. Within twenty-four hours of her husband's taking the oath, Mary was placing her dressmaking order with Elizabeth Keckly, a prominent mixed-race seamstress favored by the Washington elite, with former clients like Varina Davis, the new first lady of the Confederacy. Securing Keckly's services resolved one major problem for the anxious First Lady: trying to negotiate her way up the slippery slope of fashion within the capital.

On his last day in office, March 4, 1861, Buchanan stopped signing bills at the Capitol around noon, before riding in a barouche to pick up Lincoln. The two men, president and president-elect, were thronged by crowds along their route to the Capitol. They arrived together for the ceremony, where Chief Justice Roger Taney administered the oath of office to Lincoln. Then the new president was escorted to the White House where well-wishers by the thousands arrived in hopes of shaking Lincoln's hand.

The invitational inaugural ball was held in a large tent dubbed the White Muslin Palace of Aladdin, where five thousand were on hand to inspect and rub shoulders with the First Couple and their entourage. Lincoln's secretary complained that the White House was overcrowded with Mary's visiting gaggle of Todd and Edwards relations, which included two half sisters, Mrs. Clement B. White (Martha Todd) of Selma, Alabama, and Mrs. Charles Kellogg (Margaret Todd) of Cincinnati, Ohio. Mary Lincoln glided into view wearing blue silk, bedecked with pearls, gold, and diamonds. She dazzled the crowd and wanted to savor the feeling. While Lincoln left at midnight, his wife stayed on—dancing polkas and schottisches into the night.

Despite the boycott of the event by secession partisans, Mrs. Lincoln surprised Washington snobs to become the belle of the ball. Elizabeth Ellet, an influential cave dweller, commented on Mary Lincoln's "exquisite toilet," complimenting her "admirable ease and grace." The *New York Herald* weighed in: "She is more self-possessed than Lincoln and has accommodated more readily than her taller half to the exalted station to which she has been so strangely advanced from the simple social life of the little inland capital of Illinois." She wore the pearls her husband had bought her at Tiffany's, which became such a hit that copies were made by a Washington jeweler for purchase in his store.

But like the proverbial Cinderella after the ball, she had wicked stepsisters with whom to contend, in some cases *literally* with her Confederate kin. Most elite Southern women did not attend the inaugural festivities, turning their backs on Mary Lincoln.

Once her husband was installed as president, Mary struggled to establish a domestic routine for her family while learning the ropes of her new town. The Lincoln family could only find privacy on the second floor of the Executive Mansion in the nine-room family suite. Abraham Lincoln had seen the White House kept under lockdown during Polk's administration, while he was a Congressman, and did not want to risk a closed-door policy. As a result, his expansive welcome mat put a considerable strain on his family, his staff, and him.

1861 inaugural fashions

Visiting women relatives turned into a godsend when both of Mary's younger sons contracted measles during March and were confined to sickbeds. Her cousin Elizabeth Grimsley was drafted into nursing duties. Tad and Willie's illness brought Lincoln some relief; he would wander into the sickroom, take a cup of tea, and read aloud from the Bible or recite some poems, enjoying the respite until "recalled to the cares of state by the messenger."

During these earliest weeks in Washington, Mary was pleased that, after their recovery from measles, her sons became inseparable from two boys who lived nearby, Bud (Horatio) and Holly (Halsey) Taft. Along with their older sister Julia, they were frequent visitors to one another's homes. Once, the Tafts' father noted, "The Lincoln boys have dined with us every day this week." The hospitality was joyfully repaid. Mrs. Lincoln was pleased that her boys were so happily situated, and she warmed to Julia as a substitute for those absent younger women whose company she missed.

As the wife of the president, Mary threw herself into the required social obligations and held her first levee. Custom required that the White House levees, which anyone might attend, be hosted weekly. In the Lincolns' case, they were held on Tuesday evenings.

The great parade of humanity that appeared was so colorful that William Stoddard, one of Lincoln's secretaries, recommended that an attendee would "see more varieties in one evening than Broadway would furnish in a week."

The Lincolns' first White House public reception, on March 8, 1861, was described by one paper as a "monster gathering" and a great success. Wearing her magenta watered-silk dress and a wreath of red and white japonicas in her hair, Mrs. Lincoln looked regal. A visitor described the occasion with humor and delight: "We are rewarded with a propulsive movement in the rear, which nearly precipitates our party of five into Abraham's bosom. Our ladies blush with shame and indignation; but promptly recover their self-possession, they are introduced to 'Old Abe' who shakes their hands cordially, smiles graciously, addresses them familiarly, and we pass on to Mrs. Lincoln, who, nearer the center of the room, maintains her position with the steadiness of one of the Imperial Guard."

Mrs. Lincoln's steadiness was even more remarkable because she was thrown into the marathon of exhausting, emotionally draining ceremonies where every aspect of her appearance was critically examined. She knew observers were watching every move and gesture she made. This time, which was allegedly for celebration, was in fact a baptism by fire for Washington's newest hostess.

Lincoln shook hands with the parade of well-wishers for over two hours as hundreds turned out, including Charles Sumner, who graced the White House for the first time in several years. (Sumner, an impassioned champion of abolitionism, had been the victim of a brutal attack on the Senate floor by a proslavery Congressman, and had boycotted the Buchanan White House.) By the end of the evening, Mary's cousin Elizabeth Grimsley admitted, "We must confess to a sigh of relief when we heard the marine Band strike up 'Yankee Doodle,' the signal for retiring." This was the last event in the Lincoln White House where Southern and Northern representatives from Congress would mingle freely.

Mary was required to welcome rounds of guests to the president's home, and she conducted weekly at-homes between two and four on Saturday afternoons. Despite all this company, White House protocol hampered her by limiting her access to the drawing rooms of Washington's elite.

Mrs. Lincoln's houseguests were issued social invitations to which Mary was not privy. Ironically, invitations were not extended because the wife of the president was forbidden by protocol from accepting them. She could only learn about William Seward's glittering dinner parties and Mrs. Charles Eames's Sunday salon from her relations, not from her own experience. All month long, she kept messengers busy delivering fresh-picked bouquets from the White House conservatory, with compliments and invitations attached. This bounty of flowers, however, did not bring about the desired results.

From the beginning, letters to Mary had been screened. In this way, she was protected from most of the hate mail flowing into the White House. She was particularly prone to hysterics where threats were concerned. Lincoln's secretaries were forced to dissemble to calm Mary's frequently aroused fears, most especially about her husband.

Mary's voracious interest in politics continued to pit her against her husband's phalanx of advisers. She was caught between the proverbial rock and hard place when she attempted her husband's policy of reconciliation early on. In some ways, this policy was not difficult at the outset, since she had Southern relatives living in the region, such as Appelline Alexander of Versailles, Kentucky, who married politician Frank P. Blair, who would become a Union general. Lincoln's wife was hailed by the couple as Cousin Mary, but this familiarity with Southern sympathizers rankled many staunch abolitionists. Also, when Mrs. Lincoln honored the wife of Stephen A. Douglas and other wives of Democratic leaders by having them on her receiving line, she "gave great offense to many Republicans." Slights and insults abounded in response.

Mrs. Lincoln was brought up to be a proper Southern lady, with the impeccable manners of her region and culture. When she tried to demonstrate hospitality, however, her actions were constantly being challenged, her motives misconstrued. Her attempts to be gracious and cordial, her desire to bring people together, were thwarted by angry critics. This vilification only increased over time, as her detractors hammered away at any effort to follow her husband's example—to be conciliatory toward those pulling away. While Lincoln would later be applauded for his mercy toward his enemies and his "malice toward none," his wife suffered stinging attacks for her charitableness.

Mary Lincoln muddled onward, wading into the swamps of Washington's social scene. She hosted her first state dinner at the White House on March 28, where Kate Chase called even more attention to her beauty and wit by being one of the few women in attendance. The paucity of women at White House social gatherings continued. One journalist reported that during a reception on March 30, "there were only two or three ladies in the drawing room when I arrived." Women were snubbing the new First Lady.

By April, the political cold shoulder was offered to the new president, as well. Republicans were flocking to town in droves, but the powerbrokers were slow to roll out the welcome wagon. Even Yankee newspapers took savage aim at Lincoln's administration. An antiabolitionist journal, New York's *Evening Day-Book*, slammed the newcomers: "The Great Black Republican 'Wigwam,' in which Abe Lincoln was nominated for President, was to have been sold at auction on Saturday last. The country under his administration is rapidly 'going' the same way."

There was a seismic social shift in the District of Columbia. Virginia families who had ruled Washington society for decades disdainfully withdrew, and many Marylanders demonstrated their loyalty to the secessionist cause as well. Whether it was a political act or snobbery, the Lincolns were being treated like pariahs by mainline Washington elites, and as one observer complained, "Both the President and his wife were mercilessly lampooned, and yet Mrs. Lincoln was the peer of any woman in Washington in education and character."

Mary might have likened herself to a bird in a gilded cage, denied the social butterfly role to which she long aspired. But the cage was not exactly *gilded*. Her visitors were shocked by the shabby, run-down condition of the president's residence. The furnishings in the Red Room, which the Lincolns used for private callers, had pieces left over from the Madison era, nearly forty years before. There were only ten matching place settings in the White House china collection. Her Springfield friends thought the place resembled a second-rate hotel with threadbare carpets and chopped-up drapes—during the midcentury period, many citizens would cut off a piece of the curtains as a souvenir. But should she have postponed her elaborate plans to redecorate?

Mary was determined to set a high standard and prove her refinement to the Washington cave dwellers. Her increasing isolation might have hastened her plans. London journalist William Howard Russell discovered that, even after a month, "The Washington ladies have not yet made up their minds that Mrs. Lincoln is the fashion. They miss their Southern friends, and constantly draw comparisons between them and the vulgar Yankee women and men who are now in power."

Mary decided she would have to make a splash to prove herself. In April, she looked forward to some respite from social pressures, because she had given her last levee before the summer. She wrote to a friend that once Congress recessed, "the crowd will be gradually leaving the city, and henceforth, we may hope, for more leisure."

But events intervened. Following the attack on Fort Sumter on April 12 and Lincoln's call to arms on April 15, her new home became the nerve center for a divided nation. She opened her drawing rooms to soldiers, as recruits marched into the East Room, where, "under the gorgeous gas chandeliers, they disposed themselves in picturesque bivouac on the brilliant patterned velvet carpet." A remarkable vortex of events kept the Lincoln White House both under the microscope and within dangerous crosshairs.

The military advised Lincoln to evacuate his family, but Mary stubbornly refused. This resolve appeared a good one for the First Lady's public relations, but not all of her decisions had such happy outcomes. The declaration of war launched a particularly stormy series of crises for Mary Lincoln, setting her on an unfortunate course.

As debates over the war began to heat up in the public rooms of the Lincoln White House, gloom invaded the family's private domain. Mrs. Lincoln was forced to cajole her husband from his spells of melancholy. During this period, it was his wife, as one of Lincoln's friends recalled, who "was the only one who had the skill and tact to shorten their duration."

She struggled against the odds to keep her husband and family on an even keel. Trying to enforce leisure was an uphill battle. Lincoln's secretary explained that the family were "beginning with an effort to keep Sunday . . . they are trying to be private citizens once a week, but the circumstances are against them." Mr. Lincoln accompanied his wife to services at the New York Presbyterian Church, and their sons enrolled in the Fourth Presbyterian Sunday School, where they might attend with friends, the Taft family. Tad and Willie might not have been happy with sitting for long hours in services, although Willie was much more devout than his younger brother, Tad, who became restless and challenged the need to attend. However, they both enjoyed the egg-rolling on the White House lawn, which took place on Easter Monday, April 1, 1861.

The Lincoln boys took their baskets full of hard-boiled eggs, gaily colored, to roll down the hill, competing with other children. Tad vowed to get a cast-iron egg so he might win the race the following year. The Lincolns attempted to keep their boys diverted from the more serious matters swirling around. But even Mrs. Lincoln's talents could not withstand the pressures brought to bear as the war approached. As William Stoddard would later remark, "War has no Sunday, no day of rest, no hour that is sacred above the others."

The threat of armed conflict set up a dramatic hullabaloo. Washington received daily bulletins of impending disasters: The navy yards of New York went unprotected; troops in Virginia might march to capture Washington; the rails from Annapolis to the District had been torn up by Rebel sympathizers. Soldiers in Jim Lane's Kansan Frontier Guards filled up the East Room of the White House, while Cassius M. Clay's Home Battalion billeted at Willard's Hotel. Troops summoned from farther north moved too slowly for frayed nerves in Washington. The White House was turned into a fortress, as Washington girded itself for attack.

A Massachusetts regiment passed through Baltimore and was waylaid by a mob on April 19. Scores were injured and a handful killed. News of battles to the north of Washington caused a panic, and many deserted the city. Ominously, post and telegraph communication with the North was cut off. The District became a ghost town as public buildings closed and hotels shuttered. At the same time, prices shot up as beef and salt were hoarded in expectation of a flood of troops.

Alarmists redoubled their efforts to safeguard the president and his family. But military strategists were stymied: How could they set up any protective circle around the District, to hold off an attack on this city built on a swamp with dozens of routes of access, when no troops were available and Southern states surrounded it? The tension as the city awaited the arrival of troops—enemies from the South or reinforcements from the North—was unbearable. Citizens turned out in droves, with relief and elation, when New York's Seventh Regiment marched into town. One member of this elite corps described the occasion, as "[some] what [as] Mr. Caesar Augustus must have felt when he had crossed the Rubicon." They paraded in front of the White House and shouted their huzzahs for the First Family, and their cheering embrace was returned in full. More regiments followed, and by April 27, Mary Lincoln reassured a Springfield friend that there were thousands of soldiers nearby, so she felt secure and was hoping for peace.

Perhaps Mrs. Lincoln, like her husband, cloaked her anxieties with bravado. She was pleased when Abraham offered one of her Kentucky brothers-in-law a commission. They were both extremely fond of Ben Hardin Helm, son of a former Kentucky governor, who had married Mary's favorite half sister, Emilie.

Lincoln thought of him as a younger brother, and pressed him to become paymaster in the U.S. Army—with the rank of major, quite a plum for a thirty-year-old. Mary especially urged him to return to a position in Washington so she and Emilie might be reunited. But when the Lincolns bade him farewell on April 27, it would be the last time they saw Helm alive: He declined Lincoln's offer and died fighting for the Confederacy.

Kentucky's divided loyalties caused Lincoln and his wife deep angst. The president's concerns were not sentimental but strategic, as the loss of this border state would have a domino effect. If Kentucky seceded, then "we cannot hold Missouri, nor Maryland." A friend reported, "The President would like to have God on his side—he must have Kentucky."

A favorite native son from Illinois, E. Elmer Ellsworth, proved more eager to capitalize on Lincoln's patronage. Ellsworth was only seventeen when he left his Midwest home for Chicago—where he trained volunteer military companies and rose to the rank of major by the age of twenty. He was an accomplished fencer and skilled rifleman. Ellsworth met Lincoln during his visit to the state capital in December 1859, coming to town at the invitation of the Springfield Greys to prepare the militia for an exhibition drill.

Lincoln urged the charismatic young man to move to Springfield to study law, and Ellsworth relocated in September 1860. He was a favorite in the office, although not particularly gifted in his law studies. Yet Elizabeth Grimsley described him as "a magnetic, brilliant young fellow, overflowing with dash and spirit." With Lincoln elected, Ellsworth's stock rose.

Ellsworth became part of the honor guard accompanying Lincoln on the train eastward. One day after his inauguration, Lincoln wrote the secretary of war to urge him to find Ellsworth a permanent military appointment. Two weeks later, he sent a strongly worded follow-up letter requesting that Ellsworth be given the post of adjutant and inspector general of militia for the United States (complete with an office, clerk, messenger, and a raise in pay grade). When war erupted less than a month later, Lincoln issued his call to arms and penned a note to his protégé, explaining that he wanted to find him the "best position in the military" possible, without incurring the wrath of "older officers of the army." Ellsworth was flattered by Lincoln's unflagging support, and raised a regiment of Zouaves, pledging undying loyalty to Lincoln and the Union.

By the time May rolled around, Washington was reeling from wartime's quickening pace. Besides band concerts and drilling militias and forced gaiety, the town was coming to grips with the onset of armed conflict. Treason and death were on the horizon. Thousands of family members, like Mary and most of her Todd kin, were walled apart with secession's geographical divide. Many stood alienated at this ideological crossroads, fearing the abyss. Even if family members were united in purpose, an abyss still beckoned, as households were forced into tearful good-byes when men went marching off to war—and unknown fates, with women left behind.

In spite of all this gloom, business was booming in the District—and war fever ran high. Liquor stores and brothels flourished. Other fevers erupted, with overflowing gutters and fetid alleyways on the rise. The *Evening Star* asked, "Are we to have pestilence among us?"

Mary Lincoln decided to take an excursion. She was not trying to elude fever (or she would have taken her children along), but she did hope to escape the fishbowl. One journalist recalled, "If she but drives down Pennsylvania Avenue, the electric wire trills the news to every hamlet in the Union."

Again, this was a misstep, as her May 1861 trip north only increased unwanted attention. The newsmen hounded Mrs. Lincoln on her journey. It was a difficult time for the president and the nation. Many called for the resignation of the Cabinet, and demanded that the commander in chief step down as well. One political wag said that "those who wish to set a house on fire begin with the thatch." So Mary Lincoln became an easy target.

Elizabeth Grimsley, Mary's companion on this infamous junket, had no idea of the trouble brewing when they set out northward on May 10. She said that they intended to stop over in Manhattan and pick out a carriage on the way to visit Robert Lincoln at Harvard. Yet they became the target of celebrity gawking at a variety of places: in church, at Laura Keene's theater, at a local hotel.

She was shadowed at every turn: "Mrs. Lincoln . . . busily engaged in 'shopping' the greater part of yesterday [May 13]. In the morning she inspected a number of carriages at Brewster's manufactory, and later in the day visited Stewart's and other dry-goods stores, purchasing quite extensively. . . ." The ladies later enjoyed a soiree hosted by New York merchant Alexander Stewart, also chronicled in the paper. While in Boston, Senator Charles Sumner arranged for Mary's entertainment by local luminaries. Any meetings with dignitaries were overshadowed by attention paid to Mrs. Lincoln's wardrobe in press reports.

At home, Mr. Lincoln, also, garnered notice, and not just for his famous call to arms on May 15. Any move a Lincoln made was considered good copy by the newspapers, even sitting on the White House piazza to listen to the Marine Band or attending the wedding of a friend's son.

Mary Lincoln scheduled her trip to New York after Congress had appropriated $20,000 for White House refurbishing. This might seem like a princely sum for the period, but not only was it inadequate to the task, it was paltry compared to other Congressional appropriations for District expenditures at the time. It was also considerably less than the $125,000 Andrew Johnson's family would be given for redecoration a little over four years later.

The new Congress designated generous amounts to renovate District public parks and statuary, and to update heating and cooling systems for the federal government's ten dilapidated buildings. It authorized the purchase of artwork for the Capitol, with its unfinished dome. Indeed one single painting bought to hang in the Capitol building cost nearly as much as the entire sum designated for the White House renovation. Thus economy and finesse would be necessary to complete the formidable task of turning the White House into a showplace. Mary Lincoln believed she would be well served to allow the commissioner of buildings for the District, William S. Wood, to advise her. He volunteered to accompany the First Lady to New York City.

William Wood was a former hotel manager and tour organizer who had been drafted by New York party boss Thurlow Weed to assist the Lincolns on their trip from Springfield. During this time together, he became an intermediary for patrons seeking influence. For example, he presented Mrs. Lincoln with a pair of horses, anonymously donated. His combination of shameless flattery and smooth manner had ingratiated him to the president-elect's wife. Trusting him contributed to Mrs. Lincoln's drubbing by the press, which worsened with Wood's eventual downfall. She championed him for office during the spring of 1861, but when later investigated concerning misconduct, Wood was forced to resign. Naturally, critics painted his relationship with the First Lady as darkly as possible. Gossips intimated that their connection had been both illicit and sexual—a rumor that was vindictive and maintained lethal longevity.

Nevertheless, Wood proved himself a cad. Instead of safeguarding Mrs. Lincoln during her excursion to Manhattan as First Lady, he continued to wheedle for a permanent appointment, peddling his influence and pocketing kickbacks. In May 1861, one particularly outraged New York reporter voiced his disdain, "She is expending thousands and thousands of dollars for articles of luxurious taste in the household way that it would be very preposterous for her to use out in her rural home in Illinois." The fact that she was the First Lady refurbishing the tattered and decaying symbol of the nation, and the fact that her own Illinois home was irrelevant, did not lessen the sting of this attack. Lizzie Grimsley was adamant in her cousin's defense that Mary "did not indulge in one hundredth part of the extravagance with which she and I were credited."

Mary might even be commended for eventually selecting for state occasions a fine Haviland dinner service with purple-and-gilt decoration and with the seal of the United States on each piece. Historian Jean Baker suggested that the entwining of a gold border with two lines is "signifying the Union of North and South." However, what is remembered is neither Mrs. Lincoln's taste nor her patriotism, but that she ordered a second set, at a cost of $1,100, with her own initials emblazoned on this china. Grimsley insisted that this "was not paid for by the district commissioner, as was most unkindly charged." Although a payment of $1,106.73 was withdrawn in November from Lincoln's account at the Riggs & Co. Bank (presumably to cover the purchase billed in September), scholars have suggested that the Lincolns did not pay the cost themselves. Was this an honest mistake? A clerical oversight? A deliberate tactic introduced by Wood? Or Mary Lincoln's willing complicity? Grimsley complained that "Mr. Lincoln was not in this respect 'worldly wise.' "

Mary Lincoln may have been compromising ethical standards, standards to which her husband subscribed. Certainly she was overspending her budget and desperate to find ways to cut costs. However, there is no evidence to suggest she participated in a criminal conspiracy.

This debate will continue, as it has from the first days when merchants' invoices were submitted to Secretary of Treasury Salmon Chase, and rumors began to circulate—about her extravagance, and about irregularities. Chase apparently regaled his daughter with tales of Mrs. Lincoln's avarice and overspending. And "scandalous reports" about Mrs. Lincoln (and Lincoln himself) reportedly "emanated from the Treasury Department" and its "special agents." Newspaper stories full of allegations fueled the controversy, and gossips had a field day. Such rumors, which erupted during her first weeks in Washington, dogged Mrs. Lincoln long after she left public life.

Again, the custom of presenting personal gifts to the president was long established. Mary could have turned a blind eye to the fact that favors might indicate repayment, and that the bills—literally as well as figuratively—would come due. She was aware of her own financial limitations. While a New York reporter accompanied her wardrobe shopping, she told him that she could not afford the cost of some luxury goods in cashmere and lace with which merchants tempted her, and she was "determined to be very economical." But when luxury items ended up in her personal trunks, when she appeared in a parade of new fashions, an avalanche of bad press followed. One New York reporter fumed that Mrs. Lincoln's china would match "the mulberry colored livery of her footmen." This attack rings shrill, and arguably false.

Such a suggestion—that Mrs. Lincoln bought china to match servants' livery—proves even more mean-spirited when contemporary descriptions of the Lincoln White House reveal quite the opposite. A diplomat's wife noted: "There were before the White House no sentinels, not even a porter . . . no crowd of gorgeous liveried footmen was to be seen." Prince Napoleon was taken aback when he arrived at the White House on an official visit and there was no one to greet him. So the idea that Mary Lincoln would pick out china to match liveried footmen was not just fanciful, but purposely misleading. These libels were published and circulated even before she arrived back in Washington from her trip in May. The First Lady found herself the center of attention, but not the renown of which she had dreamed.

9.

Dashed Hopes

So much of promised usefulness to one's country, and of bright hopes for one's self and friends, have rarely been so suddenly dashed as in his fall.
—ABRAHAM LINCOLN *on Col. Elmer Ellsworth*

Mary Lincoln swept back into town in May 1861 and found Washington on tenterhooks. She eagerly anticipated the arrival of her purchases for her new home: the beautiful carpets, mantel ornaments, and other furnishings ordered for the Executive Mansion. But her shopping expedition sparked controversy, even though Mary Lincoln seemed unwilling to smell the smoke.

The president's wife was released from her years of longing for a perch at the top of the social ladder. But in Washington, she found herself by turns frustrated and flattered by all the attention her every move garnered. After yearning for recognition of her husband's potential, his sudden elevation was such gratification that Mary Lincoln was unwilling to delay. During her first few weeks in the White House, she charged full steam ahead, letting her critics be damned.

From childhood onward, she had endured long harangues about the interference of the federal government and the rights of slaveholders. There had been duels aplenty among her father's political cronies over just such issues. While this ruckus was being resolved, why shouldn't she make the White House a home of which to be proud? Why not make it a proper setting to reflect her proud Bluegrass heritage? Mrs. Lincoln might have sensed the easy target she made, but she dismissed the press corps as some kind of pesky cloud following her around. That a horsefly could bring down a horse was folk wisdom that would have appealed to the president, but it was ignored by his wife, entranced with her new status.

Back at the White House, Mary Lincoln delighted in being reunited with Willie and Tad. Think of what a luxury it was for her to *miss* her children—and then to see them again, knowing they were taken care of, protected, and having the time of their lives. Before her husband's election, only illness would excuse Mrs. Lincoln from the domestic duties associated with home and family. Once her son Robert was born nine months after the wedding, the marathon of maternal duties began. All of the menial aspects of housekeeping had weighed her down, prior to the hiring (and firing and rehiring) of servants. Regardless of household help, Mary Lincoln had spent nearly two decades as a devoted homemaker, and her primary responsibilities were to maintain the household for her growing family.

For decades, her sister Elizabeth had enjoyed the comforts of wealth and a prestigious social position. As a young bride, she moved in with her father-in-law at the governor's mansion to play hostess when she first arrived in Springfield. She then resided in a large home on a hill, mixing with the aristocracy and looking down on most of Springfield from this lofty position. Sisters Frances and Ann made equally comfortable matches, while Mary's husband entered into marriage saddled with family debt.

Mary (who bore her first child in a boardinghouse) got her first home only through her father's largesse. On her first visit to Washington when Lincoln was in Congress, she had to shuffle off with the children when things did not work out for the family in a Capitol

Hill boardinghouse. Finally, she became the proud resident of the most famous home in America. After so many years of domestic scarcity, imagine how she enjoyed this exalted status.

It was perhaps particularly pleasurable for Mary that she was able to travel with a Todd cousin, to enjoy an agenda of which she had only dreamed while marooned in Springfield. This early flush of elation allowed her to float in bubble of near oblivion as the nation inched closer and closer toward military Armageddon.

She found that her new duties in the White House were more ceremonial than onerous. For example, the First Couple reviewed the troops during a dress parade of the Seventh New York Regiment on May 23. Mary discovered that, during her absence, Springfield's own Colonel Ellsworth and his New York Zouaves had taken the city by storm. Ellsworth had journeyed east with the Lincolns, even moving into the White House (sharing a room with Robert Lincoln before he returned to college), and had been given a commission with the War Department. In April, Ellsworth resigned his commission to raise recruits in New York for a regiment. He administered the oath to his volunteers en masse on May 7. They were the first group to sign up with a "pledge not

for thirty days, not for sixty days, but for the war."
His men wore colorful uniforms and drilled with
jubilation—passionately committed to their youthful
commander. The Zouaves won even more plaudits following heroic rescues during a fire near the Willard
Hotel on May 15. Their pluck and patriotism buoyed
the citizenry during what became, with each passing
day, a sense of fearful uncertainty. Mary had faith
that her husband could address the current crisis,
but many seasoned politicians, especially skeptical
Washingtonians, feared what lay ahead.

All of Washington was in an anticipatory panic:
What would happen with this new Confederacy? How
could Lincoln hold the center while the Deep South
slipped away? What might happen along the border
states? But most of all, when, and especially where,
would the Confederacy attack? How could Washington
protect itself from rebel marauders? Ellsworth wanted
to keep morale high, and seized every opportunity to
prove his bravado.

For weeks, Washingtonians seethed at the sight
of a Confederate battle flag unfurled over a tavern in
Alexandria—across the river, but within striking distance of the White House. Until Virginia officially
took a position, this flapping in plain sight would have
to be tolerated. The Old Dominion had not formally

broken with the Union, so Virginia farmers crossed the Potomac to bring goods to District markets—as things remained business as usual. But all this would change when the state voted to join the Confederacy on May 23. This date sealed Virginia's fate, and the Union stood prepared to strike.

Ellsworth's men were poised to invade and cut telegraph lines to capture the town. Shortly after dawn on May 24, the colonel moved into Virginia with his men. He led a small contingent directly to Marshall's Hotel, where the offending flag still waved from on high.

Colonel Ellsworth in 1861

Ellsworth dashed up three flights and ripped down the Stars and Bars. As he came charging down the stairs, captured flag in hand, he confronted the hotel owner, James T. Jackson—who blasted him with a double-barreled shotgun. Ellsworth immediately slumped to the ground. His fellow Zouave, Captain Brownell, returned fire—shooting Jackson in the face—and then stabbed him with a bayonet. Ellsworth's chest wound proved fatal; he could not be revived and died at the hotel. Although telegraph wires were cut, news of Ellsworth's death spread like brushfire. The tolling bells in firehouses confirmed the rumor of his demise. This first casualty sank the capital into a miasma of despair.

Lincoln was with visitors in the library of the White House when future Assistant Secretary of the Navy Gustavus Fox (who became a Lincoln favorite), arrived with news of Ellsworth's death. The president found himself overwhelmed by his emotions and unable to speak. Mary and her husband went to the Navy Yard in the afternoon to view the body and pay their respects. Lincoln also returned later, to mourn alone beside the corpse. The president's gloom was shared by so many who knew Ellsworth.

Ellsworth was too close to Lincoln's oldest son's age for comfort. He was the flower of youth that Lincoln now realized would be crushed under the boot heel of war. This was difficult to absorb in the abstract, but tragic to experience with the loss of someone as close as Ellsworth had been to the Lincolns. Like some youthful knight, he had paid homage to the Republican White House, representing the best it had to offer, lifting spirits with his youthful bravado. His early and violent end signaled the horrors to come, and provided the First Family with a symbol of the war's human costs.

The next day, Ellsworth's body was moved to the East Room of the White House, where thousands filed by his coffin as he lay in state. The Ellsworth funeral on May 25 signaled the Lincolns' first personal contact with wartime loss. Willie and Tad were somber, and the family was comforted by the arrival of their oldest son, Robert, home from Boston on May 27.

Mrs. Lincoln was deeply touched by this death. She was presented the Confederate flag, the very one that led directly to Ellsworth's death. She also requested that Julia Taft, the sister of her sons' favorite playmates, learn to play "Colonel Ellsworth's Funeral March," composed by Septimus Winner, on the piano. Although Julia was a shy performer and did not like recitals, she confessed, "When I played [it] for Mrs. Lincoln, she would stand beside me to the last note, turning the leaves of my music. Somehow I never minded playing for her."

But the "first blood shed on secession soil" was even more personally grievous to Mary Lincoln, because the man who shot Ellsworth was the brother of Dr. John Jackson, a physician in Mary's Kentucky hometown. She was one of the first to appreciate the double mourning the war would impose. Mary Lincoln deeply regretted the loss of Ellsworth and proudly supported his cause. But she could also empathize with the loss of a beloved brother by the Jacksons of Lexington, the grief that might engulf them. This very particular intersection of loss and pain caused Mary Lincoln dread. These fatalities signaled more deaths to come—deaths that would divide not just homes and families, but drive a wedge into the soul of the nation.

Yet mourning did not suspend the continuous demands of life in the White House. Mary rose to varied occasions: the Lincolns hosted a levee on the May 28 for dignitaries, and the Marine Band began to play concerts on the estate grounds when they were opened every Wednesday and Saturday to the public. National grief intervened, once again, when Stephen Douglas died on June 3. The president closed government offices for the senator's funeral on June 7.

Mary found her duties multiplying exponentially during the onset of summer. On June 27, during one among the seemingly endless round of military reviews, an army colonel broke a bottle of champagne over a carriage-wheel to christen as "Camp Mary" the tented field where his men were bivouacked, to honor the First Lady.

Mrs. Lincoln earned praise for her continued devotion to the military men stationed in Washington—first in her own home, then in camps, and soon in hospitals. She requisitioned guns from the War Department to donate to efforts in her home state, writing to a Union colonel in Kentucky, "Please accept, sir, these weapons as a token of the love I shall never cease to cherish for my mother State . . . and the confidence I feel in the ultimate loyalty of her people, who, while never forgetting the homage which their beloved State may justly claim, still remember the higher and grander allegiance due to our common country." Mary angrily invoked the sacrifice of her ancestors and the loyalty of her fellow Kentuckians. She wanted to hearten her countrywomen and remind them that she was loyal to the Union above all.

On June 21, while visiting soldiers' encampments, the pole of her carriage broke, and after the driver was thrown from the vehicle, only the quick wit of soldiers saved Elizabeth Grimsley, Mary Lincoln, and her two young sons from injury. The *New York Times* reported that members of the New York Twenty-fifth Regiment witnessed the runaway carriage and came to the rescue, and that "Mrs. L. clung to her youngest boy and leaped to safety."

Despite occasional complimentary snippets from the press, the vast majority of journalists were content to repeat gossip and print partisan reports about the First Lady. One Washington insider commented that "the women kind are giving Mrs. Lincoln the cold shoulder in the City & subsequently we Republicans ought to Rally."

During this period, Mary Lincoln begged Hannah Shearer, her Springfield friend who now lived near Philadelphia, to come for a prolonged stay at the White House. She had seen her during the train ride to the inauguration, but she missed her good friend, and wanted to be reunited. Mrs. Shearer was pregnant, and not feeling up to it. Thus Mary changed her plans and invited Hannah to join their party later in the summer for an extended excursion to Long Branch, New Jersey. She promised rest and quiet, as she had been offered accommodations at the best shoreline hotels, with sea bathing available. Mary even proffered her guest railroad passes, noting that "the trip will cost nothing, which is a good deal to us all these times." She closed the invitation, "If you love me, give me a favorable answer. I have set my heart having you with me."

While Mary tried to sort out excursions, the city was awash with intrigue. Rose Greenhow, the widow of a Virginia politician and a leading Washington society hostess, continued to fill her parlor with a mix of guests. When war was declared, Greenhow's decidedly Southern sympathies made her suspect to the new administration. She entertained politicians from both sides, and her glittering social gatherings allowed her to collect intelligence and tease out valuable information to pass to Confederate couriers. Greenhow would later be arrested for spying and confined to the Old Capitol Prison.

Amid Fourth of July celebrations and reviews of marching militia, Lincoln sent Congress a message of war. Troops were being funded, war was being waged, and the confiscation of Confederate property was authorized, as restoration of the Union by force was well under way.

As troops collected outside Manassas Junction, Virginia, residents in the capital enthused over the prospect of upcoming battle. Every carriage, wagon, and hack available for hire had been secured in advance. Caterers were raising prices to keep up with requests for hampers. Civilians traveled to the battlefield in high spirits with picnic blankets, eager to watch as events headed toward a reckoning on the afternoon of July 21. Sadly for the Washington sightseers, the battle turned into a Union disaster, as 32,500 Confederate soldiers under the command of General P. G. T. Beauregard sustained less than 2,000 casualties, while the over 35,000 Union troops suffered the loss of nearly 3,000. The Rebels routed the Union invaders and drove them into retreat—and this Great Skedaddle was the first major military loss of Lincoln's generals.

The first reports back to Washington were wildly contradictory, but soon the War Department confirmed the worst. General Winfield Scott appeared at the White House at two a.m. on the morning of July 22, insisting that Mrs. Lincoln and her sons evacuate Washington to make their way north to safety. As bedraggled, defeated soldiers tramped back into town, reality set in.

The horrors of war were again reinforced when the Lincolns began the first of what was to be a regular parade of hospital visits—together and separately over the next few years. In the wake of this disaster, Mrs. Lincoln insisted on staying put, and delayed her trip to the seaside until the middle of August. She wrote to Hannah Shearer that she wanted to meet her soon, because "I have passed through so much excitement, that a change is absolutely necessary."

Ostensibly, Mary Lincoln was finally willing to leave the city because of undesirable weather, but also because workers would begin renovations on the White House. She hired John Alexander, of the dry-goods emporium located three blocks away along Pennsylvania Avenue, to oversee the renovations and redecorating. Lincoln apparently approved of this activity; he had insisted that work on the Capitol continue, believing it would suggest to the viewing public "that the Union will go on." Although the president might have approved of repainting and refurbishing, he certainly expected his wife would keep to the budget Congress appropriated—even if Mrs. Lincoln had other ideas.

Lincoln was less inconvenienced than he might have been, because Mrs. Lincoln had taken Buchanan's advice and visited the Soldiers' Home, across the district line in Maryland, which had become a summer White House. President Buchanan had furnished Anderson Cottage with castoff White House furniture and moved to this residence from July 5 until cool weather prevailed in the fall. On March 6, Mary Lincoln had inspected the property—surveying her domain. She recognized that this would be a suitable place to escape the dust and disorder, the fumes and crates that would descend on the White House in her absence.

Under her direction, the White House was given a complete exterior paint job, and the interior was scheduled for extensive renovation—with new carpeting, canvas floorcloths, and straw matting installed throughout. She concentrated on the public rooms and the bedroom reserved for guests. The most impressive of the redecorated rooms was for the guest quarters, the "Prince of Wales Room." Most striking was "the Lincoln bed," as it has come to be known:

> the head surmounted by a gilded half coronet emblazoned with the American shield. The coronet was suspended from high on the walls . . . from it hung curtains of purple satin trimmed with yellow-gold fringe over long, full panels of glistening gold lace drawn back with cord and tassels. Bolster and spread were of figured satin in purple and gold; deep purple and gold fringe formed a valance along the sides of the bed.

Mrs. Lincoln also installed pale purple wallpaper, and gold curtains with purple fringe.

Mary Lincoln delayed her vacation to host visiting royalty, staging an elaborate state dinner for Prince Napoléon III on August 3. She was annoyed that Secretary of State William Seward, a thorny rival whom she had not been able to unseat, had suggested that *he* might host the formal reception for Prince Plon Plon, as the French royal was nicknamed. But Mrs. Lincoln's will prevailed.

Mary Lincoln organized what she hoped was a grand fete when the French emperor graced her home. Her cousin Elizabeth Grimsley was the only other female in attendance, and "the French tongue predominated." This grand occasion boasted members of the Cabinet, the son of George Sand, Senator Charles Sumner, and Generals McClellan and Scott as distinguished guests. General Scott confided to Lincoln at the close of the evening, "I have dined with every President since Jefferson, and in my mind, the last should be first." The press was equally complimentary.

Mary was flush with victory and celebrated by borrowing the complete works of Victor Hugo (in French) from the Library of Congress. When the bills from the dinner were submitted to Seward at the State Department, the secretary claimed that as the prince was traveling as "a private citizen" and not in any official capacity, the cost of the extravagant evening must be borne by the president, not the nation. Perhaps not even Seward's machinations spoiled Mary's memories of triumph.

Bills came due in other ways as well. Commissioner William Wood was not working out, although he was reappointed on August 12. Despite Mary's initial enthusiasm for Wood, she became convinced he was a liability and needed to be replaced. She and her husband interviewed Benjamin Brown French to take over Wood's duties. The replacement appointment would not be made official until September, but having this suitable candidate in place allowed Mary Lincoln to leave the city, with renovation activities in full swing.

Elizabeth Grimsley was more than ready to pack up and go to Springfield. Both homesickness and money worries pulled her homeward to Illinois. By August, the White House social whirl became tiring even to Mary, who floated at its center. Both summer weather and the war took their toll on the First Family, especially on Mr. Lincoln. His wife tried to counter the negative effects of stress and overwork by demanding a regimen of regular meals and fresh air. Lincoln's Cabinet as well as his assistants began to require more and more from their leader—the Tycoon, as he was dubbed by private secretaries John Nicolay and John Hay.

The tug of war between family and work was intensified during these weeks of tension and political dramas. While reports of riots and battles, defections and demands piled up in the executive inbox, Mary Lincoln and her sons stubbornly resisted the boundaries imposed.

Lincoln found himself more and more hemmed in. Fears for his safety increased the number of doormen (most in civilian clothes with concealed weapons) at the White House, including members of the newly formed Metropolitan Police. Uniformed sentries ringed the grounds. Within the mansion, although halls and corridors were crowded with those seeking appointments, Lincoln was often closed off from the crowd, taking the service stair more often than the grand staircase or business stairs. This back stairs had its own separate hall and guard, allowing the president access to any room in the house, plus a means of egress through the basement.

Upstairs at the White House, the Lincolns turned the Oval Room into a family parlor and library, where they might read in the evening, or where Mary might wait for her husband to join her as she did some sewing. The Lincolns slept in adjoining bedrooms (west of the library), with the children's sleeping room and a guest bedroom across the hall. Modern convenience had come to the White House in 1861 when Buchanan had installed running water and marble washstands in all the upstairs rooms but the library.

Mary Lincoln found it more and more difficult to try to maintain a healthy schedule for her husband. She expected him to accede to her wishes, as he had pretended to in Springfield, but she found concession a thing of the past. For example, during early months of the first administration, "Mrs. Lincoln instituted the daily drive, and insisted upon it, *as her right* [emphasis added] that he should accompany her." Many have suggested that Mary Lincoln was selfish and demanding, and she might well have been. But in this case, charges of her pettiness and indulgence are unfair.

Mary insisted that her husband accompany her on a carriage excursion daily. Yes, she was stealing him from the nation, but she was also forcing him to take time out from his hectic schedule. She might insist that he listen to her everyday news and cares—but these domestic and family details would allow relief from the official business that ground him down.

Clearly, her husband needed to conserve his strength, to learn to pace himself—especially in light of previous patterns involving his overextending himself. In order to prevent collapse, Lincoln needed diversion from the woes of war. This was crucial with the weight of the Union on his shoulders. Lincoln never conceived of his own comfort and health as a priority during perilous times.

Mrs. Lincoln had to insist on excursions, and, as her cousin confided, "this was the only way in which she could induce him to take the fresh air which he so much needed." That her requests became both nagging and annoying was legendary among friends and staff. Lincoln and his wife often displayed a dance of wills. Once when first the butler and then one of their sons tried unsuccessfully to escort the president from his study to the dinner table, the First Lady finally appeared:

Lincoln, without the slightest show of displeasure . . . quietly arose and passed across the room to Mrs. Lincoln. . . . As Mr. Lincoln approached her she partly turned as if intending to leave the room with her husband. On reaching her side, Lincoln took hold of both Mrs. Lincoln's arms just above the elbows and slowly moved forward, gently pushing her before him, until she was through the door. He there released her arms, and stepping back, closed the door between them and, locking it, quietly returned to his seat at the table without a word of reference to the incident . . . the episode was ended.

The two played out these domestic dramas regularly. Mary had an equally legendary knack for being able to draw her husband out of his infamous blue periods. For example, Lincoln was frequently called for meetings in the early morning, holed up in the Cabinet room where Mary would have to send along his coffee. Knowing he would sometimes be so busy that he might forget to eat, Mrs. Lincoln might invite old friends to have a meal with them, and then request that her husband come greet their guests. Once again, this might be interpreted as interference with weighty government matters, but Mary viewed her tactics as necessity. Mrs. Grimsley described the "tonic" of guests at breakfast: "Mr. Lincoln would come in looking so sad and harassed, seat himself, and with a bare nod of recognition, saying, 'Mother I do not think I ought to have come.' . . . Presently Mr. Lincoln's mouth would relax, his eye brighten, and his whole face lighten, as only those who have seen the transformation would believe." Even Mary's detractors would have to admit: "If left to himself, when he was very busy he would skip several meals and not be aware of it."

Indeed, the notion that Lincoln couldn't dress himself was perpetuated by the sloppiness of his appearance. His casual indifference toward his wardrobe became notorious, as cronies noted that they could tell when Mrs. Lincoln was away by the "disorder in Lincoln's apparel." Benjamin French called on the White House and spotted four men, including Lincoln, but suggested that no one could have guessed the president was among them, as he "was dressed in gray woolen clothing, and had upon his head a most ordinary broad-brimmed slouch." However, French indicated, Lincoln moved with an identifiable and "peculiar gait."

After nearly twenty years of marriage, Mary clearly understood her husband's haphazard style, and though it might still madden her, one close friend believed "they understood each others peculiarities." Lincoln provided a constant challenge, as Mary sought to enforce regularity to lessen his burdens.

Lincoln weathered his wife's absences, especially when she took the children away for a refreshing holiday by the sea. And meanwhile the White House was being invaded by a horde of workers: upholsterers, painters, and others charged with renovations. Lincoln vowed during August 1861 that he would not abandon the capital while the country was at war, which was a promise he kept in spirit if not to the letter, with only a handful of excursions during his nearly fifty months in office.

Mary was eager to escape the heat and take the children to the seaside. So she picked up Hannah Shearer in Philadelphia on the way to the Jersey shore, where son Robert would join his mother and brothers. John Hay was delegated to escort the party of women and children. They stopped off in Manhattan for a courtesy visit with Princess Clothilde, Prince Napoleon's wife, who had remained on shipboard in New York harbor rather than venture to Washington in August.

Mary Lincoln's large party made for a welcome addition to Long Branch, the New Jersey resort where they all converged: "Mrs. Lincoln's presence here at the Branch, whatever may be its ultimate influence, has had the immediate effect of crowding the hotels to suffocation." Hannah Shearer was extremely ill during her first few days at the shore, which resulted in Mary's personal and constant attention. She was very solicitous of her pregnant friend, who had pledged to name her next child William, after Mary's son, should the baby be a boy.

Although balls given in her honor were diverting, Mrs. Lincoln's nursing duties continued when Tad caught cold, which delayed the group's departure for upstate New York until August 26. By then, Hay had deserted the party for duty and confided, "I impressed her [Mary Lincoln] with the belief that the delay of another day would break the blockade, recognize Davisdom, impeach the Cabinet and lose the Capitol. Like a Roman matron she sacrificed her feelings to save the Republic." His departure meant Mary Lincoln would have to fend for herself, among the swarm of journalists shadowing the First Lady.

Mary stopped off on her way to Niagara Falls to pay a call on Frances Seward—surprising in light of her generally cool attitude toward the Secretary of State. But Seward actually accompanied Mary and her boys from Albany to his home in Auburn. On the train ride back home to Washington, Mrs. Lincoln spent time shopping in New York, where she purchased new china—190 pieces for the presidential mansion— at a cost of $3,195. No doubt the lack of proper place settings for the dinner for Prince Napoleon added to the urgency of this particular purchase.

But again the purchases prompted unwanted comment that was relentless and unreasonable. The *Chicago Tribune* editorialized near the end of Mrs. Lincoln's summer tour:

> . . . if Mrs. L. were a prizefighter, a foreign dan-
> seuse, or a condemned convict on the way to

execution, she could not have been treated more indecently than she is by a portion of the New York Press. . . . The whole editorial corps of the New York *Herald* pounced upon her like buzzards . . . they pursued her from the blue room to the boudoir. They divided into squads and platoons, part of them alighting upon her milliners, part on her flunkies and part on her hostler, while the main pack flapped their unseemly wings around Mrs. L herself from morning to night. . . . No lady of the White House has ever been so maltreated by the public press.

The paper scolded, "We implore the tribe of water toadies to let the wife of the President alone." Sadly, they did not, as persistent and pervasive attacks continued.

Perhaps the onslaught of bad press caused Lincoln to suggest, just days after her return home, that Mrs. Lincoln prompted the release of a man sentenced to be shot for falling asleep on duty: "Mr. Lincoln came this morning to ask me to pardon a man that I had ordered to be shot, suggesting that I could give as a reason that it was by request of the Lady President."

During this difficult time of press battles and real battles, Mrs. Lincoln struggled to keep her new house afloat, firmly tethered to the principle of family first, focused on the comfort of her husband and sons. Her devotion to her children was notorious. The Lincolns' lax discipline was a hallmark—as Mrs. Lincoln was fond of saying, "Let the children have a good time." Yet both the Lincolns also loved learning and introduced their sons to the world of knowledge. Mrs. Lincoln wanted her boys to have the best possible education, and had already sent Bob off to prep school and then to Harvard.

With all the schools in the District shut down, and with her little Tad barely literate, Mrs. Lincoln decided not to ship the boys away or separate them, but rather to improvise. She drafted the Taft boys as schoolmates, ordered chalkboards, and hired a tutor, a Scotsman named Alexander Williamson. Williamson was detailed to a paid government job for half his workweek, and employed the other half as a White House schoolmaster.

Such bureaucratic shuffles were quite common in the White House—as the work mounted for the president's secretaries, their salary was defrayed from the Ministry of the Interior. Soon all the male White House staffers were given military appointments and put on a military payroll. At the same time, any of these shifts to benefit the First Lady's White House budget were viewed as suspect and provoked smear campaigns against Mary Lincoln's "crimes."

Despite every effort to regain some semblance of normalcy, the White House was not an easy place for the Lincoln family to maintain equilibrium. Just as the boys were getting used to the loss of their hero, Colonel Ellsworth, another White House favorite died tragically—this time on the field of battle. Colonel Edward Baker was an old friend of the Lincolns from Springfield, a popular politician who was sent to Congress in 1844 and succeeded by Lincoln. Baker was so esteemed a friend that the Lincolns named their second child after him—Edward Baker Lincoln, born in 1846, who died at the age of three. Edward Baker migrated to California in 1852, and next was elected senator from Oregon. He was a staunch supporter of Lincoln and enlisted in the Union army shortly after the call to arms. Baker was killed in action at Ball's Bluff on October 21, 1861.

The day before his death, Baker had breakfast at the White House. An officer passing the Executive Mansion spied him and the president in a pastoral moment: "Mr. Lincoln sat on the ground leaning against a tree, Colonel Baker was lying prone on the ground his head supported by his clasped hands." Baker shook Lincoln's hand when he bade farewell, and gave young Willie, playing nearby, a farewell kiss.

Baker's funeral on October 24 was another emotional family ordeal. Mrs. Lincoln wore a dress in a shade of purple instead of the traditional black. This Victorian fashion of wearing purple in mourning was not universally recognized, and allegedly offended some ladies, which caused quite a stir. Mrs. Lincoln was embroiled in layers of scandal; more than just a tiff over dress code threatened to engulf her.

While the work of refurbishing continued, Mary Lincoln was forced to deal with escalating political fallout. Meanwhile, there were servant problems to address, as both the British usher and the French-speaking butler were replaced. In the fall of 1861, she sent home Ellen, the maid she had brought with her from Springfield. Instead, Mary Ann Cuthbert, a former seamstress and lady's maid, became the executive housekeeper in the mansion, and Mrs. Lincoln began to rely more heavily on the services of seamstress Elizabeth Keckly.

The Lincolns welcomed Benjamin Brown French as the new commissioner of public buildings, but William S. Wood, smarting from his dismissal, did not go quietly. Mary wrote to Secretary of War Simon Cameron on September 12 to thank him for appointing John Watt to the cavalry and asking that he be detailed to the White House. She warned that Wood had made "false charges" against Watt, and asked Cameron to "settle Major Watt's business this morning." In an even more candid letter to French, Mary complained about his predecessor, "He is either deranged or drinking," but Wood continued to circulate nasty tidbits. The Lincolns paid French a visit on September 13 to discuss his upcoming confirmation testimony in front of a Congressional committee.

During this period, Mrs. Lincoln was supportive of Watt. She volunteered her opinion—"I know him to be a Union man"—to the chairman of the House committee investigating loyalty within government departments. She went on to discredit Wood as an "unprincipled man" whom the president only renominated to save Wood's family from disgrace, with the promise that he would resign.

Her defense of Watt might have been a genuine belief that he was wronged on this question of political loyalty. But detractors have suggested that her spirited support of him was bribery to ensure that he would be a staunch ally—if not an accomplice—in her campaign to squeeze every available dollar from the government to fund her White House projects. The gathering clouds of this scandal would continue into the fall and new year.

Watt would finally get the ax when a portion of the president's annual State of the Union message appeared in the *New York Herald* in December, before its delivery to Congress. Mary Lincoln was implicated in this tempest, and her most vituperative critics suggested she was having an affair with Henry Wikoff, a well-known man-about-town, known as "the Chevalier Wikoff."

Wikoff was charming, sophisticated, and sycophantic. He had served as a spy and confidant in European capitals, working at different times for both Lord Palmerston and the Bonaparte family before settling in Washington, D.C. The chevalier's campaign of ingratiation bewitched the vulnerable First Lady. When Henry Wikoff was arrested on suspicion of spying in the White House, Mary Lincoln discounted these claims of his criminality. And after he was released from jail, he continued to worm his way into Mrs. Lincoln's parlor. Lincoln finally banished the chevalier from the Executive Mansion, and William Watt left as well.

The suggestion that Mrs. Lincoln and the chevalier were involved in an affair seems a contrived falsehood. Allegedly this dalliance led her to hand over Lincoln's private papers to Wikoff. Charges of treason and adultery were part and parcel of the smear tactics employed to bring down the controversial Mrs. Lincoln.

British journalist William Russell confided to his diary in early November, "The poor lady is loyal as steel to her family and to Lincoln the First; but she is accessible to the influence of flattery, and has permitted her society to be infested by men who would not be received in any respectable private house in NY."

Tongues were wagging when Mrs. Lincoln held her first reception of the winter season on December 7, but in the spirit that the show must go on, she held her head high and won the admiration of many. One of her severest critics, writer Mary Clemmer Ames, nearly always found fault with the Lincoln White House, but was forced to praise Mary at the sight of the newly carpeted East Room, because the "ground was of pale green, and in effect looked as if [the] ocean, in gleaming and transparent waves were tossing roses at your feet."

She strove to please with a transformed White House, hoping that after her redecoration, visitors might not recognize the place. Yet still the carping continued. As a journalist noted, "The ladies in Washington delight to hear or to invent small scandals connected with the White House."

Benjamin Brown French became Mary's confidant while the refurbishment budget crisis bubbled during the closing weeks of 1861. She summoned French to the White House on December 14, where during an interview she begged him to use his influence with her husband—to get the president to approve bills to cover cost overruns. She also asked him not to let Lincoln know about their conversation, essentially luring him into her web. He recorded the conversation in his confidential diary, but when French met with the president later that day, things did not go well.

First, Lincoln had just returned from listening to eulogies of his old friend Senator Baker, and was emotionally drained after confronting the question of the war's "costs"—human and otherwise. Second, he was acutely embarrassed at the ammunition this kind of fiscal miscalculation would give his enemies, and exclaimed, "It would stink in the land to have it said that an appropriation of $20,000 for furnishing the house had been overrun by the President when the poor freezing soldiers could not have blankets, & [he] swore he would never approve the bills for flub dubs for that damned old house!" Lincoln went on to suggest that he would prefer to pay for these expenses himself, rather than to ask Congress for more.

Apparently, Mrs. Lincoln's plots had some effect, as Congress appropriated $6,000 more in January to cover costs—which Lincoln approved. Even with these additions, the White House budget was hardly out of proportion, as Congress had already doled out thousands for upgrading public buildings in the District. The initial White House renovation appropriation was equal to the commission for a single mural for the Capitol building.

So for a short time, French and Mrs. Lincoln became a team, and during the Christmas season of 1861, he reflected, "She is an admirable woman. She bears herself in every particular like a lady and say what they may about her I will defend her." And indeed, critics would continue to say what they would about her, to such a degree that the editor of the *New York Herald* complained in late October about the abuse heaped upon her, and Mary replied with thanks and to extend an invitation to him and his wife to visit the White House. But Mary would remain at the center of controversy.

Watt was accused of submitting false bills for flowers and manure—which would clearly have given rise to rude nicknames for this White House scandal. By October, the secretary of the interior confided that the president was being dragged into this dilemma. Watt argued that it was Wood who had made the corrupt bargain and that "he assured Mrs. L [Lincoln] that the transaction was right & legal and that she had no idea that anything was done which was not authorized by law."

Mrs. Lincoln tried to ride out these storms, and she was able to bring her husband's closest friend from early Springfield days, Joshua Speed, and Speed's wife, to the White House for a traditional Thanksgiving celebration on November 28. The beautiful meal was supplemented by champagne at each place setting. A visitor noted that a glass was by the president's plate, but he did not partake. Whenever friends sent a stock of liquor to the White House, which became a Republican custom, Mrs. Lincoln would send bottles to be distributed at the hospitals. Her generosity largely went unheralded, but her fans continued to champion her.

Christmas at the White House in 1861 was a joyful affair despite the looming concerns of the Southern rebellion. Her renovations had turned the public spaces of the Executive Mansion into show rooms, while she had fitted out the private rooms of the household simply. Mrs. Lincoln might fondly celebrate the holiday with her three sons and husband gathered round the family fire. But, once again, Washington was visited by scenes of horror when a fire broke out in government stables, and hundreds of horses were let loose into the city streets, giving the town a very apocalyptic look. Little did the Lincolns and their fellow Washingtonians know that this drama during the 1861 holiday season would pale in comparison to future tragedies.

10.

Grand Designs Gone Awry

When Mrs. Lincoln rang in the New Year with her family after her first Christmas in the White House, she must have experienced a stir of emotions. Despite obstacles, Mary took delight in Christmas holidays with her menfolk gathered round—the family circle reconnecting. Her sense of accomplishment had grown with each passing month: She had managed to bring the house up to an admirable standard and had found allies to help her with pet projects. Bob came home from Harvard, and Willie and Tad enjoyed the Executive Mansion as both schoolroom and playground.

Lincoln's work multiplied exponentially, especially after the nation erupted into war in April. He was perched atop an enormous pyramid, and it was Mary's job to try to reduce his vertigo. His wife's ability to distract and exert influence over his moods waned, however. Mrs. Lincoln was increasingly supplanted as her husband's domestic adviser and sounding board, elbowed aside by both his official Cabinet and a retinue of young, ambitious men on staff.

She was disappointed to be elevated to the role of chatelaine of a house that was considerably de trop. She was equally distressed to have her every move scrutinized by a pack of hungry journalists, with domestic dramas magnified a hundredfold on the national stage.

As First Housewife, Mary was enmeshed in the everyday workings of the White House. She had sent her serving girl Ellen back to Springfield and hired Mary Ann Cuthbert as her chief housekeeper. When Mrs. Cuthbert replaced Ellen, Mrs. Lincoln engaged Elizabeth Keckly as her personal seamstress. She hired two other African-American women, Rosetta Wells and Hannah Brooks, to do her "plain sewing."

By 1862, only four staff members remained from Buchanan's administration: John McManus and Thomas Stackpole, doormen, John Watt, the gardener, and Edward Burke, the coachman. The large staff consisted of chambermaids, valets, cooks, and couriers, the majority of them African Americans. As the daughter of a slaveholder, Mary was used to the pres-

The First Lady in a Brady portrait

ence of blacks in the household, finding such circumstances both familiar and favorable. Meanwhile, the
onset of war and the issue of slavery's future hung in
the balance.

Conditions in the capital were difficult for all, but
especially bleak for refugees. A Washington society
woman described desolate conditions, with outbreaks
of smallpox, typhoid, and measles increasing worries,
especially among "the poor neglected contrabands
[the name given to runaway or captured slaves]."

African Americans within the District viewed employment for the First Family as both a sign of arrival and a means of entrée. If it remains true for whites today, then why not for blacks of an earlier era? As one historian confided, "The colored people who worked in the White House were considered the cream of Washington's colored society," and he goes on to explain, "Nearly all were members of the 15th Street Presbyterian Church, one of the leading colored churches of the day." Thus the ebony elite had its own institutions and strong networks, side by side with the whipping post and auction block, true throughout most slaveholding urban centers, such as Charleston and New Orleans.

African Americans had been a familiar sight in the White House since the opening years of the century. Some presidents had even brought slaves to Washington with them—a practice unofficially suspended after John Tyler's tenure. Tyler had brought James Hambleton Christian with him to the Executive Mansion to serve in the White House from 1841. Christian later ran away after his mistress died, and he shared some unkind confidences about his former master: "Tyler was a very cross man, and treated the servants very cruelly; but the house servants were treated much better, owing to their having belonged to his wife, who protected them from persecution." Christian also revealed that he was born his master's son, which would have made him a half brother of the First Lady—to whom he had been given upon her marriage to John Tyler.

The free blacks serving in the White House perhaps had the best vantage point to observe the racial views of the First Family. Clearly, the Lincoln boys brought with them the prejudices instilled by life in a prairie town in mid-nineteenth-century America. Yet the African Americans who worked for the Lincolns, at times living in the same household, assessed the two boys—especially Tad—as relatively color blind when it came to playmates.

The black children of the White House servants were not shunted off into dark corners but were part of the gang who roamed the halls, enjoying their children's perspective on the rambling Executive Mansion. They might have joined in on the pony rides, or sat enthralled during plays put on upstairs by the Lincoln boys, or generally enjoyed the glee. Yet segregation was also observed, as none of these free African-American children were invited to join the makeshift schoolroom.

William Slade, a light-skinned black employee (described in a biographical sketch as "a Virginian of distinguished southern ancestry"), played a key role in the running of the White House. Slade was not just the keeper of the keys, but took charge of supervising all the black workers in the Executive Mansion, as well as the arrangements for all public and private functions involving food and service. It was no accident that Hannah Brooks, a cousin of Slade's, secured a position sewing for Mrs. Lincoln.

William Slade knew all the significant callers—the diplomats and statesmen, the military figures, and those whom the president might summon to his office. Slade would spend many evenings alone with Lincoln, at times listening to him talk into the wee hours of the morning or, if necessary, summoning others to come to the president's office in the middle of the night. Slade's role was confidential and increasingly essential to Lincoln.

Slade and his family lived on Massachusetts Avenue NW, between Fourth and Fifth streets. His three children, Katherine (Nibbie), Andrew, and Jessie were often invited to play at the White House. They became such close and constant playmates that Mr. Lincoln sometimes took Tad to *their* house for an afternoon.

Mrs. Lincoln clearly found herself at ease among the blacks on the White House staff, and reminiscences about her from these former employees offer a distinct and favorable perspective. Rosetta Wells recalled that the President's wife "had her ways, but nobody minded her, for she would never hurt a flea and her bark was worse than her bite." Another former servant confided that "Mrs. Lincoln could put on her apron and cook as well as Cornelia [Cornelia Mitchell, White House cook] herself." This intimate insight stands in stark contrast to any portrait of her as the haughty and domineering matron painted by detractors.

From her years in Lexington, Mrs. Lincoln was well versed in the politics of color within the black community—she knew about the snobbery of skin color that had permeated the African-American slave quarters in Kentucky. But she and her husband were disturbed by events that developed when they tried to add William Johnson to the black staff at the White House. Johnson was the only African-American servant who had been brought to Washington from Springfield. The Lincolns intended that William would shave Mr. Lincoln, and serve as a handyman and messenger for them. However, his dark color set him apart from the other African Americans in the White House, and this created a rift.

Indeed, William began looking for a new place as early as March 1861. Lincoln penned notes, seeking an alternate position for him, and he finally landed a spot in the Treasury Department on November 30, 1861, with a salary of $600 per year. William Johnson remained close to the Lincolns, as he was linked with their home town. At one point, Lincoln collected William's pay for him, to tend to his bills when he was down with smallpox.

The black with perhaps the most influence in the Lincoln White House was Elizabeth Keckly, who had moved from a role as modiste to confidante within a few short months. Keckly rented rooms from Walker Lewis, a leading Washington caterer. She lived in and maintained her dressmaking establishment on Twelfth Street, but she also became a frequent visitor to the White House.

After Mary's confidante Elizabeth Grimsley departed, pointedly no Todd relative was willing to take her place. Hannah Shearer was preoccupied with her new baby, William Lincoln Shearer. Mrs. Lincoln found herself feeling considerably alone—and turned to Keckly for companionship.

Elizabeth Keckly was a remarkable woman who made her mark among Washington society ladies shortly after her arrival in October 1860. She was born the same year as Mary Lincoln, in 1818, and they both had slaveholding fathers. But this overlap—and their personal ambitions—were all they shared from their

Southern upbringing, because while Mary's mother had also been from a prominent white aristocratic clan, Elizabeth (Lizzie) was born the daughter of an enslaved woman. Her mother had been forced to serve as the concubine of a white master.

When she was not yet twenty, Lizzie herself was coerced into a sexual liaison with a white man in her Hillsborough, North Carolina, community. This union produced her only child, a son, who became the focus of her life—with her passionate obsession for emancipation for herself and her child. As an industrious young woman, Elizabeth eventually became a seamstress in St. Louis, and in 1855 was finally able to buy her own freedom and that of her son, George.

Through the loyalty of her patrons, she was able to move to Washington in 1860 and establish her reputation there as an invaluable couturiere. She was so highly recommended that Mrs. Lincoln summoned her to Willard's Hotel on the morning of the inauguration, and then Mary asked her to appear at the White House the next day.

Naturally, Elizabeth Keckly's hopes faltered when she arrived at the family quarters to find three rival seamstresses cooling their heels as well. But when she made her way into the interview room, Mrs. Lincoln eagerly engaged her services and promised, "If you do not charge too much, I shall be able to give you all my work." This was quite a coup for the black dressmaker.

Over the next few weeks, Keckly made herself indispensable to Mrs. Lincoln—coming to the White House for fittings, and soothing Mary when her feathers were ruffled, which was quite often. These fitting sessions frequently took place in Abraham Lincoln's presence, and thus endeared her to the president. Her ability to make exquisite gowns for her client—more than a dozen during her first four months in the White House—enraptured Mrs. Lincoln. Over the weeks and months, the relationship became more personal, more intimate, so much so that news of her would be included in letters back home to Elizabeth Grimsley in Springfield.

Mary wrote to her cousin on September 29, "I know you will be sorry to hear that our colored Mantuamaker, Elizabeth, lost her only son & child in the battle of Lexington Mo—She is heart broken. She is a very remarkable women [sic]." Her simpatico, soothing qualities were going to be taxed in the months and years to come, as troubles continued to break in waves over the Lincoln White House.

Mary became increasingly dependent upon Keckly, and confided in her as in no one else, indeed "more

than she did on her own kinfolks." When a deep tragedy struck the family circle in the early months of
1862, she would need to rely on the woman she came
to affectionately address as Lizabeth or Lizzie.

On New Year's Day in 1862, Frances Seward, who
was visiting her husband, the secretary of state, from
their Albany home, observed that "the carriages are
rolling along the streets as they used to do in old times."
When the doors to the Executive Mansion opened, as
many as a thousand paraded in. Many gossiped about
the scandals brewing over government corruption.
When Lincoln replaced Simon Cameron with Edwin

Elizabeth Keckly

Stanton at the War Department later in the month, a new regime was under way.

Stanton made it clear to Mrs. Lincoln that he was no Cameron. Her requests for appointments would no longer be honored during his tenure. Even so, by February, Mary Lincoln was riding a wave of social success. Although complaints about refurbishment would continue to plague her, she seemed more confident about her White House business. She was so buoyant that she was willing to challenge both protocol and custom by suspending the practice of expensive state dinners and biweekly receptions. The Buchanan White House offered weekly dinners with forty or more guests, which forced Lincoln's predecessor to dip into his own pockets to supplement the White House budget.

Once Mrs. Lincoln saw what the costs would be to maintain the elegant style to which she aspired, she decided to revise her protocol for entertainment. She proposed to the president that they drop the customary state dinners, and not simply for reasons of economy. She suggested they substitute large receptions, because they would be "more in keeping with the institutions of our country." When she first broached the subject, her husband was skeptical, but her arguments won out. One of Lincoln's secretaries, John Nicolay, proclaimed, "La Reine has determined to abrogate dinners." And La Reine got her way.

She would continue her own at-homes on Saturday afternoons, and the weekly public receptions. "Levees will be held at the mansion every Tuesday evening during the remainder of the session of Congress," the papers announced. These social occasions were obligatory, and many of the staff found them wearying, as Nicolay confided: "They are both novel and pleasant to the hundreds of mere passers by who linger a day or two to 'do' Washington; but for us who have to suffer the infliction once a week they get to be intolerable bores." A congressional wife complained of the president standing "gaunt and careworn." Perhaps to relieve the tedium of these occasions, the First Lady introduced the practice of inviting performers to the White House, from magician Herrmann the Prestidigitator to Polish pianist Alexander Wolowski.

Indeed, the Lincolns were the first residents of the White House to see the potential of making their home a national stage. They introduced the practice of bringing artists and performers into the Executive Mansion for social occasions, as part of their White House entertaining. Lincoln's favorite singers, actors, and others might be singled out for recognition— even one of P. T. Barnum's most famous acts would be extended an invitation, as Mrs. Lincoln recognized the public-relations potential of this practice.

She decided to throw a very large ball in early February and was in the thick of her plans by the end of January. Her lavish gestures and grand manner invited criticism. Mary decided to issue seven hundred invitations and planned to funnel all these guests into the East Room—shades of her party for her remodeled house back in Springfield. Not only the work of such an event, but the worries associated with it became immediate and acute to the Lincoln secretaries, who were by this time openly hostile to Mrs. Lincoln. They referred to her as Hellcat behind her back. Surely their banter and nicknames reached the servants' quarters and perhaps echoed back to Mary's own boudoir, but these acolytes were confident of their indispensability to the president, whom they had dubbed the Tycoon.

The ball's guest list caused quite a ruckus within Washington society: "Half the city is jubilant at being invited while the other half is furious at being left out in the cold." William Stoddard, Lincoln's secretary, was put in charge of the list and had to disabuse several dignitaries that they might impose on Mrs. Lincoln for some extra tickets for gentlemen of the press.

There were, of course, detractors on all fronts. Mary was firmly convinced that diversion was an absolute necessity. She ignored Senator Benjamin Wade, who wrote indignantly: "Are the President and Mrs. Lincoln aware that there is a civil war? If they are not, Mr. and Mrs. Wade are, and for that reason decline to participate in dancing and feasting."

But feast the others did, as heaping plates of partridge, quail, duck, turkey, foie gras, beef, and the president's favorite, oysters, greeted guests, as well as an elegantly turned-out Abraham Lincoln with his wife at his side. A cake in the shape of a fort, as well as elegant spun-sugar desserts amused the throng. The Marine Band played the "Mary Lincoln Polka," a piece composed to honor the First Lady. The rooms were not overcrowded, because only about five hundred showed up, but the *Washington Star* pronounced it the "most superb affair of its kind ever seen here." Not even Kate Chase's lavish soiree the next evening (with songs by a political singing group, the Hutchinson Family) could detract from the occasion as a social and political success for the Lincolns. However, dissenters composed a counter ditty, "My Lady President's Ball," which offered mournful verses written from the perspective of a wounded soldier bemoaning these follies at the White House. Satires and lampoons appeared in several journals.

One thing nearly caused Mrs. Lincoln to postpone the event, however—a "bilious fever" suffered by her beloved eleven-year-old son, Willie. Living in the White House less than a year, this favorite child, William Wallace Lincoln, represented the dearest hopes of both his parents. He was a studious and affable boy, and an especially good example for his brother Tad. A frequent visitor to the White House commented, "His self-possession—aplomb, as the French call it—was extraordinary." His resemblance to his father was common praise, and because of this, it was often assumed he was his father's favorite child.

During those first months in the Executive Mansion, Willie had proven an excellent companion for Mrs. Lincoln, and might be found perched next to her of an evening in the family's oval sitting room, book in hand, curled on a chair or lounge. Mary came to depend on him, and was deeply fearful when he fell ill. When he didn't improve rapidly, his mother wanted to do away with the scheduled ball. Lincoln prevailed on her to wait and see, and called in Dr. Robert Stone, who pronounced that the boy was "in no immediate danger." They were able to set aside their fears and leave Willie in the capable hands of Mrs. Keckly.

Elizabeth Keckly served as Mrs. Lincoln's dresser the night of the ball, and she recalled an intimate moment when Lincoln saw his wife in her white satin dress with a long train and a low neckline, and re-marked, "Whew! Our cat has a long tail to-night." When this failed to get a response, he went on: "Mother, it is my opinion, if some of that tail was nearer the head, it would be in better style." Neither Mary Lincoln nor her dressmaker paid any attention to his jesting complaint, and Keckly remarked, "She [Mary] had a beautiful neck and arm, and low dresses were becoming to her." This easy banter demonstrates that they were not in the grip of anxiety, but had set aside their fears, leaving Willie to Mrs. Keckly's care.

Lincoln had not been willing to cancel the event, but he did veto dancing. Also, the couple absented themselves from the party to look in on Willie several times during the evening. Under watchful eyes, his breathing grew labored. By the next day, his situa-tion had deteriorated. The papers reported Willie's dire condition over the next several days. Lincoln canceled a Cabinet meeting, and Mary omitted her regular White House levee to stay by Willie's bedside. She was gripped by fear, remembering little Lincoln Smith's death, about the same age, just two years before.

Intimates reported the terrible trials of the family during this fortnight; a Cabinet secretary confided to his diary that the president was "nearly worn out with grief and watching." Willie's dear friend Bud Taft attended the sickroom, trying to help him to rally. The president would find the Taft boy asleep on the floor, fearful to leave Willie's side. When Tad became sick as well, confined to another bedroom, Lincoln's official correspondence included a reference to his "domestic affliction."

Two days before the president's birthday (February 12), with Willie and Tad's health hanging in the balance, Henry Wikoff was hauled before the House Judiciary Committee to testify concerning the president's purloined annual message to Congress in December. Wikoff was sent to jail overnight for refusing to testify, then he confessed that the White House gardener, James Watt, had leaked the material. The whole sordid story was dragged into the headlines by midmonth, interspersed with bulletins on Willie's health.

Although the government was moving ahead with plans to celebrate Washington's Birthday with an illumination of public buildings, the Lincolns remained in grim shape. The president was distracted, while his wife was paralyzed. She sat next to Willie's bed day by day, trying to erase memories of the sickbed vigil she spent during winter a dozen years before, when she sat by her little Eddie, praying for his recovery before he died.

Finally his mother's constant prayers and his father's nightly visits failed to rouse Willie from the permanent slumber into which he drifted. The boy breathed his last on a long, gray winter afternoon, dying February 20 around five p.m. As the light slipped away, so did their boy. His parents were inconsolable.

Elizabeth Keckly washed and dressed the body, and watched as the president looked down on his dead child, confessing, "My poor boy, he was too good for this earth. God has called him home. I know that he is much better off in heaven, but then we loved him so. It is hard, hard to have him die." Lincoln sobbed over the small, frail body laid out on the bed.

Mrs. Lincoln's grief was volcanic, as she gave into hysteria and convulsions. Tad collapsed, physically felled by sorrow. Mary was so distracted that she keened long into the night, ignoring her husband and other sons. Robert, called home from college, struggled to control his grief. Robert was very like his mother in his appearance, but was determined to distinguish himself in his conduct. He struggled to project dignity and reserve, in stark contrast to what he felt were his mother and younger brother's sentimental displays.

The White House was shaken by the loss of this boy. William Stoddard suggested that malaria had felled him, while some papers reported that typhoid had killed him. Whatever the cause of death, the loss was a heavy burden for the entire Executive Mansion. Attorney General Bates sadly remembered supplying the boy with marbles when he visited the White House.

The president had gone immediately from Willie's deathbed to Tad's sickroom and recognized that something must be done for his ailing boy in the wake of this calamity. He knew his wife wouldn't be up to

Willie at age eleven

tending to Tad. Rebecca Pomroy, a nurse working in a Washington hospital, was ordered to the White House. Pomroy did not want to leave the military wards, filled with scores of bedridden soldiers. Yet she was touched by the president's relief when she arrived at the White House, and decided she was desperately needed.

The days following Willie's death were an incredible strain on the president. His wife was too stricken to even get out of bed. News of federal military victories flooded in—the capture of Fort Henry in Tennessee on February 6 and surrender of Fort Donelson on February 17—the latter after Ulysses S. Grant's famous ultimatum: "No terms except unconditional and immediate surrender can be accepted."

Hundred-gun salutes competed with the tolling of bells. Mrs. Lincoln wrote to Mrs. Taft to ask her to keep her boys at home during the funeral. The president overruled his wife, to invite Bud to have one last visit with Willie before he was put into the casket. After the funeral, Mrs. Taft waited in vain for a summons that her sons might come to the White House to visit Tad, a call that never came. The Taft boys went north with their sister to attend school, and Tad never got to say farewell to his playmates.

Willie was laid out in the Green Room. The family planned a funeral service in the East Room, to be conducted by Dr. Gurley, the pastor of Mrs. Lincoln's New York Presbyterian Church. Mary joined Robert, Tad, and Abraham for a final, private farewell. She refused to subject herself to the public services. White House mirrors were covered in crepe, and bouquets surrounded the body. Members of Congress and the Cabinet, as well as generals and diplomats, crowded in to pay their last respects. Lincoln went with the casket to a vault at Oak Hill Cemetery in Georgetown. The day of the funeral, a dramatic storm seized Washington, blowing out the skylights at the Library of Congress, toppling church steeples, and collapsing buildings throughout the city. It was as if the heavens opened in protest.

Lincoln had his own mourning ritual—locking himself in the Green Room on the Thursday after Willie died. He wanted to be alone to meditate on the one-week anniversary, as well as for several Thursdays thereafter, in the very room where his son's body had lain before being shut into a casket.

The president was occasionally called out of meetings by Tad to give him his medicine. He would stubbornly refuse to take a dose from anyone else. Lincoln would often pick him up and carry him along to meetings. He ignored convention and counsel, indulging the boy shamelessly. Perhaps it was a way of putting his mourning over Willie into the background by putting Tad in the forefront.

While the two remaining Lincoln sons struggled to help their father cope with their mutual grief, Mary retreated. She took to her bed and drew no practical relief from visiting friends' religious platitudes. Only the trappings of mourning seemed to rouse her. Victorian mores dictated that a mother might mourn for her child formally for a year, while a widow should mourn for two and one half years (and widowers only three months). These prescribed rituals were clung to like masts in the stormy seas of grief that washed over Americans during these terrible times.

Lincoln feared that his wife might never regain her equilibrium. Elizabeth Keckly recounted Lincoln's taking his wife to the window of the White House and pointing out St. Elizabeth's, a mental hospital in the distance, and warning her she might have to be sent there if she could not recover.

A gush of obituaries reopened her wounds, as when she read Nathaniel Willis's tribute to Willie: "He retained his prairie habits, unalterably pure and simple, till he died. His leading trait seemed to a be a fearless and kindly frankness, willing that everything should be as different as it pleased, but resting unmoved in his own conscious single-heartedness."

Robert Lincoln summoned his Aunt Elizabeth Edwards, who arrived at the White House the day after the funeral. She found her sister shut up in a room, prostrate with grief, and her nephew Tad weeping that he would never see Willie again. By February 25, Tad had improved under Rebecca Pomroy's care, but he could not yet sit up. Elizabeth finally persuaded Mary to get out of bed and dress, and attend church services. Elizabeth believed that "my presence here has tended very much to soothe the excessive grief."

As her husband and son Tad became closer, Mary let anxieties engulf her, wrapping around like wraiths of fog refusing to lift. But the good counsel of women like the wife of her husband's secretary of the navy, Mary Jane Welles (who would lose six children), and nurse Rebecca Pomroy, as well as other spiritual advisers, supported Mary during these initial weeks of grief.

Tad could not abide being left alone. Losing his constant companion of a brother was a shattering blow. He and the president began to spend more time together, as a kind of mutual therapy. The little boy could be found shadowing his father by day, in and out of his father's office all day long. Since Lincoln kept such late hours, Tad also might often be found curled up asleep next to his father's desk of an evening. Many nights the father would carry his son into his own bed, and the two might try to keep one another from nightmares.

Mrs. Lincoln spent much of March in seclusion, trying to weather the clouds swirling around her. Politicians contributed to the rumor mill linking Mrs. Lincoln to the leaking of the State of the Union message in December 1861, before it was delivered to Congress. Although White House gardener James Watt (who had by now left for Scotland) had taken the fall for passing on Lincoln's text to Wikoff, scandalmongers suggested that Mary was the one who had committed the deed. As these wild stories circulated, Mrs. Lincoln was much abused.

Meanwhile, debates over the abolition of slavery in the District of Columbia gathered steam. Many ladies in the District resented the changes they foresaw. Racism reared its ugly head and exacerbated tensions among the society women of Washington. The president's enemies played on these fears, especially the charismatic and clever Kate Chase.

One day, when an African-American teacher came to call on Mrs. Lincoln at the White House, having been invited to tea, she was escorted around to the kitchen entrance by the doorman. Mary was infuriated by this slight. The First Lady became especially solicitous during tea with her guest in the Red Room. Afterward, her black guest enjoyed the First Lady's promise to bring the cause of African-American education to her husband's attention. Then she was cordially escorted to the formal entrance, where Mary pointedly shook hands with her while bidding the woman good-bye. This gesture was observed by both the Chases, who just happened to be driving up to the Executive Mansion at that very moment. Naturally, Chase's daughter spread the story—willing to use it to her father's political advantage by portraying Mrs. Lincoln as someone who was "making too much of the Negro."

This jab at Mary Lincoln was particularly hypocritical coming from Kate Chase, whose father was a staunch abolitionist. Chase had long promoted emancipation as part of his political agenda, and criticized Lincoln for being weak on abolition. But when his wife made a gesture toward racial equality, the Chases made political hay by broadcasting Mary's liberality among unsympathetic listeners.

The Chases also maintained a relentless campaign against the First Lady in other ways, portraying Mrs. Lincoln as someone whose financial improprieties lowered the tone of the White House. This was also hypocritical, because Kate Chase accepted extravagant gifts from Jay Cooke, a Philadelphia financier who sent lavish tokens of his esteem—including an expensive open carriage. Cooke received contracts from the Treasury Department during this era, which made these presents, as well as Cooke's "investments" on Chase's behalf, compromising of the high ground the Cabinet secretary and his daughter claimed.

To add further fuel to the fire, Mary, as the sister of Confederate soldiers, was openly disparaged as an enemy spy in the White House. Mary's enemies were not bothered in the least by the inherent contradictions in these competing charges. At a time when Mrs. Lincoln might have been gaining some sympathy from the public—locked up and weeping over her lost boy—the long knives were drawn out in the press, charging her with improprieties at best and treason at worst. While she was imprisoned by her grief, critics in the outside world mounted new offensives.

Some even hinted that Mary Lincoln's being laid low was her own fault, attributing her fall from grace to her risky, high-flying social ambitions. Did this kind of cruelty cross a line? Did her unpopularity not echo the vilification of Marie Antoinette? Mary Lincoln, dubbed the Republican Queen in the pages of *Leslie's Illustrated Weekly*, became engulfed by intrigue that rivaled that of Versailles.

Perhaps most unnerving were the courtiers nattering on the second floor of the Executive Mansion—the president's secretaries rarely managed any sympathy for Mrs. Lincoln during this period. John Hay wrote to John Nicolay on March 31: "The 'enemy' is still planning campaign in quiet. She is rapidly being reinforced from Springfield. A dozen Todds of the Edwards breed in the house." A few days later he confided, "Madame has mounted me to pay her the Steward's salary. I told her to kiss mine." On April 9, he snarled, "The Hellcat is getting more Hellcattical day by day."

Surrounded by this bubbling cauldron, Mary Lincoln must have felt much comforted by having her sister Elizabeth with her, and sought to postpone her departure. Lincoln certainly wanted his sister-in-law to stay, recognizing Elizabeth's positive influence over her emotionally overwhelmed younger sister. Elizabeth Edwards wrote to her family back in Springfield in April, "Your Aunt Mary wonders if Mary Wallace [Frances's daughter] will not feel like coming on, when I am ready to leave . . . she feels since her trouble that she cannot again be alone." It was an uphill battle to convince Frances to send her daughter Mary. Elizabeth reminded her sister that the summer season requires "a very simple wardrobe" and told her, "Aunt Mary says that she will be only too glad to defray her expenses in coming, which you can repeat if that objection is urged." But she coaxed in vain, because Mary Wallace did not come.

The situation worsened when gossip flared following the death of her half brother Samuel B. Todd, a Confederate officer who died at the Battle of Shiloh on April 7—though it was not public knowledge until later. A banner in the *National Republican* trumpeted: "Mrs. Lincoln's Brother Killed in Battle."

The president had no time to slow his pace in the weeks following Willie's death, as he threw himself into ongoing debates over slavery and getting the Emancipation Bill through Congress—an abolition measure for the District of Columbia that included compensation for slaveholders. He was preoccupied with what to do about General McClellan's seemingly endless hesitation and increasing calls from Congress for his resignation. He followed war news closely, including the naval battle between the *Monitor* and *Merrimac*. His visiting sister-in-law observed, "Mr. Lincoln looks very anxious and careworn." It wasn't just the loss of his son that wore him down: By May, the press reported nearly 75,000 troops sick, disabled, and missing.

The president resumed his theatergoing, one of his few remaining diversions from the cares and routines of the White House. After his entrance into the theater, however, all the opera glasses would be trained on the presidential box rather than on the actors. Lincoln had a fondness for Shakespeare, both for reading the plays and attending performances. He invited actors to the White House to discuss the fine points of the text. Throughout his troubled presidency, he sought escape from his life-and-death decisions with a night at the theater.

Three months after his son's death, Lincoln was reading aloud a passage from the third act of *King John*—Constance grieving over the death of her son: "That we shall see and know our friends in heaven / If that be true, I shall see my boy again." He asked a nearby army officer: "Did you ever dream of some lost friend, and feel that you were having a sweet communion with him . . .? That is the way I dream of my lost boy." Lincoln tried to embrace the sense of loss, to cherish those mementoes that Willie left behind, while his wife tried to avoid troubling remembrances.

The president, mindful of his wife's precarious mental condition, did not dwell on their child's passing in her presence, but took comfort from recollections among a small circle of friends.

Mrs. Lincoln struggled to ease herself back into some semblance of normalcy. She ordered books and clothes, and made excursions to Mt. Vernon and to the Navy Yard, among other outings. She recommenced her custom of sending out bouquets from the conservatory. She wrote solicitously to her minister's wife in mid-May, sending along raisins and figs. She began supplying Mrs. Pomroy with delicacies for soldiers back at the sick wards. Her own young son continued to require care, but Tad found it easier to turn to his father for solace.

Two months after Willie's death, Mary began to resume former routines, which lightened the president's burdens. She realized how difficult the war demands had become, and forced herself to stroll with friends on the grounds or take a ride in the carriage. She might drag her husband to a concert, because otherwise his fondness for music might not outweigh his aversion to the "swallow-tailed coat and kids" [gloves] required for such an outing.

Mrs. Lincoln may have been able to handle social obligations, yet still she struggled to regain her equilibrium. Her roiling emotions unnerved her, while her younger son "bears up and teaches us a lesson, in enduring the stroke to which we *must submit*." As Washington began to fulfill the promise of a gorgeous spring, Mary confessed that although "our home is very beautiful, . . . everything appears a mockery, the idolized one is not with us . . . and we are left desolate." These words and feelings would haunt her.

11.

Struggling Against Sorrows

By May 1862, motherhood gave Mary the distraction she needed to overcome her paralysis, as she focused on her two remaining sons. The First Lady, heartened by her younger son's recovery, knew Tad was both lonelier and even more mischievous—without his brother Willie to function as companion and restraint. Mary Lincoln longed for her older son's return from Harvard: "Robert will be home from Cambridge in about 6 weeks and will spend his vacation with us." She added with maternal satisfaction: "He has grown & improved more than any one you ever saw."

Mary took special pride in Robert's current educational accomplishments. When he first did so poorly on his college entrance exams, it had seemed the end of the world to her. But once he began his undergraduate studies, Robert advanced rapidly. During his first year at Harvard, the president's oldest son flourished, attending courses in Latin, Greek, composition, mathematics, elocution, religion, and history. In addition—his mother's boy—he studied French. Although Robert's grades were mediocre and placed him near the middle of his class of a hundred young men, Mary exulted at his accomplishments. She glowed with pride over the fact that her son succeeded on a campus with such outstanding faculty as Oliver Wendell Holmes and Louis Agassiz. Many years later, Robert's professor James Russell Lowell would recall his pupil: "I don't think he

Robert at Harvard

distinguished himself as a scholar—perhaps that may have been my fault as much as his—but he was always an honest fellow with no harm in him and I was sure he would turn out well."

Robert became involved in the Delta Kappa Epsilon fraternity (a group founded at Yale in 1844). He lived on campus in Stoughton Hall, and became vice president of the Hasty Pudding Club during his senior year. As the namesake of his Kentucky grandfather, as the son of a president, Robert Todd Lincoln was his mother's pride and joy. She might not have been amused that he was dubbed the Prince of Rails (a reference to his father's nickname, the Railsplitter), but he remained princely to her, even though he had taken up the habit of smoking, of which his mother disapproved. She would not let him smoke in the house.

Robert became even more precious following Willie's death. Thus newspaper reports on May 30 stating that "last Monday evening, seventy-five students from Harvard University immediately offered themselves as volunteers" sent chills through her. Pressure was mounting, as the *National Republican* reported in July 1862 that "a son of Secretary Seward has just enlisted. . . . A Son of Governor David Tod, of Ohio has done the same thing. All over the States the best classes of young men [are enlisting]." She remained adamant that Robert not join the army, and father and son continued to humor her on this matter.

Even the promise of having this cherished son at home for the summer could not relieve Mary's discontent. Everywhere she looked, she saw traces of her beloved Willie. Mary had barred Tad from any contact with his former playmates—the Taft boys. She would not risk the emotional quakes that such a reunion might trigger.

Meanwhile, Tad protected his own feelings by shipping off to Springfield some of Willie's favorite toys, including two railroad cars, for his cousin Edward Lewis Baker Jr. (Elizabeth Edwards's grandson, called Lewis). But still, the memories lingered on, making a prison out of the Executive Mansion. The president continued his vigil, moving back and forth from the War Department to the White House, trying to hold together the dissolving Union.

With her husband absorbed in duty and Tad still frail, Mrs. Lincoln devised a plan: She would move them to the Soldiers' Home. Mary had first seen this place over a year ago, making her first visit by carriage two days after the inauguration. Lincoln first inspected the site the day after his wife—but this second home had lain fallow. By the end of May 1862, the plan had been set in motion to move the household, to relocate to

Anderson Cottage, Soldiers' Home

their country cottage. Mrs. Lincoln convinced herself that a change of venue was beneficial, if not essential.

She wrote to a friend that she planned to "go out to the 'Soldiers' Home', a very charming place 2½ miles from the city," explaining further, "quiet is very necessary to us." The papers reported that the Lincolns would summer at Buchanan's old place after it was renovated. Once the plan was announced, gawkers drove by to inspect the new presidential headquarters, and one confessed, "I have seen nothing in Washington that has charmed me more than this quiet and beautiful retreat."

After all, fashionable ladies in the District were fleeing the city in droves on account of the heat, which often rose to over 100 degrees. John Nicolay complained about the bugs, "The air is swarming with them . . . in countless numbers, they are buzzing about the room." The papers warned, "If Congress should remain in session long enough the coming summer they would have a good chance to become its first victims." One woman complained that the unhealthy climate in the District was exacerbated by the rotting corpses of dead horses littering the city. Spending the summer outside the dense urban center and moving to higher ground attracted Mary to her new home, but she was primarily seeking isolation when she relocated on June 16. Benjamin French reported that her removal put her in "excellent spirits."

The summer residence was on the grounds of the Soldiers' Home, a military asylum established as a place to care for disabled veterans. The facility included a group of cottages nestled on a hilltop next to the larger wards, with pathways all around for resident soldiers to take the air. A cemetery had been dedicated after the First Battle of Bull Run the summer before. The largest of these smaller dwellings was known as the Anderson Cottage, a stuccoed, gabled structure built in the 1840s as a summer house for banker George Riggs. The government had also built Scott Hall (the main building, erected in the 1850s and named after General Winfield Scott, who had been a founder of the project), and the rather unimaginatively named Quarters 1 and Quarters 2, as well as Corlisle Cottage (located in a nearby grove). Buchanan, a lifelong bachelor, had occupied Quarters 1 during his summer stays. The Lincolns selected the Anderson Cottage for their residence.

This cottage suited the entire family. Lincoln found it a comfortable distance from the Executive Mansion, where he arrived around ten every day, departing for the Soldiers' Home about four. In addition, he sometimes sent a carriage to summon friends to breakfast meetings at the cottage, as he did on June 18, when Senator Orville Browning brought along New York store magnate Alexander Stewart (which must have pleased Mary Lincoln) and New York Justice Henry Hilton. Later that same day, Vice President Hannibal Hamlin dined with the Lincoln family, then apparently discussed with the president the possibility of a general military proclamation of emancipation. Residence at the cottage did not reduce Lincoln's administrative business, but it allowed him to channel his energies productively, rather than having them monopolized by zealous office seekers.

With Willie gone, Tad sought playmates among the soldiers, many not much older than he. The young men on the hospital grounds proved a boon to Tad, who, like his brother Willie, was crazy about martial affairs. The whole Lincoln family found the soldiers' kindness touching. Tad was "adopted" by many of the military guards who surrounded Lincoln; he sometimes took his "rations" with the troops, and was given the unofficial rank of "third lieutenant."

Lincoln enjoyed being surrounded by men in uniform. Secretary of War Edwin Stanton was assigned a residence near Lincoln's. Stanton rarely spent the night at his Soldiers' Home cottage, sleeping more often at his house on K Street, nearer to the War Department. But he moved his family to the cottage for the summer in 1862. The Stantons' young son James became ill, and took a turn for the worse at the end of June. Ellen Stanton's sickbed vigil created a terrible dilemma for Mary. She had been out at the cottage less than a month when the Stanton boy became ill. After all, she had come to escape memories of her own son's ordeal. It was particularly stressful when a messenger summoned Stanton from the War Department on July 5, saying, "the baby is dying."

The war had also taken a turn for the worse. Lincoln departed for a visit to the Virginia Peninsula on July 7, following McClellan's defeat during the Seven Days' Battles (June 25–July 1). From his days in the Illinois militia during the Black Hawk War, Lincoln enjoyed the company of common privates. Now, riding home on horseback to the Soldiers' Home on the afternoon of the Fourth of July, 1862, Lincoln encountered ambulances overflowing with the Union injured returning from Virginia battlefields.

The president listened to stories about conditions on the peninsula and decided to undertake the twelve-hour journey to see the front for himself. While able to rally the troops and confer with McClellan, he confirmed for himself that it would take something dramatic to expand the army and win the war.

With her husband away, with raw memories engulfing her, Mary packed her bags and took Tad and Robert to New York, arriving on July 9. The very next day, the Stantons' child, less than two years old, died of his illness, brought on by smallpox inoculation. This was a tragic reminder of Willie's death, which both Lincolns were trying to put behind them.

On the way back from the front, Lincoln's boat ran aground on the Kettle Shoals, and he and his entire party took a swim in the Potomac. The president returned to Washington reinvigorated, and decided he would take a gamble and introduce the military proclamation of emancipation, which he had discussed with the secretary of state and the secretary of the navy as the men rode together in the funeral procession for Stanton's young son on July 13. Mary had not yet returned from New York.

Four days after the Stantons' child was laid to rest, Mary came back to the Soldiers' Home with her two sons. Her time away had certainly been a study in contrasts. During her two previous visits to Manhattan as First Lady, her every move had been reported and dissected with partisan zeal in the papers. But this time the press seemed to take a pass, perhaps out of respect for her loss of a child. During this tour of New York, her proclivity for fashion was not an issue. Mrs. Lincoln was swathed from head to toe in black, hidden by her mourning garb when she ventured out, and her only recorded purchases were books by Shakespeare and Scott for the presidential library. The only public outing that attracted press attention was a visit to the Broadway clinic run by the New England Relief Association. Following this stop, she headed for the Park Hospital, where the First Lady was escorted through the wards, greeting wounded soldiers, who "fully appreciated her kindness and womanly sympathy."

When Mary Lincoln returned to Washington in mid-July, she also seemed renewed. She confided that she had left with much reluctance but found her visit "more pleasant than I could have hoped." Mrs. Lincoln next embarked on a program of ministering to soldiers—which her location, close to several hospitals, favored. By the summer, nearly eight thousand soldiers in the District were sick and wounded, and their care was a critical concern to the women in Washington. Mrs. Lincoln threw herself headlong into this new vocation—and garnered some modest gains, such as the $1,000 donated by a Boston merchant for the military hospitals.

Mary conducted a rigorous campaign of regular visitation. Journalists reported, "A noble example was set by Mrs. Lincoln, yesterday, in her visit to Mt. Pleasant Hospital." Popular novelist Ann Stephens accompanied Mary Lincoln during several of these tours of duty, "walking for hours through the wards to say cheering words of hope and encouragement to the wounded and sick; laying fresh flowers on their pillows and offering them delicacies brought from the White House." During their time together, the First Lady confided to her companion that if it were not for these visits, her heart would have been broken by the loss of her child.

Mrs. Lincoln wrote letters and performed mundane tasks for those in sick bay, as well as handing out wagonloads of fresh flowers and baskets of fresh fruit. By the fall, the number of federal casualties more than doubled, with nearly eighteen thousand hospitalized in and around Washington. Volunteers in medical wards were critical to the Union cause.

Visitors commented that during this period Mrs. Lincoln regularly lost control of her emotions. Shortly after Willie's death, the great opera singer Madame Patti appeared in Washington and was invited to the White House to perform. The president had heard her sing nearly ten years before and wanted to hear her again. When Patti entered the parlor, she saw the First Lady, "almost regal in her deep black and expansive crinoline, only an outline of white at throat and wrists." Her singing brought tears to listeners' eyes.

Laura Redden, who earned distinction as a journalist despite being deaf as well as female, stopped in at the summer cottage in August 1862. Mrs. Lincoln greeted her warmly. But when Mary "burst into a passion of tears," Redden was so moved by this display that she broke protocol and reached out to embrace the distraught First Lady. She put her arm around the bereaved mother who "could neither think nor talk of anything but Willie."

Of course, fresh torrents of grief flowed freely when her young half brother Alexander Todd died of wounds received at the Battle of Baton Rouge on August 19. Mrs. Lincoln never publicly acknowledged the deaths of any Confederate relatives, but privately mourned these additional losses. Indeed, "baby Alec" had been a particular favorite of Mary's, even though it was his birth in 1839 that contributed to her leaving her Lexington home. But Alexander's youthful death, struck down by friendly fire, was one of those everyday tragedies that cut to the quick in wartime.

Mrs. Lincoln threw herself more ferociously into the cause of dying soldiers, making sure fresh fruit arrived at the hospitals. The Northern papers were full of misgivings about all the death and dying. A "letter to the President" by Lydia Maria Child appeared on August 22, complaining, "Surely the women of America have the right to inquire, nay demand, whether their husbands, sons and brothers are to be buried by thousands in Southern graves." She demanded that Lincoln free the slaves and put black soldiers into the ranks, demonstrating the widening of the debate over slavery and the war. While Abraham Lincoln was caught up in demands for black progress, Mrs. Lincoln felt swept off into a parallel sphere, increasingly separated from her husband and his concerns.

Struggling against her overemotional nature, she found herself enthralled by an increasingly popular pastime in Civil War America: spirit circles. These were gatherings organized by mediums who practiced spiritualism, communing with those who had "crossed over"—talking with the dead. This belief in contact with the dead was one of the fastest-growing movements in nineteenth-century America, accelerated by the mounting Civil War death lists.

The attempt to reach out to those taken proved irresistible to Mary in her weakened state—and indeed to hundreds upon thousands before and during the war. The American penchant for belief in the spirit world predated the nineteenth century. It was popularized especially among the Methodists, who believed their leader, John Wesley, had been an early pioneer of the practice of "rapping." In 1726, Wesley suggested that his childhood home had been haunted by "raps and knocks, footsteps and groans," and the family had even nicknamed the family ghost, which Margaret Fuller featured in her review of Wesley's life in the 1840s. The early decades of the century proved a fertile period for the flowering of spiritualism.

With the appearance of *A Report on the Mysterious Noises Heard in the House of Mr. John D. Fox* (1848), a pioneering spiritualist publication, curiosity shifted into a growing craze, a movement that swept thousands of Americans into its sentimental thrall. This "report" told of the "Hydesville Rappings" of April 1848. Mysterious spirits allegedly inhabited the home of John and Margaret Fox, spirits that communicated with the couple's three daughters: Leah (who was older and married), Margaret, thirteen, and Kate, only eleven.

Sometime in early 1848, at their home near Hydesville, a village near Rochester in upstate New York, the Fox family heard rapping noises, which appeared to come from the walls and furniture. The three daughters said they believed that they might be in communication with the supernatural if they could decode these rapping noises. Their ideas gained fame, and within months, the Fox sisters were a sensation—primarily Margaret and Catherine, as Leah, the married older sibling, remained in a kind of managerial mode. The two younger girls began to give public séances in America and soon were shipped off to perform in Europe as well. The Fox sisters came along at a time when Americans were receptive to the idea of communing across time and space.

By the 1850s, the American public burned with spiritualist ardor, as magnetism and legerdemain also became popular. The ideas of the popular eighteenth-century Swedish philosopher Emmanuel Swedenborg were also in vogue. When the Lincolns lived at the Globe Tavern in 1843, Alfred Bledsoe was a practitioner of mesmerism (hypnosis), which captured Lincoln's imagination. During this fertile era, the country became suffused with a wide range of scientific and popular areas of interest, including sociology, phrenology, and anthropology. As a keen intellect, Lincoln was curious about many of the claims of these new sciences and wanted to test some of them by observation—as did many Americans.

The public eagerly flocked to spiritualist performances, and the Fox sisters became the subject of intense debate. Margaret and Kate were booked into a hall in Rochester, New York, in 1849 to withstand intense public scrutiny. Onstage they were forced to stand on a pillow, then on glass, while threatened with electricity, but no matter how they were bound or maneuvered for observation, the rappings persisted.

After the public attempts at unmasking, a Committee of Ladies disrobed the girls in a private room following their performance, searching for devices. The young women were virtually imprisoned, crying and begging for release until Rochester feminist Amy Post rescued them from this strip search. Post and her community of radical reformers formed a tight-knit circle of supporters. The debates persisted, as did spiritualism's popularity.

The Fox sisters were followed by "trance lecturers," including girls such as Emma Jay, not yet eighteen, who spoke with simplicity, logic, and clarity beyond her years. In 1858, seventeen-year-old Cora L. V. Hatch held audiences spellbound at the Broadway Tabernacle, responding eloquently to such questions as "Was Jesus of Nazareth divine or human?" The *New York Tribune* reported that Hatch was a slender belle with "a profusion of sunny ringlets and a fresh youthful complexion [that] gives her an almost childish air." One of the most famous of these trance speakers, Cora Richmond, became fictionalized as the charismatic Verena Tarrant in Henry James's novel *The Bostonians.*

Horace Greeley knew how to stir up controversy and sell papers. Greeley's personal life had been sadly adrift due to his terrible marriage—described as a "domestic hell" by intimates. The Greeleys lost several children. After his beloved son Pickie succumbed to cholera at the age of five in 1849, Greeley wrote to his friend Margaret Fuller, "The one sunburst of joy that has gladdened my rugged pathway has departed." When his wife prodded him, Greeley attended the private séances of Charlotte Fowler Wells, a phrenologist who initiated weekly meetings in December 1850. Wells led the cream of the Manhattan elite in communing with their dead. Her brother, Edward Fowler, performed "automatic writing" when he went into a trance state to commune with spirits, and he became a popular medium.

Kate Fox was hired by the publisher of the *Christian Spiritualist,* who paid her an annual salary of $1,200, at a time when wages for a textile worker were less than $5 a week. Naturally, rival sects erupted, discrediting one another in an atmosphere of bitter denouncements and partisan feuds. Congress introduced a bill in 1854 calling for an official investigation of spiritualism, labeling it an "occult science." (The bill was defeated.)

Some spiritualists branched out from spirit circles to venture into what was termed magnetic healing. At midcentury, the practice of mesmerism was in its infancy. However, the laying on of hands, a practice involving touch as well what we would now call hypnotic suggestion, had been around for centuries; spiritualists were eager to broaden their scope. One celebrated case in particular helped ally the two: After two years of lying flat in a darkened bedroom after slipping on ice and suffering paralysis, Olivia Langdon was desperate and sought nonmedical treatment. A spiritualist healer was summoned to the girl's Elmira home, and he prayed over her. After the visit, Olivia rose out of her bed, completely recovered. The spiritualist attributed his miraculous success to the "form of electricity passing from his body to his patients." (Olivia went on to marry Samuel Clemens, later famous as Mark Twain.)

The rise of Victorian sentimentality blended nicely with spiritual philosophy. Abby Sewall explained,

" 'Tis sweet to call to mind the loved of earth who have passed away, and think how soon we shall be like them, free from sorrow and care . . . the thirst for immortality is not implanted in our hearts in vain. . . ." Sewall suggested that the church, the state, and medical experts could not provide succor for the "disease of a starved heart." Naturally, many took advantage of these starved hearts.

Spiritualism went on trial in 1857 with the "Cambridge investigations," sponsored by the *Boston Courier*. A panel of Harvard professors—Louis Agassiz, Benjamin Pierce, Eben Horsford, and B. A. Gould—cross-examined leading spiritualists, including two of the three Fox sisters. The debates revealed the depth of the cultural standoff. Oliver Wendell Holmes reflected on the spiritualist culture wars in the *New York Tribune*: "While some are crying against it as a delusion of the Devil, and some are laughing at it as an hysteric folly, and some are getting angry with it as a mere trick of interested or mischievous persons, Spiritualism is quietly undermining the traditional ideas of the future state which have been and are still accepted." Men of learning remained skeptical about this new field that was claiming to be a science, while others, such as Judge John Edmonds of New York, became convinced of this new movement's transcendent significance. Edmonds resigned his position on the New York State Supreme Court to become a medium.

One practitioner argued that "when a sympathetic cord is established between two spirits," thought might flow between the two "from mind to mind, as electric fluid on the telegraphic wire." It was a kind of fearful symmetry that the telegraph (along with the concept of electrical forces) premiered in 1848, just when the Fox girls burst into the headlines. It was no accident that their talent was described as a "spiritual telegraph," which allowed them to communicate with those who had "passed over."

Women disproportionately flocked to spirit circles. As one rapturously described her vision of the spirit world, "The gentle whispers fall upon our ears, while angel hands smooth back the damp hair from our brows, press their angel-lips to ours, and sing us to heaven." The language of this movement was deeply and unabashedly suffused with gender. Spiritualism was "the only religious sect in the world . . . that has recognized the equality of woman." The popular spiritualist writer Cora Wilburn heralded a movement whereby "the medium may be man or woman—woman or man—but in either case, the characteristics will be *feminine*."

Spiritualism caught fire because it coincided with the deepest needs of American women at midcentury. At a time when nearly half the deaths in New York State were children under five, it is no wonder that grieving mothers sought escape with dreams of Summerland, the spiritualist name for heaven. Novelist Elizabeth Stuart Phelps's portrait of burial grounds in her influential novel *The Gates Ajar* reflected the spiritualist ideal that the graveyard must be transformed from doleful stone gardens into a "site designed for the contemplation of happy memories and heavenly reunions."

Harriet Beecher Stowe (like her sister Catharine) had a healthy distrust of mediums. Nevertheless, she was obsessed with the idea that she might be able to talk to her dead children. She was not alone. The more than two million Americans subscribing to spiritualist beliefs in 1850 would triple their numbers by the summer of 1862, when Mary Lincoln, like so many American mothers, faced grief over the loss of a child. The Civil War fueled interest in spiritualism.

When Mary Lincoln sought the comfort of spiritualists in the summer of 1862, she was too needy to be cautious. Her involvement became the hottest topic among Washington gossips. Dabbling in the spirit world, not her good works among the hospital wards, stimulated rumors in 1862 and into 1863. Mary Lincoln's association with shady characters continued to tarnish her reputation.

Betrayed by her susceptibility to flatterers, Mary Lincoln was taken in by Charles J. Colchester ("Lord Colchester"). This stout, suave British gentleman, who claimed to be the illegitimate son of an English duke, became a familiar sight in wartime Washington society. He conducted his affairs out of a suite of hotel rooms, where he would advise "women who had lost their blue-eyed children, young men in love, and suspicious gentlemen, who felt their relatives had robbed them by faulty wills." His reputation as a medium led to an introduction to Mrs. Lincoln. She invited him to hold séances at the Soldiers' Home in the summer of 1862.

Lincoln became suspicious as a result of gossip in the wake of Colchester's appearances at the Anderson Cottage, and asked Dr. Joseph Henry, head of the Smithsonian Institution, to investigate him. Henry delegated Noah Brooks to attend one of Colchester's spirit circles, where Brooks was able to disrupt the proceedings and unmask fakery. Colchester tried to use blackmail—threatening to expose his relationship with Mrs. Lincoln—to extract a favor from Brooks. The scheme backfired. Colchester lingered in Washington after Brooks dealt with him, but he did not trouble Mrs. Lincoln anymore.

In 1863, Mary confided to Senator Orville H. Browning during a carriage ride that while visiting a medium in Georgetown (Mrs. Laurie), she had been in contact with her son Willie. The number of séances Mrs. Lincoln attended following Willie's death (with or without her husband), the number of séances held at the summer cottage or in the White House, the president's attitudes toward spiritualism, and the depth of his wife's feelings continue to be a source of debate, unresolved by verifiable evidence.

The memoirs of spiritualist Nellie Colburn assert that she knew both of the Lincolns from spiritualist circles in wartime Washington and that Lincoln was a spiritualist. Colburn and her fellow believers derived their claims from Lincoln's willingness to attend séances to appease his wife in the months after Willie's death. The allegations that Lincoln adopted the Emancipation Proclamation at the behest of the spirit world or that he climbed on a piano and was levitated by spirits demonstrate how a cloak of overstatement is fashioned from thin threads of truth.

Lincoln's presence in the Lauries' Georgetown parlor does not indicate much beyond his indulgence of his wife. Lincoln did take meetings with spiritualists— even without Mrs. Lincoln present—but this may have reflected his extreme curiosity and courtesy rather than any affinity.

By contrast, Mrs. Lincoln acknowledged her participation in spirit circles and confided that these sessions allowed her to communicate directly with her son Willie. She even confessed that she had been in contact with her son Eddie, as well. Her belief that her dead boys appeared brought Mary great comfort—a comfort sought by thousands of mothers across the nation during those dark days.

The Lincolns appeared to be having more marital difficulty during this period of extraordinary strain. Some have made much of the friendships that Lincoln struck up during this period, in particular with one of the soldiers assigned to presidential guard duty.

Mary Lincoln had become so alarmed about her husband's safety in the weeks after Willie's death that she insisted that a cavalry group escort him on journeys from the White House to the Soldiers' Home. The duty was assigned to a New York company. At the same time, a Pennsylvania company was detached to the Soldiers' Home to guard him there as well. Lincoln disliked military escorts and their "clatter of sabers and spurs." When Mary was with him in the carriage, he complained the two of them couldn't hear themselves talk.

He resisted protective measures, and often he would send home the soldiers dispatched to guard his cottage. Spying his sentries near midnight, Lincoln might pull rank as commander in chief to order the guards on duty back to their own quarters. But he eventually grew to accept his assigned wardens, and even came to appreciate the company of two of his special guards: Captain Henry Crotzer of Company D and forty-four-year-old Captain David D. Derickson of Company K. Derickson quickly became a favorite shadow after September 7.

Derickson was often a dinner companion, and his closeness to the president caused comment. He not only shared meals with Lincoln, but, during Mary's absence from Washington, was said to sleep in the same bed. His presence at the Soldiers' Home cottage became a "joking matter," and one fellow officer recalled Derickson's "making use of His Excellency's night shirts!" Virginia Fox (wife of Assistant Secretary of the Navy Gustavus Fox) repeated gossip in her diary in November 1862: "'There is a Bucktail soldier here devoted to the President, drives with him, and when Mrs. L. is not home, sleeps with him.' What stuff!"

Some have taken these fragmentary comments and spun them into an erotic friendship replete with sexual content and sweeping significance. Any assertion of this relationship as some full-blown sexual affair overwhelms the slight and suspect evidence on which it precariously rests.

This interpretation also overlooks the evidence historian Matthew Pinsker unearthed that in interviews with a Company K soldier, Ida Tarbell found a soldier who recalled that when he was forced to awaken the president in the middle of the night, he also disturbed Mrs. Lincoln, who was sharing her husband's bed. This confirmation of the couple's intimacy has been pointedly neglected by those who want to paint a different portrait of the marriage, but it is drawn from the same well of evidence. This intimacy also reflects why there would be reports in the press that the First Lady was pregnant (reports that proved false). Lincoln's behavior during this period is complex and his relationships were equally intricate, with no overarching argument about his personal affections based on his own testimony.

Nevertheless, Derickson did have a special relationship with the president, one he treasured, proud to call Lincoln "a friend and acquaintance." He must have fulfilled a need, even if what that need was remains a matter of heated debate.

Most scholarly discussion over Lincoln's male intimates has become inflammatory. Lincolnistas have squared off with such ferocity that it seems unlikely to be resolved by weighing the evidence. Interpretations of Derickson's role will doubtless continue to produce controversy—as does scholarship over Joshua Speed.

Even if Derickson was a frequent bedmate of Lincoln's, the ease with which contemporary accounts and former soldiers' recollections reveal this sleeping arrangement demonstrates its lack of sexual undertone. Pinsker has suggested with some persuasion that what *did* raise eyebrows was that Lincoln bestowed such favor on a mere captain. This was unusual but demonstrated Lincoln's disregard for rank and hierarchy, perhaps stemming from his own Black Hawk days as a soldier.

What has taken many modern readers by surprise is that sharing a bed did not convey the same meaning during Lincoln's lifetime as it does today. Nineteenth-century soldiers frequently "spooned," sleeping in close proximity to keep warm. This practice did not indicate sexual preference and rarely evoked sexual innuendo among the men who practiced it or, more crucially, those who didn't.

Abraham Lincoln had a ready supply of fuel to warm his rooms, but perhaps he needed company to take the emotional chill off his evenings, especially during his wife's absence from the Soldiers' Home. Derickson's companionship for her husband was acknowledged by Mary Lincoln, who had won Derickson's approval by being "one of the best rebel haters that I met during my stay in Washington." Mary, in turn, recognized Derickson's influence during his daily rides with her husband to the White House. She even attempted to

have the captain use his powers of persuasion to rein-
force her own political agenda in his daily conversations
with the president.

Their mutual admiration society had Mary Lincoln's
full approval, and many mornings she was sighted at
the window in her nightclothes, watching Derickson
and Lincoln head off to the city together. Mary Lincoln
appreciated how time with this affable companion from
Meadville, Pennsylvania, a man who could remind the
president of his early days in Springfield, might lift his
spirits. She viewed Derickson more as an ally than any
kind of rival.

Washington was at first flushed with triumph at
reports of the September 17 battle near Antietam
Creek, but soon the town was flooded with news of
the horrors—the Union casualties and the stream running red. This military outcome at Antietam offered
an opportunity to call a Cabinet meeting on September
22, and the next day Lincoln released his preliminary
Emancipation Proclamation. But he knew what battles
lay ahead, and took Derickson along on his excursion
to rally the troops on October 1. He wanted to force his
top general, McClellan, to mobilize more fully against
the enemy.

The cries for changes in the administration, coupled
with demands for changes in the army command, were

deafening roars by October 1862. With midterm elections coming up, the Republican Party was under siege. Mary Lincoln was deeply concerned. She not only read her husband's mail ("I hold a letter, just received from Gov. Sprague") in his absence, but she revealed the contents of the president's correspondence and contributed her own two cents to the debate: "If my good, patient Husband, were here, instead of being with the Army of the Potomac, both of these missives, would be placed before him, accompanied by my womanly suggestions."

Recognizing that this kind of meddling might backfire, she confided to James Gordon Bennett (owner of the *New York Herald*) that she had a terror of "*strong minded Ladies*," and felt it was her duty to withhold her opinion at a time such as this. She also suggested that the president was not swayed by "ambitious fanatics," especially "when his mind is made up." And by the fall of 1862, Lincoln's mind was made up—to release the Emancipation Proclamation and, acknowledging his dissatisfaction with McClellan following his inspection of the troops, to find a new general to win the war.

On September 23, a spy was discovered lurking near the Soldiers' Home quarters and sent to the Old Capitol prison. An alarm in the middle of the night on September 25 awakened the couple, as guards redoubled their efforts to keep the president safely surrounded. In the weeks to follow, Lincoln became even more dependent on Derickson. When Mary headed out with Tad on a vacation tour on October 20, she knew she was leaving him in good hands.

At the same time, these absences stirred up gossip about the Lincolns. Two Washington politicians traded tales in the fall of 1862 about the private lives in the White House. T. J. Barnett wrote to S. L. M. Barlow after Mrs. Lincoln left the city: "I suppose you have heard the on dit of Mrs. Lincoln and the 'Chevalier' Wykoff, how the old man became jealous, and taxed his spouse, and the Chevalier volunteered an explanation, and how he told the wounded and incensed consort that he was only teaching the madame a little European Court Etiquette, and how they were reconciled."

This is but one example of the countless commentaries that attempted to interpret the relationship between the Lincolns without any evidence on these matters. Idle speculation on the First Couple became popular during the nineteenth century, and continues to fascinate well into the twenty-first.

When another company was sent to replace Derickson's "Bucktail Unit" on October 31, the president sent a note to prevent the transfer, to maintain the company's status as the President's Military Guard. By early November, with colder weather forcing longer campfires among Derickson's unit, the president wrote to Mary—who had been away by then for nearly a fortnight—that he was wondering if he should return to the White House.

Mary's second excursion away from Washington since Willie's death was a more extended tour. This trip included visits with her son Robert in Boston, and she was more social and outgoing than she had been in the summer. She toured docked ships, raised donations for her favorite causes (most especially soldiers' hospitals), even volunteered to sign a petition to commute the sentence of a New York woman condemned to death for the murder of her husband, whom she had shot upon discovering his infidelity. Some critics would suggest this murderous and jealous temper was something with which Mary could empathize.

Mary missed her husband but feared he did not miss her. She could rarely cajole him to dispatch regular correspondence. On November 2, she complained, "I have waited in vain to hear from you, yet as you are not given to letter writing will be charitable enough to impute your silence to the right cause." This was two days before their wedding anniversary. Most likely they both knew that Lincoln was not feeling abandoned.

Instead, Mary's absence had allowed him to revert to his favorite atmosphere of meals around the campfire with a plate of beans and a cup of coffee sustaining him. But the cold and the press of business pushed him back into his White House web, and on November 9, Lincoln and his entourage of servants made the trek back to Pennsylvania Avenue.

By the time Mary returned from her excursions, Lincoln was tightly drawn into his circle of political advisers. They were putting their heads together to try to achieve a revolution so audacious that it would change the country, even if it meant splintering his support. The moderates were all but forced out by Lincoln's new policies. Lincoln's close friend, Senator Orville Browning, became disaffected by what he saw as his leader's shift from his former position to save the Union at all costs. Abraham Lincoln committed himself to a policy of destroying slavery, which put him on the road to revolution. And the Emancipation Proclamation became the hallmark of his presidency.

Lincoln was absorbed and engulfed with his disunited states, while Mary struggled to maintain fractured equilibrium. The Lincolns were battered and bruised by the continuing storms of life. But unlike other couples, their decisions would shape the lives of so many others. The president acknowledged this special role in his State of the Union message in December 1862, concluding with these ringing sentiments: "Fellow citizens, we cannot escape history. . . . In giving freedom to the slave, we assure freedom to the free—honorably alike in what we give and what we preserve. We shall nobly save, or meanly lose the last best, hope of earth."

12.
Gloomy Anniversaries

Pale and drawn-looking, still dressed in a black bonnet and dress, Mary Lincoln stood in a receiving line on January 1, 1863, and commented, "How much we have passed through since we last stood here." Weighed down by comparisons with holidays the year before, she had written to a sympathetic friend on December 16, 1862, "From this time until spring, each day will be almost a gloomy anniversary. My precious little Willie is as much mourned over & far more missed (now that we realize he has gone) than when so fearful a stroke as to be called upon to resign Him came."

The president strengthened his position with the midterm election outcome in November 1862. He averted a crisis with his Cabinet by refusing resignations from both Seward and Chase, confessing with characteristic modesty, "I may not have made as great a President as some other men, but I believe I have kept these discordant elements together as well as anyone could."

This political victory was mitigated by doom and gloom over the ghastly bloodbath near Fredericksburg in December. The removal of McClellan was not going to solve all or perhaps even most of the current military dilemmas. McClellan's replacement, Burnside, contributed to his own problems at the front. The pressures included demands for the president's resignation along with the "traitoress Mrs. Lincoln," as critics in the press labeled her.

Mary had been in Philadelphia on charity missions before the holiday, but on Christmas Day she and her husband visited the District of Columbia hospital wards. The president took time out from his protracted Cabinet discussions—concerning the admission of West Virginia and the forthcoming Emancipation Proclamation—to spend time with wounded soldiers.

And finally, on New Year's Day, with the Cabinet assembled, with throngs circling the White House, an official copy of the proclamation was released. Lincoln steadied his hand to make sure his signature was firm, knowing the import of this document he was issuing, aware of its significance to his presidency, to his own

The Emancipation Proclamation

reputation. He later allowed the moment to be recorded with a historic painting, as the scene was being set to create his legacy.

Crowds surrounded the president's home, especially enthusiastic groups of African Americans. Emancipation Day was a time of extreme jubilation. The Lincolns were engulfed by the celebrations that spontaneously erupted that New Year's Day. The Reverend Henry Turner, an African Methodist Episcopal pastor in the District of Columbia, grabbed a freshly printed broadsheet of the proclamation and ran down Pennsylvania Avenue to read it aloud to the blacks collected around the White House. Turner, brandishing his copy, was met with "a cheer that was almost deafening."

This celebration by D.C. blacks paled in comparison to those launched by officers in the occupied South. General Saxton in South Carolina organized an official reading of the proclamation in the Sea Islands. Anticipating the crowd of over five thousand, he ordered a feast of twelve roast oxen, bread, molasses, and other provisions. The three-hour program of prayer, presentations, hymns, and speeches was a moving and memorable occasion.

By January 1863, Mary's hospital work had become her most zealous campaign—next to protecting her husband from his enemies. She visited the hospitals two or three times a week. The Washington *National Republican* reported, "Day by day her carriage is seen in front of the hospitals. . . . The fear of contagion and the outcries of pestilence fall unheeded."

Even the sniping young male secretaries of the White House were impressed, although William Stoddard suggested: "If she were worldly-wise she would carry newspaper correspondents, from two to five, of both sexes, every time she went and she would bring the writers back to the WH, and give them some cake." Mary Lincoln, like the infamous Marie Antoinette, may have been in desperate need of better press, but she was unwilling to distribute cake to get it.

The First Lady was not above trying to ensure her husband's place in history, however. For Christmas, Mary sent a photograph of her husband to Josiah Quincy, the former president of Harvard. This likeness arrived in Boston just when the city was overwhelmed by Lincoln's Emancipation Proclamation. Eliza Quincy, Josiah's daughter, wrote to Mrs. Lincoln on January 2, following Emancipation festivities in Boston, where "it was a day & occasion never to be forgotten." These events were made even more significant by having this personal memento of Lincoln in hand.

As the anniversary of Willie's death approached, Mary Lincoln became more restless, more involved in White House social obligations. She resumed Saturday visiting hours, receiving Fanny Seward on January 17 and Wendell Phillips on January 24. Other guests included Virginia Fox, the wife of Lincoln's favorite, Assistant Secretary of the Navy Gustavus Fox. Elizabeth Blair Lee, sister of Postmaster General Montgomery Blair, recalled on January 14 Mrs. Lincoln's complaining that "there was not a member of the Cabinet who did not stab her husband and the Country daily except my Brother." The Lincolns resumed evening entertainments and began to venture out again, attending a patriotic reading in the Senate chambers by J. E. Murdoch, who recited "The Sleeping Sentinel."

Mrs. Lincoln hosted a "wedding reception" for theatrical sensations Charles Stratton and his bride, Lavinia Warren, who were known as General and Mrs. Tom Thumb. The First Lady had been urged to hold this entertainment at her home, and she rose to the occasion on February 13. Robert was home from college, and when his mother asked him to dress and come downstairs to the reception and its fifty guests, he demurred: "My notions of duty, perhaps, are somewhat different from yours."

His mother, ignoring his bad humor and his criticism, enjoyed the evening. Robert's brother Tad was charmed by the "little playfolk" in their "wedding finery" and busied himself attending to their needs. Seeing her son thus engaged, just a year after Willie's death, when he had lain so low on his own sickbed, heartened Mary. Mrs. Lincoln wore a rose-colored dress with her signature low neck and flowers in her hair. One of the earliest guests at the reception had been William Chase, with his daughter Kate, described by writer Grace Greenwood as "tall, graceful, her small Greek head borne royally, her lovely piquant face untouched by care or sorrow."

Virginia Fox called on Mary the next day and found her in high spirits. This was surprising, all the more so because the anniversary of Willie Lincoln's death was less than a week away—a date Mary Lincoln observed with deep sorrow. She sent a thank you to Benjamin French for his note of remembrance, explaining, "Our heavy bereavement has caused this to be a very painful winter to me." On February 21, she wrote another friend who had sent comforting correspondence:

My dear Mrs. Welles: Allow me to thank you for your sympathizing & kindly remembrance, of yesterday, when I felt so broken hearted. Only those, who have passed through such bereavements, can realize, how the heart bleeds at the return, of these anniversaries.

That same day, Mary greeted guests at a reception with "the affability for which she is distinguished" and hosted Dr. Anson Henry at a White House dinner.

The good doctor from Springfield might well have marveled at how far his former patient had come: He had treated Lincoln during his suicidal depression soon after his temporary rupture with Mary Lincoln in 1841. More than twenty years later, Lincoln had risen to the White House with his wife by his side, and was carrying the cares of the nation upon his capable shoulders.

Lincoln had worried when he was in his early thirties that he should make something of himself. He wrote to his dearest friend, Joshua Speed, that he would have been willing to die in 1841 except that "he had done nothing to make any human being remember that he had lived." Now Lincoln was president, transforming the nation from a slaveholders' republic to a country rededicated to the principle that all men should be free—and perhaps one day equal. He had launched a revolution by enlisting African Americans in the Union Army.

During the early months of 1863, Lincoln remained under great pressure, with mounting challenges. Yet he continued therapeutic theatergoing, and he attended the last levee of the season, on March 2, where Mrs. Lincoln looked particularly regal, in black satin and jewels. Mary Lincoln threw herself into a crowded social schedule after her year of suspended activities. She continued her hospital work and dabbled in patronage, increasing her struggles to secure positions for friends.

At an April reception, reformer and writer Jane Swisshelm was won over to Mrs. Lincoln after a short acquaintance. Swisshelm was a radical abolitionist, a journalist who defied the boundaries established by her privileged Pittsburgh upbringing. She married and had a child, but was separated from her husband and earning her own living. She was a correspondent for Greeley's *New York Herald* during the 1850s and the first woman to report on politics from the Senate gallery. During the war, she returned to Washington and secured a government job through her friendship with Secretary of War Edwin Stanton. She also served in Union hospital wards as a nurse. She remained an indefatigable champion of slave emancipation and the rights of women.

Swisshelm first met the Lincolns at a White House reception on April 2, 1863. Previous to this occasion, she had *not* wanted to meet Lincoln, and most especially not his wife. She had heard Mrs. Lincoln was a Confederate spy and therefore claimed to despise her. Imagine Swisshelm's surprise when she felt instant sympathy upon meeting both the Lincolns. When she shook hands with her, Mary Lincoln told Swisshelm not to worry that her glove might soil the First Lady's. She suggested that if it did, she would "preserve the glove to remember a great pleasure for long have I wished to see you." Swisshelm was lavish in her compliments: "I recognized Mrs. L as a loyal, liberty loving woman." The two women found they had common interests, and the unlikely pair became mutual admirers.

Swisshelm later suggested that not only had Mary Lincoln become radically opposed to slavery, but that she was the *one* who urged the president "to Emancipation, as a matter of right, long before he saw it a matter of necessity." Furthermore, "whatever aid or counsel she gave him, in her eyes, his acts were his own, and she never sought any of the credit due them." Her notion of a politically astute Mrs. Lincoln strikes a genuine note. However, modesty was rarely part of Mary's repertoire.

The First Lady's rebirth into passionate abolitionist from a slaveholder's daughter had many midwives. Her close friendship with Elizabeth Keckly in the White House revived memories of slavery's brutalities from her impressionable Kentucky years. Her deepening friendship with radicals like Charles Sumner had a powerful effect. Mary Lincoln's opposition to slavery reflected a genuine conversion, not a shift undertaken to court favor. Emancipation, rather than merely opposition to the extension of slavery, developed into a litmus test for Washington's abolitionist elite. And Mrs. Lincoln won over former critics with her sincerity: "Her expressions were strong enough to satisfy any abolitionist."

Early on, Mary Lincoln was flattered by Sumner's White House appearances, as he would stop by the presidential manse to draw Mary out. The two exchanged floral tributes, traded visiting cards, and during Sumner's frequent chats with Mary, "often late in the evening—My darling husband would join us

& they would laugh together, like two school boys." Sumner's courtly attentions to Mrs. Lincoln allowed the president to warm to this statesman and made him a particular friend.

Both the president and his wife exerted themselves for their remaining son at home, taking Tad on a water excursion on the steamer *Carrie Martin* and presenting him with a pony on April 4 in honor of his tenth birthday. The family enjoyed the company of Dr. Anson Henry, along with White House aide Noah Brooks. They camped in tents at General Hooker's headquarters near Fredericksburg, where the mood was light. While on an excursion with General Daniel Sickles, Lincoln was bestowed with kisses by the infamous Princess Salm-Salm.

The princess, née Agnes Joy, was born the daughter of a respected Vermont general (distantly related to the president himself) and was living with her sister and brother-in-law in wartime Washington. She married a titled European soldier of fortune against her family's wishes. The flame-haired princess had attracted the attentions of many soldiers, especially during her rides along Pennsylvania Avenue astride her mustang, Minnehaha. The equestrian beauty at times eschewed a saddle (claiming an Indian foremother) and rebelled against Washington's conventions. Her notorious reputation was somewhat redeemed by her marriage to titled nobility.

But after her marriage to the prince, the bride followed her husband to the battlefield. The princess nursed wounded soldiers in their campsites, again in defiance of custom. But presumably Lincoln found her spirit and patriotism refreshing. Nevertheless, he was taken aback when she entered the military tent near Aquia and "kissed him three times—once right, once left, and once on the mouth—amid considerable gaiety."

The gaiety might have become ribald on the road back to headquarters, and Mrs. Lincoln soon heard details of the incident. General Daniel Butterfield's wife reported that Lincoln made a joke over dinner the next day to melt Mrs. Lincoln's frosty manner toward General Sickles, who had been his escort. Lincoln turned to Sickles and said, "I never knew until last night you were a very pious man. . . . Mother says you are the greatest Psalmist in the army. She says you are a Salm-Salmist." This mischievous remark apparently broke the ice as Mrs. Lincoln joined in the laughter.

While Lincoln reviewed the troops, Tad had been having about as much fun as any ten-year-old could. The press had a field day painting pictures of the younger Mr. Lincoln by his father's side. Journalists were uncharacteristically low-key about the First Lady. One reporter noted, "Mrs. Lincoln's attire was exceedingly simple—of that particular style of simplicity which creates at the time no impression upon the mind, and prevents one from remembering any article of dress." This successful family outing was a media triumph, especially compared with previous dissections of Madame Lincoln's every move and wardrobe accessory.

The atmosphere was very different when Lincoln visited Hooker's headquarters a month later. Mrs. Lincoln found it impossible to keep her husband diverted, as the drumbeat of war pounded away. The Battle of Chancellorsville began on April 30 and ended on May 6. This Union defeat was called "Lee's perfect battle" because the Confederate general divided his army when facing a much larger enemy force, a gamble that won the day. The reports from the front hammered away at fraying nerves in the White House. In once instance, Mrs. Lincoln received women friends wearing just her wrapper and with her hair down, agitated as she walked the floor, exclaiming, "They are fighting at the front; such terrible slaughter; and all our generals are killed, and our army is in full retreat; such is the latest news."

As in summers past, Mrs. Lincoln became restless and made several excursions northward. She found the comfort of friends like Sally Orne, the wife of a Philadelphia carpet merchant, welcome respite from the capital, where her husband remained anchored. On June 9, the day after she left D.C. with her younger son, Mary received a telegraph from her husband with a frantic message: "Think you better put Tad's pistol away. I had an ugly dream about him." Mary sent Abraham reassurances, and after three updates from her, he finally replied, "I am very well, and am glad to know that you and Tad are so." Days later, he said he was "tolerably well," which was putting the best face on his situation, as he was under considerable duress.

The war continued going badly for the Union military. The hollows in Lincoln's cheeks were increasingly visible. Furthermore, the president was dealing with political fallout of a most unpleasant kind: His brother-in-law Ninian Edwards had been accused of using his position as Springfield commissary to line his own pockets. As if that was not enough, bulletins warned of Confederate incursions bringing the war into the North.

While staying at the Continental Hotel in Philadelphia, Mary Lincoln was advised by her husband, "I do not think the raid into Pennsylvania amounts to anything at all." Military intelligence must have warned him about Lee's push northward, which would make the crossroads at Gettysburg in a little over a fortnight so vital. In any case, Mary returned home on June 13 in preparation for the family's annual retreat to the Soldiers' Home. She organized the move for a little over a week later.

Sadly, the retreat to their summer house did not bring the peace for which Mary had hoped, and it was marred by a carriage accident. During his daily journey from the Soldiers' Home to the White House on July 2, Lincoln decided not to take his own carriage, but to ride horseback, escorted by mounted bodyguards. The plan was for Mrs. Lincoln to follow in his carriage alone.

When she did, serious mishap befell. Mary was thrown to the ground from the vehicle. She landed badly, and her skull connected with a rock. As the papers reported, "The driver's seat became detached from the carriage in some way, precipitating the driver to the ground. . . . She jumped and received severe bruises, the most severe was on the back of her head." Reports of vehicles going awry or horses bolting had appeared in the press before. And the detachment of the seat from the carriage "in some way," did not seem menacing—until investigators discovered that the "accident" was no accident.

The carriage had been tampered with, and fears over the president's safety multiplied. Mary's injury was the collateral damage of sabotage. That his wife had been injured in his stead deeply wounded Lincoln. This news unfolded over a matter of days, with the war's most dramatic events as the backdrop.

Just minutes after her fall, Mrs. Lincoln was attended by a doctor at an army hospital, who suggested that her cuts were superficial and sent her home. Nevertheless, Lincoln believed his wife would need assistance, and called on Mrs. Pomroy to watch over the First Lady— once again. This seemed only precautionary, as in the first days after her head injury, Mrs. Lincoln moved rapidly toward full recovery.

William Stoddard had been consulting with the First Lady about plans for a Fourth of July celebration. With Lee's troops on the move and military engagements forthcoming in Pennsylvania, some Washington dignitaries wanted to call off elaborate celebration plans: the Marine Band on the White House grounds, the viewing stand on the mall, regiment after regiment signed up to march alongside twelve councils of the Union League, Masons, and Odd Fellows. This spectacle of support would bolster patriotism. Congress's new measures for a military draft—scheduled to take effect in early July—were drawing fire. But with the enemy's bold advances, re-solve began to crumble.

When Stoddard shared the committees' fears with the First Lady, she replied, "Mr. Lincoln is positive Lee will be defeated. . . . Don't you stop." Stoddard prom-ised to push ahead with elaborate preparations, but he needed more flowers. She offered him her stock from the White House conservatory and asked, "Anything else I can do?" To which he responded, "No, unless you can speak to Grant and Meade and have them win their victories on the third so we can have a bigger Fourth." And indeed, prayers were answered with a double victory. As Stoddard reported, "Can anybody measure the fierce intoxication of the Fourth of July celebration, with the news from Vicksburg and from Gettysburg thundering in the ears and thrilling the hearts of men?"

While firecrackers and patriotic fervor filled the air, shades were drawn at the White House. Mary Lincoln's condition took a turn for the worse, as this head injury turned serious. Mrs. Pomroy detected infection and knew Mary's wound needed reopening. Inflammation exacerbated the situation, as the First Lady took to her bed following the surgical procedure.

Because Mary's accident had been reported in all the papers, the president initially sent a telegram to their son Robert telling him not to worry. But when her condition worsened, he telegraphed Robert in New York City on July 11, "Come to Washington." When he failed to show up, Lincoln became even more distressed over his son's whereabouts. His worries only increased when the infamous Draft Riots broke out.

Mobs attacked a draft office in New York City on Monday, July 13. Under the Conscription Act, a draftee could avoid service by either providing a substitute or paying a fee of $300, a feature that Lincoln had opposed but Congress allowed. This infuriated those who felt it was a "rich man's war, but a poor man's fight." One of the first places to be looted during the Manhattan rampage was a Brooks Brothers store. Soon riots erupted throughout the city. Diarist George Templeton Strong blamed the violence on Irish immigrants—"every brute in the drove was pure Celtic"—and anarchy reigned.

The draft riots had severe consequences for African Americans trapped in the melee during those ghastly four days. Princess Salm-Salm described their plight in her memoir: "Wherever Negroes were discovered, they were hung or either barbarously murdered." The beleaguered president was forced to send in troops to restore order.

By July 14, Lincoln was overwhelmed with worry and wrote to his absent son, "Why do I hear no more of you?" His anxiety was extreme: While much of the nation rejoiced in the victories at Gettysburg and Vicksburg, angry mobs roamed Manhattan, and his wife was fading, felled by a plot against him.

To make matters worse, Lee escaped because General Meade failed to pursue the retreating enemy. This reminded Lincoln of previous generals, causing him deep despair. He penned letters of rebuke to vent his feelings, but kept them in the drawer to keep the peace. Robert Lincoln finally did show up, and later recalled his father's deep gloom over this military blunder: He was "in tears with his head bowed upon his arms." Although nothing could be done about Lee's escape, Lincoln took solace in Mary's recovery, slow though it was.

Under Pomroy's vigilant care, Mrs. Lincoln improved, and the nurse earned the Lincolns' continuing gratitude. Mary needed a healthier environment, however, as a heat wave broke out in the capital. By July 28, she headed north—to New England, joined by both Tad and Robert. She could not suffer the tropical summer weather in Washington, especially with the malaria scare in the city.

William Stoddard described the problem at length: "We are not above the tide-water here, and there is no current to speak of at high tide . . . an ooze has

been developed which can testify its peculiar qualities to best advantage when the river is low and the
tide is out." The newspapers reported that both
men and horses were dropping dead in the streets.
Stoddard and the other young men of the White
House staff fell ill. The best advice physicians could
offer was to flee the capital, which is exactly what
Mary and her sons did.

On her journey northward to the Green Mountains,
Mrs. Lincoln did not ignore her ceremonial role. While
in New York, she paid a visit on August 20 to the
French frigate *La Guerrière*, where she was received
with a courtesy and flourish characteristic of French
gentlemen. "Mrs. Lincoln was delighted in the attentions paid to her on board."

The First Lady had selected for her convalescence a
resort in Manchester, Vermont—the Equinox, nestled
in the magnificent Green Mountains. This spot began
as a popular tavern during the American Revolution,
frequented by the legendary Green Mountain Boys. In
1853, Franklin Orvis opened the Equinox Hotel and
billed it as a premier summer resort, with its natural mountain mineral water advertised for its "health
maintaining properties." The air, the waters, and the
well-heeled clientele recommended the place to Mrs.
Lincoln.

The president was left behind, but his letters were full of charm and news, as when he wrote on August 8, "Tell dear Tad, poor 'Nanny Goat' is lost; and Mrs. Cuthbert & all are in distress about it." (Unfortunately, this letter never actually reached Mary; it was returned to the White House after going astray and landing in the possession of a soldier.)

During this period, Lincoln was clearly thriving, after what he hoped was a turning point in the war. Union victory appeared more and more a certainty, and one of the president's personal secretaries wrote, "The Tycoon is in fine whack. I have rarely seen him more serene and busy."

Mary's activities were chronicled in the papers as well—her travels to the Green Mountains with General Abner Doubleday, her toasting the emperor of Russia on a Russian frigate docked in New York, her various perambulations. Lincoln appreciated his wife's ceremonial duties as well as the good publicity they garnered, but eagerly awaited his family's return to Washington, missing especially his son Tad.

By the seventh week of Mary's absence, Lincoln's entreaties for her to return home become more wheedling: on September 20—"I neither see nor hear anything of sickness here now" and on September 21— "The air is so clear and cool, and apparently healthy that I would be glad for you to come." He also used go-betweens to try to get his family back to Washington; on the twenty-second, he wrote: "Mrs. Cuthbert did not correctly understand me. I directed her to tell you to use your own pleasure whether to stay or come; and I did not say it is sickly and that you should on no account come. . . . I really wish to see you. Answer this on receipt." Mary responded that she had called for a car to return from New York, that she was anxious to return home.

Before Mary arrived back in Washington, the president sent along news of a Confederate triumph in Tennessee, when Union General Rosecrans was bested by Confederate General Bragg at the Battle of Chickamauga. The president was subdued by military matters, and anxious for Mary's imminent return. He was devastated when black-bordered news arrived: the death of Confederate General Ben Helm, Mary's brother-in-law, the husband of the Lincolns' beloved Little Sister. Ben Helm had turned down Lincoln's generous offer for a high commission in the Federal army, and returned South to enlist with the Confederates. He had been promoted to brigadier general after Shiloh and was commanding the famed Orphan Brigade (the Kentucky volunteer unit) when he was killed at Chickamauga on September 20. When David

Davis arrived at the White House on the very day Lincoln heard the news, he found the president in a state of personal despair: "I feel as David of old did when he was told of the death of Absalom."

When Mary arrived four days later, she and Lincoln mourned in private because, as her niece confided, "She knew that a single tear shed for a dead enemy would bring torrents of scorn and bitter abuse on both her husband and herself." Mary hid her grief not from her husband, but to protect her husband.

At the same time, family was family. Thus Ben Helm's father, the former governor of Kentucky, wrote to Emilie's mother, Mary Lincoln's stepmother, begging her to contact the First Lady to help secure a pass for travel to the rebel states. Within three days, Lincoln gave written permission for Mary's stepmother to go through enemy lines to fetch her daughter and Emilie's children, to bring them back to Kentucky.

Mary was beset by worries on other fronts, as by November 1863, she was besieged by creditors—not uncommon for Washington's society ladies but a problem that Mary had to keep from her husband. Expenses were only part of the festering secrets within the Lincoln White House, as Mrs. Lincoln lapsed into a series of states of denial.

Lincoln sought diversion from his worries at the theater. On October 17, Charlotte Cushman played Lady Macbeth to James Wallack's Macbeth, in a benefit for the Sanitary Commission, which netted over $2,000. The gala not only entertained the president and First Lady, but was attended by "Master Thaddy." On November 9, the president enjoyed *The Marble Heart* at Ford's Theatre, a play starring John Wilkes Booth, a serious rival to his brother Edwin's premier status and the heir apparent to the throne of his father Junius, theretofore the greatest British actor in America. But Mrs. Lincoln found herself in a tangle of debt, worry, and social rivalries, unable to find relief from her troubles.

The First Lady pleaded illness when her husband suggested they both attend the wedding of Kate Chase to Senator William Sprague on November 12, the social event of the season. The match attracted widespread attention and even turned up in the enemy press: A notice in the *Savannah News* described the engagement ring, a solitaire worth $4,000. At the wedding, Kate Chase was resplendent in a white velvet bridal dress with a needlepoint lace veil. Lincoln knew his wife's absence would stimulate talk, so he spent over two hours at the reception to "take the cuss off the meagerness of the presidential party." But it was Tad, not Mrs. Lincoln, who genuinely took ill a few days later—causing another White House crisis.

A national cemetery had been scheduled to open at Gettysburg, and Lincoln had been invited to speak. The little town of roughly 2,500 had been overrun the previous July by more than 170,000 soldiers. The armies suffered over 50,000 casualties. The reburial of dead soldiers became a primary headache—a debt due for this glorious turning point in the war.

This festival of death in the North, on such a scale, was relatively new for the Union stalwarts, while the death of Confederates on Southern soil had become a regular part of the landscape. By war's end, nearly twenty percent of white Southern males of military age would die. The outpouring of grief would consume the Confederate nation, and particularly women. A young girl in Virginia commented on her community, "So many ladies here, all dressed in mourning, that we felt as if we were at a convent and formed a sisterhood." But this shared season of grief had visited its terrible toil on the North in Gettysburg, whose people responded to this dramatic challenge.

David Wills, a Gettysburg lawyer, led the campaign to establish a federal cemetery, involving the burial of over 3,500 Union soldiers. Elizabeth Thorn, the wife of the main gravedigger of Evergreen Cemetery (who was away in the army), dug graves herself, aided only by her aging father-in-law and hired workers. She spent the rest of the summer, including the last few months of a pregnancy, measuring off sites and interring nearly one hundred Union corpses. This burial work, which began in July 1863, was not completed until March 1864, nearly four months after the president traveled to Gettysburg. Lincoln's inspirational visit paved the way for the town to become a national shrine. Certainly one of the major events that turned this sleepy battleground into the thriving cultural heritage site it has become was Lincoln's Gettysburg Address.

The day before dignitaries were scheduled to dedicate the National Cemetery, Mary Lincoln begged her husband not to go, hysterical over her younger son's ill health. Tad's decline echoed the fatal stupor into which Willie had fallen nearly eighteen months before. As Lincoln prepared to depart for Gettysburg, Radical Republican Thaddeus Stevens, majority leader of the House, refused to go along—even to his home state of Pennsylvania. When Stevens was asked why Edwin Stanton and Salmon P. Chase were not joining the president, he replied, "Let the dead bury the dead"—a snide reference to the president's flagging political prospects. William Seward and Montgomery Blair accompanied Lincoln on the noon train on November 18, but other

Cabinet secretaries begged off. The annual message to Congress was due soon, there was another year of war to explain, and politicians were seeking cover.

Abraham Lincoln believed he had a duty as well as an opportunity to honor the dead. He wanted to highlight the nobility of patriotic sacrifice—which he certainly did during his brief, poetic speech. As the time to depart grew near, Tad became even more unwell. The morning Lincoln was scheduled to leave, the boy was too ill to eat his breakfast. Lincoln was torn between his fears for his son and his need to bolster spirits in rural Pennsylvania. He left the capital with a heavy heart. After a long day, he received a message from Stanton that "Mrs. Lincoln informs me that your son is better this evening." Lincoln was up past midnight working on his speech.

The next day, Stanton telegraphed again: "Mrs. Lincoln reports your son's health is a great deal better and he will be out today." This bolstered Lincoln's outlook even more. On the platform he sat serenely through Edward Everett's two-hour oration, which the former president of Harvard delivered from memory. Everett was the most polished speaker of his day, the one everybody had turned out to see.

Following Everett, the president rose to the podium, put on his glasses, and read out what has become the

most famous speech in American history—Lincoln's
Gettysburg Address.

Four score and seven years ago our fathers brought
forth on this continent a new nation, conceived in
Liberty, and dedicated to the proposition that all
men are created equal.

Now we are engaged in a great civil war, testing
whether that nation, or any nation, so conceived
and so dedicated, can long endure. We are met on a
great battle-field of that war. We have come to dedicate a portion of that field, as a final resting place
for those who here gave their lives that that nation
might live. It is altogether fitting and proper that
we should do this.

But, in a larger sense, we can not dedicate—we
can not consecrate—we can not hallow—this
ground. The brave men, living and dead, who struggled here, have consecrated it, far above our poor
power to add or detract. The world will little note,
nor long remember what we say here, but it can
never forget what they did here. It is for us the living,
rather, to be dedicated here to the unfinished work
which they who fought here have thus far so nobly
advanced. It is rather for us to be here dedicated to
the great task remaining before us—that from these

honored dead we take increased devotion to that cause for which they gave the last full measure of devotion—that we here highly resolve that these dead shall not have died in vain—that this nation, under God, shall have a new birth of freedom—and that government of the people, by the people, for the people, shall not perish from the earth.

Spontaneous applause following the speech was, by all accounts, warm and enthusiastic. However, no one in the crowd, least of all Lincoln, could have imagined his words would become a national gospel, a creed that transcended its context and gave him lasting and international fame.

Nevertheless, Lincoln's speech was a revelation to Edward Everett, who wrote to the president the next day: "I should be glad if I could flatter myself that I came as near to the central idea of the occasion, in two hours, as you did in two minutes." After dining with Lincoln, Everett lavished even more praise on him. Mary would have loved Everett's estimation that the polish Abraham Lincoln had acquired "may be credited to the influence of his wife." The crisis in her son's health had passed, her husband had triumphed at Gettysburg, and all would perhaps be well with the world, once the wretched war was over.

13.
Divided Houses

The strain of a divided house, which had been a metaphor for most of the war, became quite literal when the Lincolns welcomed Mary's half sister Emilie Todd Helm to stay with them in December 1863. She arrived in Washington under presidential order and protection. Her brief stay stirred up an emotional tsunami and underscored fissures within the Lincoln White House. This Southern sister's visit unbalanced the high-wire act Mrs. Lincoln struggled to maintain.

The recently widowed Emilie, whose Confederate husband was killed in battle, was trying to make her way into Kentucky to be with her mother, when she was stopped at the border for refusing to take the oath of allegiance to the United States. Emilie complained that this gesture would dishonor her dead husband's memory. Military authorities, at a standoff with the determined young widow, detained her at Fort Monroe. They finally sent a telegram to the president asking what they should do. He replied, "Send her to me, A. Lincoln."

Family members later described Emilie Helm, when she arrived at the White House, as a "pathetic little figure" with "pallid cheeks, tragic eyes," swaddled in black crepe. She was embraced tearfully by both Lincolns. Her children were comfortably ensconced, with their cousin Tad as companion. The first night, Mary and Emilie had dinner à deux, avoiding inflammatory subjects, fearing to "open a fresh and bleeding wound." Emilie appreciated tours of the grand salons, noting that the Red Room contained the portrait of George Washington that Dolley Madison, a Todd kinswoman, had saved. She settled in to the guest quarters, known as the Prince of Wales Room, with its gloomy purple hangings brightened only by yellow cords. A fire kept her comfortable despite the December chill.

The strains of White House life were extreme, as Mary felt herself "the scapegoat for both North and South." Her mail was often so hateful that Lincoln's secretaries screened all correspondence. Even Republican ally Orville Browning suggested that Lincoln wanted Mrs. Helm's presence in the house kept secret. Yet both Lincolns wanted Emilie to remain with them. Lincoln hoped the two sisters could comfort one another, as he feared his wife's nerves had "gone to pieces." After Emilie had spent some time with Mary, she understood the intensity of her brother-in-law's concern.

Emilie found Mary agitated and angst-ridden. She was alarmed by her older sister's "wide and shining" eyes during rapturous descriptions of visitations from the dead. To Emilie, this confirmed her sister's "abnormal" state. The First Lady longed to commune with her dead son Willie—not just in spirit circles, but when he came to her chamber at bedtime. Mrs. Lincoln confessed these nocturnal apparitions were transporting: "He lives, Emilie! . . . little Eddie is sometimes with him and twice he has come with our brother Alec. He tells me he loves his Uncle Alec and is with him most of the time. You cannot dream of the comfort this gives me." Mary's weepy confessions wrenched her sister's heart, as Mary continued, "When I thought my little son in immensity alone without his mother to direct him, no one to hold his little hand in loving guidance it nearly broke my heart." A belief in an afterlife might ease her suffering, but these kind of agitations unnerved Mary's intimates.

Emilie Helm, like most of her countrywomen North and South during the Civil War, was buoyed by faith in Christian redemption. She embraced the tenets of spiritual salvation but feared the effects of Mary's preoccupation with the dead. The First Lady's embrace of the afterworld, Emilie recognized, was causing an increasing imbalance—a withdrawal from the real world. Instead of being tempted herself by these escapes, Emilie recoiled. Abraham Lincoln hoped Emilie would stay as long as she could, because "it is good for her [Mary] to have you with her." Perhaps as an antidote for both spiritualists and spirits, Lincoln invited Emilie to summer with them at the Soldiers' Home.

This illuminates the pressures Mary endured as First Lady. When she lost her son Willie, she was able to lean on sympathetic friends—Mary Jane Welles, Rebecca Pomroy, Elizabeth Keckly. However, she had a limited number of fellow mourners, and many urged her to pack away her grief and get on with taking care of her family—and she struggled to do so. Mary's prolonged imbalance was a devastating hardship on Tad, coping with his own sense of loss. Mary's situation was exacerbated by her isolation from her family roots, exiled from her "old Kentucky home" and from her Southern relatives allied with the Confederacy. This sense of exile was reinforced and her isolation was accentuated by the desertion of her Springfield relations.

To further compound matters, Mary knew she was suspected of Confederate sympathies, and any gestures she made could be blown out of proportion. She was criticized for the company of her sister Emilie in the flesh, as she was for that of her stepbrother Alec in spirit. But having Emilie to talk with released Mary's pent-up feelings, emotions she had perhaps kept from her husband.

Mary lived in an age when mothers had their children recite prayers at bedtime:

> *Now I lay me down to sleep.*
> *I pray the Lord my soul to keep,*
> *But if I die before I wake,*
> *I pray the Lord my soul to take.*

During this period, Christians more openly embraced the concept that death was a part of life. Children were regularly snatched from parents, and belief in an afterworld consoled both the living and the dying. During war, with massive losses endured regularly, these fears were multiplied and accentuated. Every soldier was a mother's son, and both Lincolns mourned young lives cut down.

One of the sorest points remained that Robert was not in the army, an ongoing conflict between Mary and Abraham. During December 1863, Senator Ira Harris was visiting the president's home and suggested that Robert Lincoln should be in uniform: Harris had only one son, but he had already enlisted. At the thought of Robert in the army, Mary's "face turned white as death." Her alarm came in the middle of a disastrous evening in the White House parlor.

Senator Ira Harris and General Dan Sickles had come to quiz Emilie about conditions in the South, inquiring at first about mutual friends, then widening the dialogue. Discussion became uncomfortable when Harris boasted about "whipping the rebels at Chattanooga"— particularly insensitive considering Emilie Helm had only been widowed a matter of weeks. However, it was equally inappropriate when she brought up Yankees running away at the Battle of Bull Run to General Sickles, who had recently lost his leg at the Battle of Gettysburg. Such slights about Northern manhood were sure to provoke, and did.

Mrs. Lincoln tried to change the subject, but Harris turned to Helm to insist, "If I had twenty sons they should all be fighting the rebels." Whereupon Emilie blurted out, "If I had twenty sons, General Harris, they should all be opposing yours," and ran crying from the room. Mary Lincoln followed to comfort her, and the two wept together, once they were alone.

Sickles stormed out to complain to the president, stomping up the stairs in a rage. When Lincoln gave him an audience, Sickles repeated the verbal fireworks. The president seemed amused: "The child has a tongue like the rest of the Todds." But Sickles did not see the humor, thundering that he should *not* have a Rebel in the White House. Whereupon Lincoln replied, "My wife and I are in the habit of choosing our guests. We do not need from our friends either advice or assistance."

When Katherine Helm and Tad Lincoln were playing one day, Emilie's daughter insisted that Jefferson Davis was the president, and a shouting match between the cousins ensued. Lincoln intervened, keeping the peace. But Emilie knew her days in her sister's home were numbered, and she confessed, "my being here is more or less an embarrassment to all of us."

The breach remained an impasse: "Sister and I cannot open our hearts to each other as freely as we would like. This frightful war comes between us like a barrier of granite. . . . " At the point of her departure, Lincoln expressed the hope that Emilie did not blame him for her sorrows.

Todd family travails continued when one of Mary's other half sisters, Martha Todd White, created a scandal during her visit to Washington. At the end of 1863, Mrs. White had come to the capital to try to purchase things not available in Selma, Alabama. When she arrived, she sent her card to her sister, Mrs. Lincoln—signaling her interest in paying a call. The First Lady declined to receive or see Martha, ignoring overtures on more than one occasion. Despite this rebuff, Martha was granted a pass by the president to return to her home in the South.

Rumors circulated that Martha had transported three large trunks, allegedly filled with medicines prohibited by federal embargo. However, she had also reputedly brought back a Confederate uniform decorated with gold buttons—and the value of these buttons was reported from $4,000 to $40,000 in various press reports. Scandal erupted. Many papers abused the president for granting his Confederate sister-in-law special treatment, presuming that she smuggled goods with his connivance. Horace Greeley, editor of the *New York Tribune*, affirmed these rumors by reprinting them: "Mrs. J. Todd White, the sister of Mrs. Lincoln, did pass through our lines for Richmond via Fortress Monroe with three large trunks containing merchandise and medicine, so that the chuckling of the Rebel press over her safe transit with Rebel uniforms and buttons of gold was founded in truth."

Later evidence demonstrated that Lincoln not only denied White favors but warned her that she would be imprisoned if she did not leave without further fuss. His Cabinet heard details from the president when the incident blew up in the press. Whatever the truth of the matter, these rumors were damaging, especially Mrs. White's later intimations that she had outwitted her brother-in-law. Indeed, she was invited to the Confederate White House in Richmond as a celebrity once the scandal appeared in the headlines.

Relatives boasted that, at a minimum, she had brought back her weight in quinine, defying the blockade. The White House tried to remain above the fray, but this was a source of grievous embarrassment to both Lincoln and his wife. By late April, John Nicolay wrote a letter to the editor correcting Greeley's story, claiming that it was "from the first without truth or foundation." To his credit, after reading Nicolay's well-documented response, Greeley published a retraction, but the damage was clearly already done.

When Mrs. H. C. Ingersoll volunteered to counter charges that Mrs. Lincoln gave aid and comfort to the enemy, Mrs. Lincoln refused the offer. Instead, she thanked Mrs. Ingersoll for a patriotic piece she had already penned for the *Hospital Gazette*, but asked her *not* to defend the First Lady in print. Mary went on to describe the hard-and-fast rule that she and her husband maintained to protect her right to privacy. Ingersoll insisted that Mary would have been defended by many had it not been for this "rule of silence" imposed by Mrs. Lincoln—and by those who feared her wrath as well.

Both the Lincolns dreaded charges of Confederate sympathizing, and did all they could to quell such rumors. Emilie Helm sent a letter to Lincoln on December 20, after she arrived back in Kentucky, requesting that the president waive the prohibition against prisoners receiving clothing and other goods. Her request was ignored.

Abraham Lincoln had many more worries on his mind. His own health was precarious—he was just recovering from varioloid, a mild form of smallpox. His doctors prescribed bed rest—which gave him more time to work on his annual message to Congress. By Christmas, Lincoln felt fully recovered, and the next day proposed an outing with Tad and Mary down the river. But this third December in the White House was perhaps even more dispiriting than the year before. It was the second holiday season without Willie, and his birthday had always been a highlight, falling four days before Christmas. So it was a rather subdued holiday, another year without him, as Mary struggled to embrace the New Year.

When 1864 finally did appear, both the Lincolns realized that in addition to their regular, nearly impossible round of duties, the president would enter the race for his party's nomination, and another election. It would be an epic challenge, but he must mount his campaign or give in to pessimism. On New Year's Day, the Lincolns gamely faced their public at the traditional White House reception. The president seemed "brighter and less woebegone" according to several observers, while Mary Lincoln had abandoned her mourning garb for a dark, attractive reception dress with an impressive train.

In leaving off her mourning weeds, Mrs. Lincoln signaled a new social schedule, resuming her Saturday receptions. After the first presidential levee on January 9, the hotels began their series of balls and events, inaugurating the winter season. The whirl of socializing brought with it, inevitably, bickering. The First Lady squared off with John Nicolay over arrangements, but soon "cast out that devil of stubbornness" and acceded to his wishes for Cabinet dinners and other formal occasions.

One of her major tiffs with Nicolay was over his inviting the Spragues to a Cabinet dinner. Kate Chase

Sprague served as hostess for her father, Secretary of the Treasury Salmon Chase, who was not very deft at masking his ambition to replace Lincoln in the White House. To Mrs. Lincoln's regret, Kate continued to be one of the main attractions on the Washington social scene.

In an election year especially, Mrs. Lincoln had no patience for the magnetic Mrs. Sprague. Nicolay observed that Kate possessed the immense advantages of youth and quite unusual beauty, which proved an irritation to Mary Lincoln. Furthermore, because the First Lady was persecuted for her southern heritage, Kate, the Ohio-born wife of a Rhode Island politician, took great pride in her Yankee roots. She would take any opportunity to impugn Mrs. Lincoln whenever possible.

Mary was so wary of the Chases that she asked her husband to examine the Treasury secretary's loyalty, saying, "If he thought he could make anything by it, he would betray you tomorrow," and adding, "I am not the only one that has warned you." Dr. Anson Henry, the old family friend from Illinois, noted that Chase's minions spread false stories about Mrs. Lincoln. He observed that Chase's conduct was so disgraceful that he thought he deserved to be removed from the Cabinet, and his opinion was shared by many—most assuredly Mrs. Lincoln.

Cries of Fire! suddenly broke out on the evening of February 10, 1864. Six horses in the White House stables were consumed by the flames, including a pony ridden by the Lincolns' son Willie. Lincoln was moved to tears by this loss. A few weeks later, Mrs. Lincoln confided to Benjamin French that it was the first time she had gone into the library since Willie had died. Both Lincolns felt the continuing affliction of that loss.

News from the Confederate capital struck a melancholy chord that was familiar to the Lincolns. On April 30, Jefferson Davis's five-year-old son, Joseph, fell off the balcony at the Davis mansion in Richmond. His parents rushed to his side, but the boy died within the hour. Crowds of mourners joined his stricken parents at the Hollywood Cemetery for his burial the next day. Varina Davis, seven months pregnant at the time, became deeply forlorn, explaining to her mother that her only comfort was "to sit alone." Her personal losses were compounded by the falling fortunes of the Confederacy.

Lincoln was buoyed by the appointment of Ulysses S. Grant as his supreme commander. Although she welcomed military successes, Mrs. Lincoln did not share her husband's admiration for Grant, complaining that he was a butcher who lost two men to the enemy's one.

But she knew how important it would become to have victorious generals for Lincoln's campaign team.

As the summer of 1864 approached, Mrs. Lincoln became consumed with fears, obsessed with her husband's reelection. She politicked on his behalf, cultivating shady characters and scheming: "I will be clever to them until after the election, and then, if we remain in the White House, I will drop everyone of them, and let them know very plainly that I only made tools of them. They are an unprincipled set, and I don't mind a little double-dealing with them."

This was a dangerous game. She was out of her league when she became involved with a whole set of New York politicians, men who were skilled in the patronage and skullduggery of midcentury urban politics. During Lincoln's campaigns for reelection, he allowed supporters promising to deliver the vote considerable wiggle room to maneuver. Abram Wakeman of New York, a retired congressman who had served one term in Albany before one term in Washington, harbored lofty ambitions. He was fishing for a plum appointment in New York. He not only was a conduit to *New York Herald* editor James Gordon Bennett, but became a shrewd adviser and go-between for Lincoln. His credentials were in order, and Wakeman curried favor with kingpin Thurlow Weed. He not only wrote

Lincoln crafty letters crammed with advice, he included regards for Mrs. Lincoln. Wakeman took Mary (and most likely not the president) into his confidence about spiritualist connections. He was an opportunist in several dimensions.

As U.S. postmaster during the New York 1863 draft riots, his home in Yorkville had been burned to the ground. The wheeler-dealer hoped to secure a post as Collector of the Port—a juicy reward. Mrs. Lincoln was no great fan of Weed, even though she was not above exchanging gossipy letters with Wakeman. Some of these letters could have put her in a compromising position. In the end, the most lucrative post went to Simeon Draper, a real estate developer and bank president. But Wakeman was not out in the cold; he was given a lesser position to enrich himself by—surveyor of the Port of New York.

Draper had served as chairman of the New York State Republican Party, had championed Lincoln in 1860, and had been of critical service to the president in delicate personal matters. He had played a key role in the recovery of letters in Mary Lincoln's hand that some believed revealed her complicity in improper activities during her first year in the White House, those gone but not forgotten scandals with James Watt. John Hay asserted this hearsay as fact. Whatever the case, Draper was a valuable asset and became the new broom that swept out the old elements—especially Chase loyalists—in the New York Custom House. One reporter observed that following Draper's appointment, "an anti-Lincoln man could not be found in any of the departments." Mrs. Lincoln kept herself close to these New York titans of business, with contacts in both Manhattan and Washington.

She wrote to Wakeman in October 1864, referring to a newspaper article claiming a particular merchant had forgiven her debts, angry about such "falsehoods." Then she scribbled an ambiguous warning, "Please say not a word, to anyone not even W[eed] about the 5th Avenue business." Her clandestine and cryptic correspondence hints at duplicity. Why would the First Lady be writing to men—neither relatives nor her husband—in such a manner? She perhaps felt such dangerous games were a necessary evil, considering what she was up against. Perhaps she worried she would have to face the greatest crisis of her marriage if her husband lost his bid for reelection and discovered that she had not kept promises of economy but had run up new and extravagant debts.

Mary Lincoln had wallowed in, and nearly sunk into, emotional turmoil during the period following Willie's death. Her siege mentality, isolation, and capriciousness led to a destructive pattern of spending and denial. Pulling herself out of mourning had required a good deal of retail therapy, which she could ill afford. She often enlisted Elizabeth Keckly and the White House chief housekeeper, Mary Ann Cuthbert, to assist with her shopping needs. They would also presumably conceal the extent of her purchasing, keeping the bad news from the president. The webs were tangling during this period, and it was only a matter of time before her bad behavior might be exposed.

In late April 1864, nevertheless, she went on another shopping spree in Manhattan. The papers took sharp aim: "Mrs. Lincoln ransacked the treasures of the Broadway dry good stores." Social arbiter Mary Clemmer Ames complained of these escapades: "While her sister women scraped lint, . . . the wife of the President of the United States spent her time rolling to and fro between Washington and New York, intent on extravagant purchases for herself and the White House."

This trip came at a most inauspicious time, as the *New York Times* reported austerity campaigns among the well-heeled: Ladies were agreeing not to use any expensive imported fabrics for the duration of the war. A few days later, the Washington papers printed news that patriotic society women were organizing a "Ladies National Covenant," pledging reduction of the consumption of foreign luxuries. These developments seemed a direct rebuke of Mary's personal extravagances.

However, Mrs. Lincoln was so invested in her own desires that she demanded and received confirmation from government offices that the United States must not offend European allies by boycotting imports. She then tried to justify her actions as patriotic—bolstering the import-export market—when she indulged in foreign finery. And she knew these excesses flew in the face of her husband's wishes.

She had collected a mountain of goods on credit and by the summer of 1864 felt desperate about how she was going to pay for them. Even worse, she explained to Elizabeth Keckly, her husband had no knowledge of the extent of her debts, and she knew that "he will be unable to pay if he is defeated." She rationalized: "I must have more money than Mr. Lincoln can spare for me."

The First Lady feared both the size and style of her indebtedness. She owed over $25,000 (the amount of her husband's annual salary), the bulk of it due to Alexander Stewart's in New York. Besides her debts, there was also her seemingly limitless appetite for gifts: How would a gold enamel brooch set with forty-seven rose diamonds from her friend Mr. Mortimer look in

the cold light of newsprint? Because of scrutiny attracted by earlier financial scandals, disaster loomed. If
this string of damaging bits of information hit the headlines, it would hurt her husband's political fortunes—
and ruin her. She confided to Elizabeth Keckly that she
wanted Republican politicians to help her pay off these
debts, as "hundreds of them are getting immensely rich
off the patronage of my husband."

Perhaps redecorating bills might be footed by
wealthy merchants, happy to provide carpets and wallpaper for the government-owned presidential summer
cottage or new ottomans for the White House as a part
of patronage schemes. But why should the excessive
dress bills of the First Lady be the responsibility of
corrupt politicians?

Mrs. Lincoln had sustained a severe head injury the
year before, and her son Robert later suggested that
consequent mental impairment had triggered a growing detachment from reality. Robert himself was one
of the closest observers of his mother, and his fears
for her sanity might have begun during this period.
Robert Lincoln would later confide to his sweetheart,
Mary Harlan, "that my mother is on one subject not
mentally responsible." That subject was money, and
this massive debt hidden from her husband was a terrible burden for her. It would appear that she lost all

reason when dealing with the money she owed in the summer of 1864.

Mary Lincoln's growing panic clouded and destroyed her judgment. She began to disintegrate when she started confiding in others who she hoped would help her but would eventually contribute to her fall. As Illinois congressman Isaac Newton Arnold complained to John Hay, Mrs. Lincoln would "shed tears by the pint a begging me to pay her debts which was unbeknownst to the President."

The First Lady, in Elizabeth Keckly's words, became "almost crazy," as fear and loneliness drove her to despair. In late May, she took to her bed, prevented from visiting hospitals by severe headaches. On June 9, both the Lincolns were cheered by the news that he had been officially renominated by his party, although Mrs. Lincoln was certainly cool to the addition of Andrew Johnson as a new vice-presidential candidate. But this confirmation of his candidacy eased her anxieties.

Although Lincoln disappointed his son by not being able to attend his Harvard graduation, Mary made the trip to Cambridge on June 20 and took pride in her son's role in Class Day. Edward Everett offered the commencement oration. Robert commented on his undergraduate experience: "My life in

College has been very pleasant and has had no interruptions. I have studied enough to satisfy myself without being a 'dig.' "

Mary anticipated that the whole family would summer at the Soldiers' Home, especially since she had redecorated it—installing coconut grass for carpeting and having eight of the fourteen rooms repapered and redraped. She would not countenance Robert's request that he, at long last, enter the army. When his father acquiesced to his mother's adamant demands that Robert stay out of uniform, the Harvard graduate insisted he would stay away from the White House fishbowl. He would return to Cambridge in the fall to enroll in law school.

Even a return to the Soldiers' Home circa July 4 could not banish Mrs. Lincoln's feelings of unrest. Perhaps as penance, perhaps due to proximity and the comfort it had brought the year before, Mary Lincoln returned to her round of hospital visits. One young soldier recalled, "None were more kind or showed a nobler spirit, than the wife of the Chief Magistrate of the Nation. She called regularly, bringing wither [*sic*], by attendants, flowers and delicacies, and bestowing them with her own hand with a grace worthy the station she held." She also resumed her practice of writing letters for soldiers too ill to write. One remains extant:

Campbell Hospital
Washington, D.C.
August 10th 1864

My dear Mrs. Agen—
   I am sitting by the side of our soldier boy. He
has been quite sick, but is getting well. He tells me
to say to you that he is all right. With respect for
the mother of the young soldier.
                              Mrs. Abraham Lincoln

Yet the First Lady remained peripatetic, taking
the air in the Vermont mountains—returning to the
Equinox Hotel in Manchester, this time with both Tad
and Robert in tow in late August.

During this excursion, she stopped in New York
going up and coming back, collecting piles of packages
and bad publicity, as well as cementing her uneasy alliance with Abram Wakeman. Befriending him may have
caused her to cross ethical boundaries—dangerous in
any time, but especially during the months leading up
to her husband's reelection.

Whether or not Mrs. Lincoln was peddling appointments, she had a penchant for pretense about her
access to information and about how her relationship
to the president might lead to advancement. Strangers

and family overestimated Mary Lincoln's influence, and she continued to lead many in a merry dance—to advance her own schemes and to allow others to draw conclusions that flattered her.

By the autumn of 1864, she complained that most days she would only have a few minutes alone with her husband near the midnight hour, when he was weary from his long labors. Mary might have made this point to demonstrate her exclusive and intimate access, but it also indicated how much the president's time was devoured by his office. Mary's complaint shows that the demands of duty put her ever lower down the list of her husband's priorities. Lincoln was under siege from a new wave of office seekers trying to claim a piece of glory, and from the relentless demands of a government at war. But night after night, Mary would be there, brushing her hair and waiting to hear the news of the day.

She had been all but excluded from his circle of trusted advisers because of her troubling mood swings, her prolonged absences, and her capricious behavior. All the clever young men who surrounded Lincoln were reaping their rewards. It must have been bittersweet for Mary Lincoln to watch both John Nicolay and John Hay, who had spent so much of their time thwarting her, win diplomatic posts in Paris for their efforts.

Meanwhile, her own Todd troubles continued as her sister again was causing conflicts. Emilie Todd Helm, a Confederate widow in reduced circumstances, wanted to bring six hundred bales of her family's cotton from behind enemy lines in the fall of 1864, to be able to sell it abroad. Lincoln would only grant this privilege if she took the oath, the only basis on which Southern planters could legally avoid the blockade. Emilie resented being treated like a common enemy because she maintained her allegiance to her dead husband's cause. She wanted an exemption. She felt Lincoln should make an exception in her case—even after she saw the furor her sister Martha's scandal had caused. She had no sympathy for Lincoln's position, and need blinded her to her sister Mary's terrible dilemma.

Emilie's pleading for favors was not an isolated case. Mrs. Lincoln was besieged by people seeking special dispensations. Mary's childhood friend from Lexington, Margaret Wickliffe Preston, wrote asking for a pass to visit her husband in the Confederate army, having been denied one by local military officials. Instead of Mrs. Lincoln writing back, the president wrote a response—pleading Mary's illness. He expressed his regret that he could not grant her request, that he did not have the authority to overrule his generals. (In fact,

Lincoln was commander in chief and could do what he pleased.)

Other Confederates wrote requesting special permission to recover property. Sallie Ward Hunt, another old friend from Kentucky, claimed she was separating from her husband, who rode with Confederate raider John Hunt Morgan. She asked that she be allowed to recover some of her own items, including a piano and some furniture. Lincoln allowed Mrs. Hunt to claim her own property, provided she took the oath of allegiance.

By autumn 1864, Emilie blamed Lincoln bitterly for several family tragedies—something that he had hoped to avoid. Levi O. Todd, the single Unionist among Mary's Kentucky siblings, wrote to the president asking him to forward $150–$200 so he could continue to work on the Republican candidate's behalf in the months leading up to the 1864 election. He mentioned in passing his illness and need. But because Levi was a notorious drunkard, whose wife had divorced him citing abuse, Lincoln was not well disposed toward him, Unionist or not. Over a decade before, Mary's Springfield siblings had had to defend themselves against Levi's lawsuit alleging unfair distribution of his father's estate, leveling charges of unethical business conduct at Abraham Lincoln. Circumstances combined, and Lincoln did not reply to his brother-in-law's request—much as with his own brother's requests for handouts in the 1850s.

Levi died impoverished, a death that Emilie wanted laid on the Lincolns' doorstep. Furthermore, after her stepson was put into a pauper's grave, Emilie's mother collapsed, which sent her widowed daughter into a tailspin of recriminations. She wrote a stinging letter to the president, hoping to shame him into granting her request to sell her Confederate cotton, telling him, "Your Minnie [Minié] bullets have made us what we are."

Lincoln was quite used to attacks from both sides of the conflict. Mary Lincoln felt she was a scapegoat, but the president clearly took fire from all quarters. When a delegation of blacks visited his White House office in 1864 to lobby for equal wages for African-American workers on the government payroll, Lincoln jokingly made reference to "Cuffie" [a racial slur] during his conversation. Henry Samuels, only twenty-five at the time, responded with great dignity as well as nerve: "Excuse me, Mr. Lincoln, the term 'Cuffie' is not in our vernacular. What we want is that the wages of the American Colored Laborer be equalized with those of the American White Laborer." With Secretary of War

Edwin Stanton as witness, Lincoln replied: "I stand corrected, young man, but you know I am by birth a Southerner and in our section that term is applied without any idea of an offensive nature." Thirty days later, by executive order, Lincoln complied with the group's request. During this same period, Lincoln also granted Sojourner Truth a personal audience in the White House, as she campaigned for his reelection, and he became more sympathetic to African-American causes through his contact with Frederick Douglass.

By the time November's election arrived, Mrs. Lincoln was worn out with worry, having claimed to have gotten down on her knees to ask for votes "again and again," presumably in prayer rather than in personal supplication. Lincoln was concerned that his wife was not prepared to bear the loss if he were defeated. Despite the eerie desertion of the White House on Election Day, Hay reported a victory celebration at the telegraph office later on the evening of November 8, when reports of returns flowed in showing Lincoln ahead. Lincoln's first thought was for his wife, as it had been in Springfield just four years before, and he sent a message back to her at the White House. It was less than a week since they had celebrated their twenty-second wedding anniversary, and Mr. Lincoln was sure this would be one present his wife would deeply appreciate: a victory at the polls.

With this reprieve, Mary Lincoln could feel disaster might be postponed, if not averted. She was overjoyed by the victory, elated that her judgment day might be delayed so she could, once again, juggle with the fates. But she had her nagging doubts about burdens weighing down her husband: "Poor Mr. Lincoln is looking so broken-hearted, so completely worn out, I fear he will not get through the next four years."

14.

With Victory in View

Lincoln's win at the polls in November 1864 was due in no small part to the enlisted men who cast their ballots for their commander in chief. He remained eager for his generals to deliver him an end to the war, the victory for which he longed. Death lists and wave after wave of American youth marching off to the battlefield gave him insomnia. All the while, his generals valiantly maneuvered to bring the Confederacy to its knees. General Sherman's decisive conquest of Atlanta in September 1864 had been crucial to the soldiers' vote, and it buoyed Union confidence. Then his "March to the Sea" was capped by his delivery of Savannah to the president as a "Christmas present." These military successes energized Lincoln.

The president had outlawed Southern slavery, and he urged his generals to use whatever means they could to break Rebel resolve. At the beginning of 1865, Sherman turned northward to take his campaign through the Carolinas. Grant focused on breaking the Confederate hold on Virginia. Both the Lincolns counted on Union military triumphs to deliver them from the tunnel of war, which seemed both dark and endless.

Lincoln dreamed of enemy surrender—and even secretly attended a peace conference at Hampton Roads on February 3, 1865. The president consented to this meeting only because Confederate diplomats were willing to reconsider secession. Also his old friend from Congress, Alexander Stephens, the Confederate vice-president, would sit at the table. However committed he was to ending the war, Lincoln was unwilling to give ground on slave emancipation. On January 31, only four days before the clandestine peace meeting, legislators on Capitol Hill considered the abolition of slavery by constitutional amendment, and the Thirteenth Amendment became the first new one passed by Congress in over sixty years. Historian Chandra Manning has deftly argued that slavery's end became a fundamental crusade for the common soldier, and Lincoln was not willing to back down. Hampton Roads faded into a failed initiative.

Lincoln's commitment to emancipation would become his legacy, one that African Americans in

particular would cherish. An old friend from Springfield, William de Fleurville, warmly endorsed Lincoln's
second term for this very reason, claiming, "The oppressed will shout the name of their deliverer, and
generations to come will rise up and call you blessed."
Lincoln's former barber, a Haitian émigré, took pride
in the president's key role in slavery's destruction.

But Lincoln feared emancipation would not take
root in the South unless he could provide carrots along
with the stick. He was prepared to soften the blow of
Confederate defeat, once he got the federal government
shaped to his agenda. He was eager to begin the work
of healing and reconstruction, and had plans for charity rather than applying the boot heel once victory was
achieved.

Republicans took comfort that slavery's permanent demise would be ensured by Salmon P. Chase's
appointment to the Supreme Court. Roger Taney had
died in October 1864. Under his tenure as Chief Justice, the Court had ruled consistently against African
Americans, proclaiming in the Dred Scott decision of
1857 that slavery was legal throughout the nation and
that blacks were not entitled to equal rights. Taney's
court determined that no African Americans were
ever intended to enjoy the privileges of citizenship.
The death of this Chief Justice, in the wake of other

associate appointments, offered the president a golden opportunity to transform the Court significantly.

The promise of a Republican majority on the Supreme Court tantalized Lincoln. All that he had been working for might be undone with judicial indifference, and he needed strong leadership in place. He sought someone with impeccable credentials and political acumen who could smother rather than stoke disputes. A vacancy of this magnitude caused quite a bit of speculation and self-promotion among Republican rivals. Because there were so many candidates jockeying for the position—especially among his Cabinet—Lincoln postponed his announcement until after the election. Picking Chase, his own major competitor who had so shamelessly and consistently put himself forward as an alternate candidate in 1864, reflected Abraham Lincoln's political genius.

This appointment silenced a large chorus of Lincoln's left-leaning critics, while it also removed a fox from the henhouse. Mary was annoyed that someone so disloyal was given such a plum appointment, but by now she understood her husband's need to firm up the foundations of his emancipationist legacy. She knew what was at stake at this stage. Chase could be a superb justice in the face of peacetime challenges ahead. With her husband's second term secured, with her lease on the White House extended, Mrs. Lincoln decided to effect some changes of her own.

The First Lady imperiously dismissed the long-time White House doorkeeper, Edward McManus. As a result, in January 1865, McManus took his tales of woe—stories of the First Lady's overspending, chicanery, and ill temper—to the receptive ears of Thurlow Weed in New York. Scandalous reports began to circulate, and Mrs. Lincoln fought back with her pen, trying to exert damage control among the New York elite. By the same token, Mrs. Lincoln was willing to use her powers of persuasion to assist those most valuable to her. Extant correspondence between the reelection and inauguration reveals the First Lady initiating requests to influence patronage appointments. Also, as Jean Baker has demonstrated, her letters to Abram Wakeman indicate she was "sharing secrets with her other Abraham." Unfortunately, we have only half the story, as Mary destroyed Wakeman's letters to her—while he ignored her entreaties to burn her notes to him.

She was hoping her husband's offer to James Gordon Bennett to become ambassador to France would appease the *New York Herald* editor, whose scorpion sting was worth keeping at bay. Bennett turned down the position but softened his hard line against Mary's husband, and would later become a major champion of Lincoln's.

Mrs. Lincoln hoped to take better care of the president's health. Elizabeth Cady Stanton reported that his "shrivelled" appearance was reportedly because his wife underfed him. Mary was wounded by this idle gossip, especially such a groundless claim. Yet she knew her husband was wan, and wanted to do all she could to ameliorate the situation. Friends from his days before the White House who saw Lincoln for the first time in years were alarmed. When Joshua Speed's brother James was appointed attorney general for Lincoln's second term, Joshua traveled to Washington and was shocked by his old friend's skeletal frame and careworn visage. Despite the emotional lift of the elections, Lincoln still carried the weight of the war on his shoulders, which took a visible physical toll.

Mary insisted upon long drives daily and organized outings to the theater and other entertainments to prevent the president from continuing business late into the night. Lincoln was often up early and rarely took time to sit down for breakfast unless guests were invited. Lincoln's practice of chomping only an apple for dinner (as the midday meal was known) was legendary. Mary would often have to call him several times to supper, and worried about his diet and lack of appetite.

The First Lady might resort to cajolery. She even enlisted Tad's help to keep her husband from skipping meals. The president was especially fond of stewed chicken with hot biscuits and cream gravy, so Mary would delegate one of the kitchen staff, Alice Johnstone, to cook the meal just as the president liked it. After these delicacies were prepared, she then asked Tad to bring his father along to a room where the table would be laden with his favorite foods. Ambushed by his wife and son, the president might clean his plate and even ask for extra helpings. Mary used whatever means she could to implement regular meals.

The Lincolns' younger son, Tad, sought a variety of methods to attract his parents' attentions. His tenderheartedness had become notable among staff and regular visitors. Lincoln loved to tell the tale of Jack the turkey. Tad had been delighted when a friend sent a live turkey to the Lincolns in December 1863. Their son befriended the bird—he was a magnet for creatures great and small, animals of any kind. Unfortunately, Tad did not grasp the reason why the turkey had been sent shortly before Christmas, and became apoplectic when he discovered the bird, whom he had named Jack, was slated for Christmas dinner. He begged his father to provide a stay of execution, and Lincoln wrote out an actual reprieve to spare him. In homage, every president since Truman has issued a pardon for a turkey in a ceremony shortly before Thanksgiving, both Thanksgiving and pardoning a turkey being associated with the Lincoln White House.

The president also had to give in when Tad invited several homeless boys—street urchins—into the White House to share Christmas dinner in 1864. The spring before, Tad had dyed two dozen eggs for the annual Easter egg roll on the White House lawn. One dozen was for a boy named Tommy, whose father had gone for a soldier and died. Tommy's mother worked for the Treasury Department, and her son was lame. So Tad provided his little friend with a chair to help him lean over to roll his eggs, and made sure the president met the boy and shook his hand. A witness to this kindness called Tad "the kindest, brightest, most considerate child."

But not everyone appreciated the boy's imperious requests, and one visitor to the White House confided to Noah Brooks that the boy was "the tyrant of the White House."

Tad was sometimes viewed as a disruptive influence on the president, as he often insisted upon

Lincoln reading with Tad

monopolizing his papa's time. Perhaps he sensed that his father required distraction—or maybe he just was a needy young boy who had lost his brother and best friend, and consoled himself with his father, who felt the same. In any case, Lincoln's willingness to indulge his son rarely wavered. As a friend recalled, "There was no hilarity excepting where Tad was concerned." Then Lincoln might let the burdens of office lapse for a brief respite. This friend recalled, "Time and time again have I seen Tad sitting on his father's shoulders, while President Lincoln galloped up and down the long corridor outside their private apartments, the boy laughing with glee." Demands by Tad and Mary provided the president's only reprieve from shouldering the cares of the nation.

Tad eased slowly and uncomfortably into the role of First Child following Willie's death. He remained in fragile health and suffered a slight lisp from a speech impediment. Yet Tad was doted upon by his parents and indulged by the White House staff. A very touching image of father and son, an intimate portrait, was snapped by Mathew Brady at his studio in February 1864, as the two were looking at a photograph album. The pair appear lost in one another's presence and deeply connected—perhaps not an unusual experience for them, but one rarely shared by outsiders.

Tad had often played second banana to his poised and polished brother Willie, who had outshone him in the classroom as well as in adult company. Robert had been away at school long before the family had even come east, and he seemed a distant stranger to his brother, nine years younger. So now young Thomas Lincoln stood alone, forced into a spotlight he never sought as a solo act. Yet Master Taddie, as he was affectionately known, enjoyed hamming things up, exerting his theatrical flair—which could and did occasionally lift his father's moods.

During his years in Washington, Lincoln befriended Leonard Grover, owner of Grover's Theater. Lincoln was a regular at Grover's playhouse. During his time in the White House, Lincoln reportedly saw one hundred performances at Grover's, indicating his abiding and ongoing interest in this diversion.

Besides his great love of live performance, perhaps Lincoln attended so often because his son Tad loved to tag along. Tad became friends with Grover's son Bobby, several years younger, and the boys spent time trawling for goldfish in the White House ponds, and on other escapades. Also, Tad was fascinated with all things theatrical and enjoyed rehearsals as well as performances at Grover's. Tad and Willie had often enjoyed theatricals, and after a while, Tad decided to resume this practice. He would build sets and borrow costumes to put on plays of his own at the White House, enlisting his friends to play parts on his makeshift stage. He might draft members of his father's guard detail to take on roles or sit in the audience. He also set up a table in the public corridors to sell fruits and treats to office seekers, in a nineteenth-century version of the lemonade stand.

Tad frequently accompanied his father to a drama—although they as easily took in comedies or even a musical together. On one occasion, Tad indulged in an elaborate charade to entertain his father—slipping out of the presidential box to dress in costume and mingle with the actors onstage. When Lincoln spied his son's impromptu performance and broke into a grin, Tad was thrilled—as was the audience who witnessed this play within the play. This shared passion for the theater developed into a deep bond between father and son. Of course, Tad's mother had starred in her school plays, and was an equally enthusiastic performer, so Tad came by his gift naturally. In short, the Lincolns were regular and enthusiastic theatergoers.

Mary might forget her troubles and relax when she could look over in the presidential box, as she did on the evening of January 13, 1865, to see her husband and son simultaneously mesmerized by Edwin Forrest in the title role of *King Lear.* Tad was leaning his head on his father's shoulder to whisper in his ear. During this period, Mrs. Lincoln was pleased to be surrounded by many well-wishers from home, including Dr. Anson Henry, who always brought with him memories of how far the Lincolns had come, how well they had triumphed over circumstances.

Tad Lincoln had been mad for all things military during childhood. Living in the White House and heading out to campsites with one or both of his parents provided excellent diversion. He adored all the pomp and circumstance of martial affairs, and his mania only multiplied as the conflict continued. In the early months of the war, he was given his own outfits or uniforms, to which several photographs attest. He and his brother Willie played with toy guns, romping in the attic and wherever adventure and imagination took them—although they might be hauled off the roof when they got reckless.

After his brother's death, Tad and his father made even more frequent excursions to Stuntz's toy shop on New York Avenue. Lincoln confided: "I want to give him all the toys I did not have and all the toys that I would have given the boy who went away." But soon Tad wanted to graduate from toys.

Tad prompted his father to issue a flurry of "official requests" on his son's behalf, including a letter to the secretary of the navy to get Tad a sword, and one to the secretary of war to find him some flags, and finally one to an army captain asking this officer to locate "a little gun that he can not hurt himself with."

Once, while standing next to his father during an official review of federal troops marching down Pennsylvania Avenue, Tad mischievously had waved a small Confederate flag. When Lincoln discovered what his son was up to, he scooped the boy into the arms

Tad in uniform

of a military attaché and banished him. Tad's favorite toy soldier was dressed like a Zouave—sent to the boy by members of the Sanitary Commission of New York—a doll that he, not surprisingly, named Jack.

This toy soldier provided countless hours of fun for the Lincoln boys even though he was regularly caught sleeping at his post. The hapless doll was often "shot at sunrise," and then buried near the rosebushes, despite protests of the White House gardener. At one point, Tad burst into his father's office and sought a pardon for poor Jack. Lincoln played along, only granting clemency after a mock trial, but this reprieve was written on official paper. Apparently Jack couldn't stay out of trouble, though, and was found guilty once again and hanged from a tree at the Taft home shortly thereafter.

Tad's obsession with military matters had a darker side. Once, he became alarmed about safety at the White House, and requisitioned arms for the house staff and drilled them to stand guard duty. His brother Robert, home from Harvard at the time, was furious about this incident. But the president seemed to sense the anxiety in his younger son, and temporarily indulged this charade—dismissing the armed servants once Tad was tucked away in bed. Tad also liked to handle pistols. He had been scolded already for pointing a gun at a playmate, and had been dressed down for accidentally shooting out a window while visiting the Tafts. Yet his begging allowed all to be forgiven, and Tad was finally given his own personal firearm—a pistol.

The first month of 1865 was a fairly happy time for the family, even though Mary did not welcome the news that her husband and older son had conspired against her: Robert, at long last, was entering the military. The president finally granted Robert's wish to perform military service. Lincoln appealed directly to General Grant on January 19, saying that his son wanted to "see something of the war before it ends." At least he found him a safe harbor—far from the front, but near the war's nerve center. Grant kindly responded: "I will be most happy to have him in my Military family," and welcomed the Harvard graduate to his own staff, as a captain.

Robert had developed a serious interest in the daughter of one of Lincoln's new Cabinet members, Secretary of the Interior James Harlan. Harlan was one of the leading Republicans of his day, and his statue is now included in the statuary halls of the United States Capitol to commemorate his home state of Iowa. He served as president of Iowa Wesleyan University in Mt. Pleasant before his appointment to the United States Senate in 1856. Senator and Mrs. Harlan had four children: A son, Silas, died in infancy; their oldest, Mary, had been born in 1846, followed by a son, William, in 1852, and another daughter, Julia, in 1856. While they were in Washington, eight-year-old Julia died. Losing their daughter plunged the Harlans into mourning—something that drew Mary to this couple. The Lincolns and Harlans had each lost an infant child and then an older child. The two firstborn children, Robert and Mary, knew the burdens imposed on surviving siblings in such families, giving them something in common when they met in 1864.

Robert became enamored of Mary Eunice Harlan during the winter social season in early 1864, when she was just seventeen. The Harlans' daughter attended a fashionable D.C. ladies' academy, Madame Smith's. This school reminded Mrs. Lincoln of her own years with Madame Mentelle. Classes were taught entirely in French, and students received demerits if they spoke English while at Madame's. In addition to fluent French, Mary Harlan was taught dance, deportment, and manners, as well as literary arts. Perhaps it was at Madame Smith's that Mary Harlan became an accomplished harpist and developed her love of music.

In any case, she not only caught Robert Lincoln's eye, but his mother took notice as well. She saw Mary Harlan at the opera one evening, and sent along one of her bouquets. President Lincoln was amused by his wife's attachment to the girl, and by her proclivity for matchmaking. And so Mrs. Lincoln began to show favor by requesting that Senator Harlan accompany the First Lady to the Capitol for the inauguration ceremony.

The effect of his mother's ardent approval of Mary Harlan remains unknown. In any case, Robert escorted her to the inaugural ball. The Lincolns' older son was extremely publicity-shy and wanted to keep his private affairs private. His feelings for Harlan remained cryptic—at one point he wrote to his friend John Nicolay about "our Iowa friend." There is some evidence to suggest that when he was visiting his parents in Washington, Robert arranged to meet with Mary Harlan not at the White House, nor at her home on H Street (which would have attracted gossip), but instead at the home of Gideon Welles. The naval secretary's son, Edgar Welles, was a close friend of Robert's who later mentioned the president's son as making use of his family's home for "courtship." Perhaps Robert Todd Lincoln was more like his father than he knew.

During the early months of 1865, the president remained pragmatic about Southern surrender. The breakdown of peace negotiations in February 1865 signaled to him that the war might continue for many months, a forecast that he confided to his inner circle. He used his second inaugural not as a bully pulpit but instead to promote conciliation. His closing words at the March 4 ceremony remain some of his most quoted:

With malice toward none, with charity for all, with firmness in the right as God gives us to see the right, let us strive on to finish the work we are in, to bind up the nation's wounds, to care for him who shall have borne the battle and for his widow and his orphan, to do all which may achieve and cherish a just and lasting peace among ourselves and with all nations.

With the crowning achievement of her husband's second inaugural behind her, Mrs. Lincoln was resplendent at the official White House reception afterward. She donned a gown of white satin with a needlepoint lace shawl draped fashionably for effect. Her neck and ears were dripping with diamonds, and flowers and silver ornaments bedecked her hair. One journalist surmised her appearance gave her "greater advantage than usual." Lincoln greeted well-wishers for over three hours, as more than five thousand crushed in to shake hands.

Two nights later, at the Patent Office building, the official inaugural ball attracted four thousand guests, who had each purchased a $10 ticket to raise money for soldiers' families. As his wife glided round the dance floor with Senator Charles Sumner, Lincoln took some satisfaction in quelling speculation about his political standoff with the Massachusetts senator over the terms of Reconstruction.

The president had personally summoned Sumner with a ticket to this event and a request that he escort Mrs. Lincoln. Sumner accepted this command performance, which kept Washington gossips guessing. The pregnant Kate Chase Sprague appeared, but she stayed off the dance floor because she was reportedly feeling unwell.

Yet Mary's great social rival still attracted a crowd, as Kate did at every social gathering. Her father had stood on the platform during the inauguration. Chase administered the oath of office rather than being sworn in himself—as his daughter had hoped. But with her father's new lifetime position as Supreme Court Justice, Kate felt her star firmly anchored in the Washington firmament.

Mary basked in the pleasure of having her husband by her side and watching her son twirl Mary Harlan across the dance floor at the ball. She tried to assuage her fears, as she knew Robert must soon return to the front near Petersburg, even though his father had secured him a fairly safe berth. She knew General Grant was spoiling for a fight on the Virginia peninsula. Robert's military assignment was an anxiety, but perhaps one she was able to set aside for the evening. Mary lounged on the dais, a raised platform built for the First Couple so they might survey the glittering gathering while being viewed by the crowd. Mary would have been filled with a sense of pride—for her husband's triumph, for all the money raised for a good cause, and for her family glorying in this joyful scene. Mary and Abraham stayed on until a little after one a.m.

Despite the uplift of such an occasion, the president continued to be severely fatigued. Mary wondered how they would be able to weather the storms facing the ship of state for four more years. Lincoln's burdens only multiplied, while the elusive victory remained ever present on the horizon but never realized. She was desperate—indeed wild—for peace, especially with her own son now a soldier.

Waves of emotion crashed over the First Lady during this period. The Lincolns had heard it all before from the generals—victory was near, just around the corner—but death and destruction rattled along at a relentless pace. With Robert in uniform, Mrs. Lincoln was always worried, on edge. Both she and her husband suffered disturbing dreams and interrupted sleep. Lincoln was so unwell that, unable to rouse himself, he held a Cabinet meeting in his bedroom on March 14. Meanwhile, Mary wanted her son home and the war over—perhaps in that order.

Her behavior during this period was not just erratic. Several outbursts indicate that she was in a constant state of hyperanxiety. Trying to juggle her treacherous game of patronage, shuffling her debts, and constant alarm about her soldier son, the First Lady suffered miserably. Mary Lincoln's conduct was marked by capricious and disturbing episodes during the weeks following her son's enlistment.

Some of the most damning of the anecdotes about her emanate from this period. After Lincoln rose from his sickbed following his inauguration, he decided to take up Grant's invitation to visit military headquarters at City Point, Virginia. Since this would be an opportunity to see Robert, Mrs. Lincoln insisted that she and Tad come along. They boarded a government yacht, the *River Queen*, with the U.S.S. *Bat* detailed with them for protection.

Threats against the president were increasing rather than diminishing with the military collapse of the Confederacy. Lincoln was cautious with his agenda and telegraphed his son, "We now think of starting to you about one P.M. Thursday. Don't make public." Bad water apparently caused Lincoln to suffer indisposition onboard. Following Grant's rout of Lee's forces at Petersburg, the president recovered enough to requisition a train to take him to the front on March 25 so he could see the battlegrounds for himself. Lincoln rode horseback over the field littered with bodies. It was a sobering vigil, and one that nerved him to marshal his energies. The next day the president called for a review of troops, which Mary Lincoln and Julia Dent Grant, the general's wife, planned to attend.

During this visit to the front, undertaken at Grant's invitation, Mrs. Lincoln had not been particularly gracious. She could not conceal her sense of disdain for the general's wife, and Mrs. Grant's sister reported that the First Lady wanted to be treated like royalty, even expecting Julia Grant to *back* out of a room when taking her leave. Mrs. Lincoln was annoyed over the way James Gordon Bennett, among other prominent Yankee opinion makers, had pushed hard for Ulysses S. Grant to campaign for president the year before. She saw Julia Grant as someone eager to occupy the White House, imagining that Grant's wife harbored ambitions as intense as her own. The coolness between the women added unnecessary friction to the visit.

On their way to the military review, obstacles aplenty impeded their progress. The ladies' party encountered a very poor road, like a washboard. A sea of mud splashed back up onto the First Lady and her companions within the converted ambulance in which they were crowded.

When Mrs. Lincoln demanded that their driver speed up, the ride became even bumpier. As a result, Mary received a violent blow to her head from the inside top of the carriage. This injury and the delay seemed to unhinge her.

Adam Badeau, one of the officers riding along with the women, innocently reported that the officers' wives had been ordered to the rear—all except General Charles Griffin's spouse, who received a special permit from the president allowing her to remain. Mrs. Griffin was a noted beauty, born into the patrician Carroll family of Maryland, who would end her days as the Countess Esterhazy. Informed that a beautiful young woman was by the president's side, Mrs. Lincoln exploded.

Covered in mud, rattled into insensibility, blinded by a migraine, she had to be forcibly restrained when they got mired in the mud and she wanted to get out and walk to the parade grounds. When she did finally arrive with her party, the men had already begun their exercises, and her physical condition and mental state were shockingly bad. Her fury was out of proportion and out of control. Both Badeau and Mrs. Grant interpreted her behavior that day as evidence of mental illness, not simply temper.

It was most unfortunate that the wife of General Edward Ord—dressed in a fashionable plumed hat—rode over to greet the ladies' party. She left behind a clearly recognizable president, tall in the saddle and wearing his trademark stovepipe hat. Mrs. Ord had barely opened her mouth when torrents of abuse burst forth from Mrs. Lincoln, who dressed her down for breaking protocol. The general's wife broke into tears over this unprovoked attack. Mary Lincoln clearly snapped, lashing out at everyone, including the hapless president, in full view of embarrassed eyewitnesses.

When his wife would not be soothed or stilled, Lincoln was forced to stalk away, "with an expression of pain and sadness," as Badeau described. Back on shipboard that night, the imbroglio continued when Mrs. Lincoln, within earshot of others, suggested that Lincoln remove General Ord from his command. The president failed to give in to his wife's whining and whimsy. Her bad behavior led to a temporary break between herself and the president. She spent the remainder of her time hidden away on shipboard, claiming to be "indisposed."

After a few days, Mary decided to make a strategic retreat. She went back to Washington without Tad and her husband. Lincoln was preoccupied with military matters, and Mary left their son behind to horse around with Grant's son and keep his father company. She headed back to her home after sending several inquiring telegrams to the White House, for her husband had dreamed of fire—and considering the string of blazes they had endured this was not surprising. Mary worried whether this was a premonition or something symbolic in code. Both the Lincolns were prone to analyze their dreams. In either case, it was an excuse to escape her embarrassing situation in Virginia for the Executive Mansion.

Once again, time apart soothed the icy impasse between the couple, as their telegraph traffic during this first week of April indicated a thaw. Back in

Washington, Mary exaggerated to Charles Sumner that her husband sent her telegrams every few hours. However, Lincoln often did send messages more than once daily. His series of updates about Grant's conquest of Petersburg and the evacuation of Richmond conveyed exultation over the Confederate defeat. The two were thrilled at the military news, and Mrs. Lincoln organized an impromptu expedition to rejoin Lincoln. When the president decided to pay a visit to Richmond, Mary suggested she join him there.

She wrote to a friend, gloating, "Will you not dine with us, in Jeff Davis' deserted banqueting hall?" Mary brought along Mrs. Harlan and her daughter Mary, Charles Sumner and his French visitor, the Marquis de Chambrun, as well as Elizabeth Keckly.

Keckly accompanied the White House party, visiting a place where she had once been held in bondage. It would be a moving experience for this former Virginia slave to accompany Lincoln on his tour of hospitals in the captured Confederate town. Now the whole of the Old Dominion and its hundreds of thousands of slaves were liberated, thanks to Lincoln's firm resolve.

Only a carriage accident involving Secretary of State Seward marred these days of glory. However, Mary was able to report to her husband that his close friend was not in danger, despite his injuries, but recovering at his home in the District. When Lincoln and his wife headed back to D.C., press reports enthused, "The President is looking much better for his extended absence from the capital." He visited the bedridden Seward as soon as he returned from Richmond, to share details of Union victory. The two lay stretched out on Seward's sickbed, so Lincoln could more easily convey the news to his friend, who was barely able to move.

On the next day, April 10, the news of Lee's surrender spread, and jubilation erupted throughout the North. Perhaps, at least momentarily, Mary and Abraham could banish their troubles. Only days before, Mrs. Lincoln was requesting extra security for her husband and their home, detailing metropolitan police to the Executive Mansion to handle threats of arson, kidnapping, and other worries. But with Confederate surrender, the Lincolns joined in the jubilation. Mary rejoiced that Robert would come home, out of danger. The light at the end of the tunnel had finally arrived.

With this historic landmark, Mary found a sense of serenity that was distinctly new and uncharacteristic. After years of feeling divided by her feelings, held up to ridicule for her Union patriotism, suspected for her family loyalties, the vindication of Union triumph and her husband's victory was sweet. Furthermore, she imagined that she might be reconciled with those

Lincoln entering Richmond

alienated, that her Kentucky kin and all former Confederates might be welcomed back into the fold, as she
and her husband had dreamed.

On April 10, she called surrender "the Consummation so devoutly to be wished." This moment had
an almost holy aspect to it. She reflected to Charles
Sumner, who had been such a good friend through
the entire roller-coaster ride of her years in the White
House:

If possible, this is a happier day, than last Monday
[the day Richmond was captured], the crowds

around the house have been immense, in the midst of the bands playing, they break forth into singing. If the close of that terrible war, has left some of our hearthstones, *very, very* desolate, God, has been as ever kind & merciful, in the midst of our heavy afflictions as nations, & as individuals.

Just a few days before, Mary had been erratic and abusive, but now she had mellowed into bliss over the prospect of peace. The world seemed transformed.

Shortly thereafter, Mary invited Lafayette's grandson to the president's private quarters to hear her husband offer a speech. Lincoln addressed the crowd gathered outside his White House window on April 11, outlining his plans for Reconstruction. His remarks were not impromptu, but written out so they might be preserved. "It is also unsatisfactory to some," he said, "that the elective franchise is not given to the colored man. I would myself prefer that it were now conferred on the very intelligent, and on those who serve our cause as soldiers." One of those listening to these words, Confederate sympathizer John Wilkes Booth, took violent exception to Lincoln's suggestion that blacks might be given the vote. He vowed in anger that this speech would be Lincoln's last.

Again the Lincolns were absorbed by celebrations: the illumination of the public buildings in the capital, the riotous jubilation and parades from Chicago to Philadelphia, from Boston to St. Louis and points between. In Springfield, the whole town turned out for a parade, led by Lincoln's horse, Old Bob. Pent-up emotions erupted following the confirmation of Lee's surrender, marking the war's end.

As the Confederate soldiers began to make their long journeys home, reality began to set in. Southern rebels faced the darkening realization that all was for naught, all was now lost. Disgruntled Confederate spies within the federal capital became desperate. A small group, led by John Wilkes Booth, put a conspiracy into play. This threat had long given both the Lincolns nightmares, most especially Mary. But during these first days of peace, the presidential couple had been dwelling on positive possibilities.

Mary was especially festive following surrender, as she and her husband had endured a particularly severe blowup only days before. Unfortunately this private quarrel erupted in public, witnessed by a score of observers. The fact that it occurred so close to the Confederate surrender perhaps caused it to be over-emphasized. Most marital spats are quickly forgotten, while this incident ended up in the pages of memoirs.

Lincoln had always been extremely indulgent of his wife's moods—to such a degree that many of his closest friends were appalled by his failure to curb her ill temper. Clearly Lincoln preferred to mock her more than correct her. During one of her rants, he might only murmur, "There, there, Mother." When that didn't work, Lincoln would treat Mary's outbursts as if they were adolescent tantrums, often referring to her as his "child-bride." But by 1865, Mary was no longer a child bride, even if she ever could have qualified as one, marrying at age twenty-three.

Mary Lincoln, the president's wife, should have maintained exemplary conduct. Nevertheless, because of what she endured, because of what he knew, Lincoln forgave his wife, even at her worst. He was not indifferent to his wife's critics, yet tried to counter harsh assessments. He rarely resorted to reproach, recognizing that she would eventually be mortified by her own bad behavior without his adding to her guilt. Because of Mary's mercurial temperament and his own shortcomings as a husband, the president clearly overlooked much within the marriage. He also was aware of the sacrifices his wife had made to marry him, and even more, of what she had to put up with in more than two decades together. He knew she had been his champion, perhaps at one time the only person to believe in him.

Again and again, she brought him back from the brink during countless episodes of gloom and self-doubt.

There is no evidence of the extent of their rupture over Robert's entrance into the army, but it had a definite effect on the couple until war's end. There is plenty of evidence that Mrs. Lincoln dreaded her son's wearing a uniform. Undoubtedly Abraham Lincoln, after securing his son's commission, bore the brunt of her misery. During these tense weeks, Mary may have misdirected her anger, letting fears curdle into rage. Imagine her relief when this breach between her and her husband was resolved by the Southern surrender. She felt sweet relief that all mothers might have their soldier sons back home.

The president was eager for time in office without the war machine grinding away. While on a carriage ride on Good Friday, April 14, Lincoln proclaimed to his wife: "We must *both* be more cheerful in the future; between the war and the loss of our darling Willie—we have been very miserable." Historian David Donald suggests this might be as close as Lincoln ever came to reprimanding his wife.

The couple discussed plans for their future, imagining a life beyond the White House. Earlier in the week, the Lincolns had shared a carriage ride with the Harlans. Perhaps contemplation of his own son's future had turned the president's face "radiant," as a companion described it. Just as everyone was commenting on the president's more vigorous appearance, just as he himself could let down his defenses and believe better days lay ahead, his life was cut short.

The night at Ford's Theatre on Good Friday in 1865 has become one of America's most significant historical memories, emblazoned in popular culture and rendered totemic. Its cultural significance cannot be underestimated, and it has stimulated a daunting amount of literature.

Lincoln and his wife arrived late at Ford's Theatre on the night of Friday, April 14, 1865. Their entrance interrupted the action on stage, as the band struck up "Hail to the Chief" and the Lincolns as well as their guests, Clara Harris (the daughter of a senator) and her fiancé, Major Henry Rathbone, settled into their seats in the presidential box. Esteemed actress Laura Keene was playing a starring role in *Our American Cousin*, a popular standard for her. Having the president in the audience was a great honor—one that Keene had regularly enjoyed. The end of the war transformed Lincoln's presence into an even more exuberant public event.

Mrs. Lincoln was particularly relaxed because her husband had written her a playful note to make a date for their drive earlier in the day, a gesture that harked back to days in their youth. This had been both thoughtful and special, and perhaps a sign of things to come. Their talk during the ride was full of cheer, a quality that had eluded them too often during the previous four years. Despite her fear of an oncoming headache earlier in the afternoon, Mary decided to accompany her husband, and to make their time together a fine evening out. The president's courtly attentiveness and the young engaged couple with them enhanced Mary Lincoln's romantic mood. While watching the play, she had been clinging to her husband's arm and had teased him about what their companion in the box, Miss Harris, might think of her. To which Lincoln replied, "She won't think anything about it."

Close to ten p.m., John Wilkes Booth crept into the presidential box and fired his pistol directly into the back of Lincoln's head. Next, he attacked Henry Rathbone with a knife before leaping onto the stage, shouting, *"Sic semper tyrannis."* Whether Booth hurt his leg in the fall from the box or during his escape on horseback remains a matter of dispute, as do so many of the events that evening. Witnesses at the theater nearly all agree that Mrs. Lincoln's screams alerted the audience that the president was injured.

During the mad rush to the presidential box, Mary seemed in shock. Watching a doctor unable to find her husband's pulse and attempting mouth-to-mouth resuscitation, followed by Lincoln's revived breathing, proved traumatizing. Mary Lincoln felt helpless as soldiers lifted her husband onto a litter to carry him across the street into a private house owned by a William Petersen. Lincoln was laid diagonally across a bed as Mary stood by.

There was a decision not to let Mrs. Lincoln know her husband's condition was hopeless. Her son Robert and the Cabinet, along with dozens of concerned friends, gathered for a vigil. As the night wore on, Mrs. Lincoln would periodically dissolve into sobs and be led from the room. Elizabeth Dixon, one of the few women in attendance, remembered the First Lady's frantic grief, wailing repeatedly that her husband must "take her with him."

Secretary of War Edwin Stanton took charge of the room and began to direct operations to secure the government (because Secretary of State Seward had been attacked as well) and to launch a manhunt for the assassins and conspirators. Word of these attacks spread throughout the city as a crowd gathered in the street between the Petersen house and Ford's Theatre, waiting for news of Lincoln's condition.

Stanton kept a close watch on the president, even as he sought the whereabouts of the vice president. Telegrams and messages flew in and out of the room, and a stenographer was fetched. As the hours wore on, Mrs. Lincoln wished to remain by her husband's side, but near dawn, when she collapsed to the floor in a faint, Stanton ordered, "Take that woman out and do not let her in again."

Thus began her exile, as she was led into a parlor, away from the deathbed of her own husband. When Lincoln's breathing seemed to signal the end was near, a clergyman, Reverend Phineas Gurley, suggested a prayer, and all knelt or bowed their heads. At 7:22 a.m., Abraham Lincoln, the sixteenth president,

The room where Lincoln died

was pronounced dead. Edwin Stanton uttered his famous tribute: "Now he belongs to the ages."

But equally poignant were the words of Mary Lincoln shortly thereafter, when given the news: "Oh, why did you not tell me he was dying?" When she was told he had passed, Mary's keening could be heard throughout the house, and signaled to those waiting outside that Abraham Lincoln was indeed gone.

The news of his father's shooting had interrupted Tad's own evening at the theater, and he was escorted back to the Executive Mansion for his safety, brought home to a deserted house from his evening out. His home was by now surrounded by guards. Rumors about Seward's attack had put all of Washington on high alert. Tad was kept in lockdown, as reports of Lincoln's condition remained unclear. A kindhearted member of the staff, Thomas Pendel, whom Tad nicknamed Tom Pen, took the boy into his father's bedroom—where he had often curled up at night to sleep next to the president. Pendel was able to get the boy to sleep by lying down with him, so Tad was not awake when morning brought news of Lincoln's death.

Mary did not want to leave her husband, and stayed on at Petersen's house another two hours after Lincoln died, until Robert persuaded her to take a carriage back to the White House. With the rain gently falling, the city woke up to the gloomy news of Lincoln's death by assassination. Thus Mrs. Lincoln would face her most severe crisis. The nation would gain a new president, but Mary, left alone and abandoned to face life without her husband, would never regain the father of her children.

15.

Waking Nightmares

Lying in a darkened room was no solution. She knew she must rouse herself for her sons. They had been deserted by her husband as well. Poor Tad, left alone after Willie had gone, and poor Robert, trying to be the man of the family—when he knew his father was irreplaceable. But most of all, poor Mary: left unprotected, vulnerable, with no will to move forward—or any will to move at all, for that matter.

That first morning back at the White House—after that awful night at Ford's Theatre and the bleak vigil at the Petersen home—she could not face the family bedrooms. Instead, she sequestered herself in a small guest room. When her sons came in to see her, she convulsed with sobbing, so hard that Tad begged her not to cry. The two women who had attended her from the night before, Mrs. Dixon and Mrs. Welles, turned her care over to Elizabeth Keckly, who reported encountering a world of agony. Robert tried to console his mother, but she remained inconsolable. There is no record that Mary ever viewed her husband's body after her ordeal at the Petersen house.

The First Lady locked herself away, refusing all visitors, trying to shut out the world. Her peace was disturbed when the workmen assembled in the East Room to build an eleven-foot-high platform to display the president's coffin. Mrs. Lincoln complained about the noise of the construction of the catafalque. She lamented that every nail driven reminded her of a pistol shot. This same structure, the Lincoln catafalque, has been used for state funerals at the Capitol into the twenty-first century.

Lincoln's coffin was put on view on April 18, and nearly 25,000 lined up to pass through the East Room to pay their respects. The mirrors and chandeliers were draped with black crepe, the room totally somber. The next day, invited guests crowded into the White House to attend a funeral service. This was marked by the conspicuous absence of females: Only seven women were among the six hundred in attendance. Once again, as at the deathbed, Reverend Phineas Gurley presided. Mrs. Lincoln refused to leave her room, and Tad also could not manage an appearance. Robert alone represented the family, and the burden weighed heavily. The short service began the process of removal for burial, which was to end in Illinois several weeks later.

On April 19, a funeral procession carried the body to the Capitol, where Lincoln would lie in state in the Rotunda, only the second person (after Henry Clay) to be granted such an honor. Lincoln also was the first president with a caparisoned horse in his funeral cortege—a riderless mount with boots reversed in the stirrups—a ritual for fallen military leaders dating back centuries. His military guard brought the body to rest at the U.S. Capitol so thousands more might file by his coffin to pay their respects before his remains were returned home to Springfield.

On Friday, April 21, one week after his assassination, a nine-car funeral train carrying over three hundred passengers began the journey to escort the bodies of Lincoln and his son Willie back to Springfield, a reverse of the seventeen-hundred-mile route of February 1861. The Lincoln Special, a train with Lincoln's photo over the cowcatcher on the front of the engine, was embraced by the American public. It has been speculated that the largest proportion of Americans ever assembled for a funeral took their turns bidding farewell to Lincoln over the next few days.

The train made scheduled stops in twelve cities, and broadsides with timetables were widely distributed. The engine moved only five to twenty miles an hour, out of respect for those assembled along the rail lines.

Thousands gathered along the tracks to pay tribute to their fallen leader—with bonfires lit at night to greet the train's passage through clusters of mourners. Ceremonies in several cities demonstrated the groundswell of grief, from ten thousand in Baltimore on April 21, to three hundred thousand in Philadelphia, to the four-hour-long procession of the casket through the streets of Manhattan, lined with half a million mourners. In New York, on April 24, a lone photographer captured the image of Lincoln's open coffin, flanked by military guard. (Nearly all the copies of this last photo of Abraham Lincoln were destroyed and the image was not published until the twentieth century.)

Newspapers overflowed with reports and images of the crowds and their grief. *Frank Leslie's Illustrated Newspaper*, *Harper's Weekly*, and other publications lavishly chronicled the funeral cortege's progress. Journalists could not keep up with the public's hunger for news, as publishers struggled to produce reprints of special editions highlighting images of mourning. Americans were enthralled by this historic loss, the first assassination of a president.

Lincoln's open coffin

By the time Lincoln's body reached Cleveland, no building was large enough to accommodate the anticipated crowds, so the open coffin was placed in a specially built outdoor pavilion. In Indianapolis, driving rain forced cancellation of the public procession, which took place the next day with an empty casket. Lincoln's remains had already headed for Chicago, but the people needed their ritual. Streetcars were decorated with slogans, including a banner displaying a common sentiment: "Thou Art Gone and Friend and Foe Alike Appreciate Thee Now."

All along the way, mourners hastily patched together elaborate arches to mark the arrival of the Lincoln Special. In Chicago, thirty-six women dressed in white formed a ceremonial guard for Lincoln's hearse as the procession moved his casket through a forty-foot memorial arch on its way to lying in state.

It was both personal and poignant when Lincoln's body reached his hometown of Springfield on May 3, and the coffin was taken to the state capitol for a final viewing. On Friday, May 4, an honor guard accompanied the body, enclosed in a gold and crystal hearse, to the Oak Ridge Cemetery. Rev. Gurley offered a final prayer before the bodies of Lincoln and his son Willie were entombed together. It was a sad day for Springfield, to say good-bye to someone whom the town had figuratively adopted, cherished, and sent forth with such high hopes. While none of Lincoln's kin were present, several of Mary's Todd relatives were in attendance to represent the family.

Certainly this series of memorials and farewells, a three-week journey from the time when Lincoln was shot at Ford's Theatre to the time when his body reached the cemetery, constituted the most elaborate and momentous funeral in American history. But it also showed that the remaining trio of Lincolns were in no shape to deal with the burdens of loss that had been settled upon them. Mary, especially, was isolated and incapable of handling her situation.

During this national outpouring of grief, Mrs. Lincoln stayed sequestered in her rooms at the White House. She wished to see no one socially, but conferred with a select group. She took comfort in her sons. She welcomed Mrs. Keckly as her companion, and also admitted family physician Dr. Robert Stone, as well as her good friend from Springfield, Dr. Anson Henry, for regular visits. Perhaps Tad's needs, highlighted by both Elizabeth Keckly and Dr. Henry, finally shamed Mary into making plans—as she continued to claim that she had no interest in life beyond her sons, especially Tad, who would more keenly feel his fatherless state.

Her younger son was at a tender age, just thirteen, and he had for so long been used to both parents' utmost attention. He had not been given any proper schooling after Willie's death, spending time the past two years with indulgent caretakers who refrained from discipline. But with the loss of his father leaving him rudderless, clearly something had to be done to provide for the boy's immediate future.

Mrs. Lincoln was willing to admit select friends as advisers: Mary Jane Welles (wife of the secretary of the navy), Elizabeth Lee (sister-in-law of former Postmaster General Francis Blair), Senator Charles Sumner, and her good friend from Philadelphia, Sally Orne (the wife of a wealthy Republican carpet manufacturer). She did not busy herself with ambitious plans but tried simply to pick up the pieces.

Mary Lincoln's preoccupation with her possessions frustrated those around her, although Elizabeth Keckly recognized that the mundane tasks of sorting and packing provided periods of diversion for the distraught widow. Mrs. Lincoln spent her days and nights at the White House filling over fifty trunks.

Although she wanted to hold on to everything, she generously gave away relics of great value, offering up sentimental mementos to those around her. Lincoln's canes went to Charles Sumner and Frederick Douglass, the gift to the latter pointedly demonstrating regard between black and white. Mrs. Lincoln also gave Elizabeth Keckly several important artifacts: the cloak Lincoln wore on the night he died, as well as the comb and brush he used in the White House. Shoes, gloves, and headgear were also part of this important collection of presidential memorabilia donated to Keckly.

The African-American housekeeper, Mrs. Slade, received Mrs. Lincoln's bloodstained dress from the night of the assassination. Keckly reported also that once Mary reached Chicago with her trunks overflowing with goods she would not be able to use, she donated scores of items to charity fairs—to benefit veterans and orphans. The dispersal of goods remains a matter of discussion well into the twenty-first century.

Mary Lincoln vetoed a campaign in Springfield to build a grave site for her husband directly in the center of town. The town fathers had formed the National Lincoln Monument Association, and wanted to keep Lincoln's tomb closer to city services for the throngs of visitors they anticipated. Mary roused herself from her daybed at this development, energized by her anger at such a suggestion. She threatened to bury her husband in Chicago instead of Springfield rather than to allow this plan to move forward.

She instead demanded that the Lincoln mausoleum be built at Oak Ridge Cemetery, a beautiful graveyard on the outskirts of Springfield. She explained that this more pastoral place reflected Lincoln's own personal choice—as they had discussed their own graves while visiting a Virginia cemetery just a few weeks before his death. She told Isaac Arnold that when she was at City Point, she and her husband had been at a place where early spring flowers were blossoming on graves, and her husband had confided, "Mary, you are younger than I. You will survive me. When I am gone, lay my remains in some quiet place like this." Mrs. Lincoln laid claim to knowing her husband's preferences more than anyone else, and exerted her authority.

This right of selection of her husband's final resting place—and presumably her own—was one she refused to relinquish. In this matter, Mrs. Lincoln asserted her rights as wife, which began eroding from the very moment of her husband's assassination, indeed dramatically during the hours leading up to his death, when she had been shoved aside.

After Mary was quite literally taken from the room where he lay dying, Lincoln's male circle—rivals for his attention—began to exclude her. Indeed, this cold shoulder caused great bitterness. While she felt she should have been given support and help, many withheld kindness, and some ignored minimal courtesy. Some doubtless believed they were repaying her own vindictiveness toward them, while others maintained that she had been unworthy of her husband, and now that he was the Martyr President, she was deemed even less deserving. Mrs. Lincoln felt deliberate disrespect and knew there were men among Lincoln's political confidants eager to be rid of her.

The new president, Andrew Johnson, had been conspicuously out of the picture on the night of Lincoln's assassination, and had kept a low profile in the wake of the outpouring of feeling about the fallen leader. He kept an office in the Treasury building and lived under guard at a house at Fifteenth and H streets. Johnson generously allowed Mrs. Lincoln to stay in the White House until she was prepared to leave Washington.

Despite his gesture, Mary was upset and angry with Johnson, complaining that he never sent any written condolences, nor paid a call to show his respect. Robert Lincoln was in charge of communications during this arrangement, and he also complained of Johnson's lack of proper form. Perhaps President Johnson saw no reason to be received by Mrs. Lincoln in what had become *his* White House by reason of her husband's death, and decided to avoid what a formal social call would require.

As the days stretched into weeks, indeed longer than a month, Johnson felt Mary had abused his generosity. Meanwhile, Mrs. Lincoln was recovered enough to send along a raft of requests, if not demands, to the new president—that he might grant favors to those Mary claimed Lincoln had intended to reward with appointments. Her directives insinuated that he must carry out President Lincoln's wishes—as channeled by the First

Lady. Mary might have believed she was just fulfilling her husband's final wishes by small acts of duty,
but to Johnson, and to other Republicans who learned
of these multiple directives, it seemed as though Mrs.
Lincoln was clinging to the last vestiges of her former
status, refusing to move on.

Mary was indifferent to unfolding developments.
The capture and death of John Wilkes Booth on April
26, the arrest of other conspirators—none of this
seemed to move Mrs. Lincoln. She only could awake
herself occasionally to respond to expressions of sympathy. As Lincoln's widow, she received condolences
from around the globe, and perhaps none meant more
to her than the message penned by Queen Victoria,
whose letter is quoted below in full:

Osborne, April 29, 1865

Dear Madam,

   *Though a stranger to you I cannot remain silent
when so terrible a calamity has fallen upon you &
your country, & must personally express my deep
& heartfelt sympathy with you under the shocking
circumstances of your dreadful misfortune.*

   *No-one can better appreciate than I can who am
myself utterly broken hearted by the loss of my own*

beloved Husband, who was the light of my Life—my
stay—my all, —what your sufferings must be; and
I earnestly pray that you may be supported by Him
to whom Alone the sorely stricken can look for
comfort in this hour of heavy affliction.

With the renewed expression of true sympathy,
I remain, dear Madam

Your sincere friend Victoria

Queen Victoria had been widowed in 1861, and her prolonged mourning would become a hallmark of her reign. Mary penned a note to the wife of a diplomat, Madame Berghmans, on the last night she spent in the White House: "I do not have the least desire to live. Only my extreme agony of mind has prevented my receiving yourself & other friends. God alone knows the agony of this crushed heart." Thus she felt her grief was shared by European aristocrats and royals who might appreciate her dilemma.

Religious faith was minor comfort to Mrs. Lincoln, because even if it meant the promise of greater glory, she deeply resented having her loved ones taken from her. Tad might remind his mother there were "Three of us on earth, & *three* in Heaven" (referring to Eddie, Willie, and his father), but this was something Mary found nearly impossible to bear.

Mary Lincoln could only manage to function minimally with the constant assistance of Elizabeth Keckly, whose steadfast care provided somewhat of a continuum, a link between before and after. Mary also leaned heavily on her Springfield confidant, Dr. Anson Henry, who was trying to coax the widow from the depths of despair, and to lead her back to the Midwest.

Lincoln's close friends, Judge David Davis and Orville Browning, counseled Mary to move back to Springfield to the home she still owned, at least while her husband's estate was being settled. Lincoln did not have a will, so Davis was handling the president's business affairs. (These affairs ranged from evaluation of his property back in Illinois to depositing his pay, for Lincoln's legendary indifference to accounts meant he had several uncashed salary checks tucked away in his desk drawer.) Although he died intestate, Mr. Lincoln's property would go to his wife and children. Mrs. Lincoln and their two sons would each inherit a third of his considerable estate, roughly estimated at $85,000 in 1865.

Although such an inheritance would allow her the financial wherewithal to reside comfortably in her former hometown, Mrs. Lincoln believed Springfield was no solution. She had burned too many bridges. She had become the subject of tittle-tattle, and so it would remain. The granddaughter of a Springfield merchant told the story that Mrs. Lincoln would "buy an entire bolt of yard goods from him, so that no other woman would have a matching garment." Such stories continued to swirl around her name, and much bad blood remained. In addition, her dearest confidante from the days before Washington, Hannah Shearer, was now settled in Philadelphia. And Julia Trumbull was no longer on speaking terms with Mary, permanently estranged.

As for Mary's own blood, two Springfield sisters, Frances and Ann, had been conspicuously absent during Mary's years in the White House. Her older sister and surrogate mother, Elizabeth, had been attentive following Willie's death. But Mary's uncharitable sentiments about Elizabeth's daughter (her niece) Julia—including indiscreet and waspish remarks in correspondence—had frayed their relationship. So if her own Springfield siblings were so indifferent to her plight, she could not expect a warmer response from others. Perhaps the former First Lady sensed that without Lincoln's mitigating presence, she could not endure what awaited her in Springfield.

Yet her husband's grave—a place where she imagined she might take comfort regularly—would be located in Springfield. Thus she required a residence nearby. So Mrs. Lincoln finally settled on Chicago—

the city of Lincoln's first grand political triumph, his nomination at the Wigwam in 1860. This was a community where many of Lincoln's oldest and strongest political supporters remained, the town where he had been launched as a presidential candidate, and a place where Mary hoped both her boys might receive the kind of consideration and attention their mother felt they deserved. After several false starts, mother and sons departed from the White House on May 23.

When Lincoln's body had left Washington over a month before, the world had turned out to watch. When his widow took leave, the scene could not have been more different. Her paid companion on the journey home, Elizabeth Keckly, wrote, "There was scarcely a friend to tell her good-bye." Mary, Tad, and Robert were accompanied on the train by Dr. Anson Henry, as well as White House guards Thomas Cross and William Crook, detailed to escort them back home to Illinois. Crook wrote sympathetically about Mrs. Lincoln's impaired condition, that she seemed in a stupor during the entire fifty-four-hour trip. (Considering the liberality with which laudanum and other opiates were prescribed, there is every reason to believe she was heavily drugged for the journey.)

When the small band arrived in Chicago, there was no welcoming party. Out of a sense of entitlement,

Mrs. Lincoln decided to head for the Tremont House, one of the most fashionable of Chicago addresses. She wanted only the best to be at her disposal. No sooner had she settled her sons into the sumptuous quarters at the Tremont, than she soured on the experience and drafted her older son Robert to find less expensive accommodation. Her sense of privilege was only outweighed by her fear of financial embarrassment, as obsessive money fears regularly overtook her.

Robert found the family "summer quarters" in the Hyde Park Hotel, roughly seven miles from the city center, on Lake Michigan. This move signaled more than a measure of economy, as Robert saw it not only as a sign of reduced circumstances, but as a symbol of abiding misery. He complained bitterly to Elizabeth Keckly at the Hyde Park that he would "as soon be dead as be compelled to remain three months in this dreary house." This reflected his own impairment at the time, as it was Robert himself who had selected this accommodation, yet he scorned it.

Robert was twenty-one, in love with Mary Harlan, and eager to get on with his law studies. His pursuit of his career and his own personal fulfillment was hampered by his sense of responsibility for his younger brother, coupled with fears about their unstable mother. Tad remained a sunny boy, despite all that he had been through. As the Lincolns' firstborn, Robert was expected to fulfill the promise his parents, and especially his father, had always presumed.

Mrs. Lincoln hid herself away at their shore hotel, allegedly because of frequent migraines—but there were other issues. Mary found it easier to cope within a contained environment, rather than confronting new realities. There is evidence that she began to medicate herself—with doctor's remedies as well as perhaps with alcohol, to dull her pain. After years of being at the center of attention—demanding deference while indulging material desires—Mary Lincoln was now shunted aside, facing debt and doubt alone.

This dramatic shift resulted in her clawing out for more—she craved admiration. With her husband gone, Mary felt even more emotionally bereft than she had when he was preoccupied by the war. Her grown son, Robert, had neither the temperament nor the inclination to substitute as his mother's comfort. Quite naturally, Mrs. Lincoln turned to her younger son, Tad, for her emotional coddling.

When she moved to Chicago, she had perhaps anticipated an outpouring of sympathetic and favorable attention, of being cosseted as Lincoln's widow. She had hoped that both she and her sons would be enfolded into welcoming arms. Instead, Mary found that those who came to their wealth and position during the rise of a Republican president were quick to forget that president's widow. She seemed unable or unwilling to grasp the full extent of her marginalization.

The status she had come to expect was now virtually gone, and nothing would replace it—especially in the tangible forms she had hoped for. Even if callers made their way to her hotel, they were systematically turned away. Mary was embarrassed to greet visitors in her humble circumstances. She perhaps thought some guardian angel would insist on installing her in suitable lodging, all expenses paid. But no offers materialized. When none of her husband's wealthiest supporters jumped to the rescue, Mrs. Lincoln deliberately refrained from socializing.

Mary limited most of her social contact to formal letters. She spent hours penning personal epistles to confidants, or lengthy responses to formal condolences. But the bulk of her correspondence during this period focused on her financial plight. The loss of her husband initiated a severe downward spiral, emphasizing her crippling sense of impoverishment. This was squarely in conflict with her inexorable demands for finery and expense, her obsessive need for new things. This friction plagued her to the end of her life.

Mary still owed thousands of dollars, personal debts run up while First Lady. The government was unwilling to step in and offer assistance, even for materials that she had purchased to decorate the White House. She did not present her bills, but instead made elaborate pleas that she be given government compensation in the wake of her husband's death. First and foremost, she requested the remainder of Lincoln's salary—all four years, because one year's salary ($25,000) would barely cover her expenses to one store account, A. T. Stewart's. By this time, Stewart's New York emporiums had made him a millionaire several times over. Shortly after the war's end, he was reputedly earning nearly one million dollars per year from his commercial empire. Even so, Mary wanted to pay her debts to him because he had been such a close friend during the White House years.

While Republicans were raking in profits, Mary Lincoln decided she would have to mount a personal campaign to restore her family's financial health. She was mortified when the extent of her debts became known to her son Robert and her husband's executor, David Davis. She insisted that these obligations would be honored—and determined to somehow bargain her way out of her financial hole. She regretted that appeals for the family of the slain president did not produce the large sums she had anticipated, but she was determined to prevail.

Mary decided to direct appeals toward some of the rich and powerful men who had grown fat off their appointments by President Lincoln—contacting them, reminding them of their indebtedness, looking for financial handouts. This strategy would eventually backfire.

Shortly after she settled in Chicago, the press reported Lincoln's estate was worth over $75,000. The information about the size of the president's inheritance came as an enormous shock to the American public, schooled on Lincoln as the Railsplitter, forgetting his years as a corporate lawyer. News of the amount splashed into the headlines while most Americans were reeling with their own wartime losses. Mrs. Lincoln had difficulty painting her portrait as needy, and the press would not let her forget her privilege.

Thus campaigns to raise funds for the president's widow were foiled by reports of her substantial inheritance. Of course, the money was tied up in Lincoln's estate, not readily available. The inheritance was slated as well for division with her sons, one of whom was still a dependent minor. Moreover, as income, she would only be given the interest on these sums, not the principal. Yet all of this was lost in the gossip. Talk about

Lincoln's estate dampened the general public's impulse for charity toward Mrs. Lincoln. Still, she hoped for help from the small but powerful group who had been made wealthy by her husband's patronage and who would view a $75,000 inheritance as inadequate.

Mary retreated into an unrealistic bubble, trying to resolve her problems in a vacuum. She was obsessed with restoring her status at a time when her children and her books became her only companions. Mrs. Lincoln was able to conduct her business only fitfully, as Elizabeth Keckly described:

She places her hand upon her forehead, as if ruminating upon something momentous. Then her hand wanders amid her heavy tresses, while she ponders for a few seconds—then, by a sudden start, she approaches her writing stand, seizes a pen, and indites a few hasty lines to some trusty friend, upon the troubles that weight so heavily upon her. Speedily it is sent to the post office; but, hardly has the mail departed from the city before she regrets her hasty letter, and would give much to recall it. But, too late, it is gone.

Even worse, these plaintive appeals failed to produce desired results.

While waiting for answers—which mostly did not arrive—Mrs. Lincoln contemplated treasures she had moved from the White House, which were sitting in a warehouse nearby. The number of trunks she had shipped from Washington was reported as well, so many people let their imaginations run wild about the "sacking" of the White House. Mrs. Lincoln was subjected to another wave of bad press.

During the weeks following the president's death, thousands filed through the public rooms and the Executive Mansion furnishings were pilfered and desecrated. Souvenir seekers went on a rampage and were not beaten back by experienced staff. The dismissal of the longtime doorkeeper, Edward McManus, in January, left the staff ill equipped to handle the deluge of visitors descending in the wake of Lincoln's assassination. They were unable to maintain order, and when the Johnsons moved in, they found the public rooms blighted and half-emptied. Despite the obvious cause, gossip laid the blame squarely on Mrs. Lincoln's excessive baggage.

During this era, all goods and services donated to the president and his family were completely acceptable and considered the personal property of those to whom items were given. It was not viewed as corrupt or even inappropriate to receive such things. Accepting such gifts was customary and familiar to all Americans, especially Southerners. Both Mary and her husband were born into this Southern tradition, steeped in elaborate rituals of gift giving. At the same time, receiving *lavish* gifts caused qualms concerning taste—then as now.

Public perception of Mrs. Lincoln's proclivity for gifts—that she was always thrusting an open hand to those in a position to seek her good-will—was sharp and negative. And perversely, when she did not take handouts, when she paid for goods and services she sought, then her spending priorities came under fire—even by her own husband, who raged over her "flub-dubs."

However, claims of her looting were vicious and part of a continuing press campaign against her. Mary Anne Clemmer, no fan of Mrs. Lincoln's, defended her against charges that she took what was not her own. Clemmer claimed the "vandals" who roamed the White House were "sufficient to account for all its missing treasures." So as Mrs. Lincoln sat pining in Chicago over her neglect, the public salivated over the worst kind of libels against her.

At first Mary was pleased by the way Robert was stepping into the breach, trying to help manage family affairs—consulting with David Davis and her husband's former circle, organizing their residence, and handling other details too upsetting for her.

But this pleasure was mitigated when Robert came under Davis's influence, as the judge was advising Mrs. Lincoln that she could live comfortably on her income, especially if she were to return to Springfield, where she owned a home. She adamantly rejected this notion, claiming that people misunderstood the "particulars" of her situation. And she would not let anyone to whom she wrote forget "*from whence*, we have just come."

Mary Lincoln's letters from this period were laced with self-pity and despair, as when she complained to Elizabeth Blair Lee that when she stared out into the waves of Lake Michigan from her window, she felt she would like to slip under them. She did not withdraw into spiritualism and religious devotions as many wartime widows did. Instead she launched a campaign of such enthralling intensity that it would in many ways become the crusade that kept her alive despite suicidal depression—a crusade to change her reputation, to recapture the status she coveted, to devote herself fully to her husband's memory. Indeed, memorializing Abraham Lincoln became Mary's battle flag, to be raised at any and every opportunity.

While Lincoln was alive, she had been forced to share him with rivals, but now that he was gone, she wanted things changed. As his widow, she believed her sacrifice should be acknowledged by the government and the people. She devoured the newspapers to track the progress of her husband's legacy, and filled her missives with sharp and current political observations. She took umbrage at any honors or charity bestowed upon others, harping on the theme that any former general or prominent businessman owed his bounty to her dearly departed husband. Mrs. Lincoln wanted to restore herself through the status of "First Widow," and she railed against those who failed to pay deference in this regard.

First Widow

Mary's letter writing took on the quality of a publicity campaign. Her resolve is reflected in references to unflattering press and in demands for action, fired off with regularity and intensity. She couched repeated demands for assistance in identical florid prose: "Roving generals have elegant mansions showered upon them [referring to General Grant being offered homes by private subscription], and the American people—leave the family of the Martyred President, to struggle as best they may!" Her constant laments repeated a familiar theme: "The family of the man who sacrificed his life for his country—barely able to meet expenses, in a boarding house—A generous Nation, will certainly not permit this."

She asked friends to write to members of the House of Representatives for her, or to use their influence on her behalf with influential politicians. Her begging was a relentless, self-pitying ploy. She scribbled and dispatched, devising plots within plots, as when she instructed one of her agents, "Please do not mention that Sen. Sumner wrote *me* lest *Seward* will suspect, *he*, advised me *not* to let the letters be published. . . ." She mounted her campaign with great intricacy and fury, tossing request after request into the winds.

Mary felt abandoned again when her beloved Lizzie, Elizabeth Keckly, headed back east—as had been agreed upon at the outset. It was nevertheless a difficult parting when she bade farewell to the woman who had been with her in the White House from the beginning. Mary Lincoln promised to meet up with Lizzie again soon, to arrange a rendezvous in the near future.

Mrs. Lincoln suffered another setback with the loss of Dr. Anson Henry in July 1865. Henry had provided one of the few links with Springfield she treasured. In the weeks following Lincoln's death, while Mary was holed up in the White House, Henry's solicitous care shielded her. When he was reported lost at sea during the summer, Mary was completely distraught.

Absorbed in her own campaign, Mary seemed to take no interest in her husband's killers. If she had any thoughts about the trial of the Lincoln conspirators in June 1865 and the execution on July 7 of those condemned to die on the gallows for their part in the plot—Mary Surratt, Lewis Powell, David Herold, and George Atzerodt—she kept them to herself.

In August 1865, Mary moved with her boys back to the Clifton House, a respectable residential hotel in central Chicago, where Tad enrolled in a local school and Robert continued his apprenticeship at the offices of Scammon, McCagg & Fuller, a leading firm of Chicago attorneys. By this time, Tad had learned his letters and was stubbornly working on his reading skills. He seemed finally keen to make up for lost time, having suffered years of academic neglect.

The younger surviving Lincoln son was a bright child who demonstrated his talents only when it suited him. Perhaps his lisp and the shining exemplars of Willie and Robert had discouraged him from competing in the classroom. His intellectual laziness could and did translate into his being mistaken for someone far less gifted than he was. Casual observers might take him for being slow or mentally handicapped. But Tad seemed to thrive once he was given tutors, discipline, and regular classroom instruction. He was able to find his own niche among the sons of the Midwestern middle class with whom he mingled.

With both of her boys studying away, Mrs. Lincoln continued her schemes to restore the family fortunes. She concocted the idea of ridding herself of unwanted items from her former wardrobe, as those piles of clothes she had transported from the White House were of no use to her now. She had pledged to remain cloaked head to toe in black bombazine, her mourning armor. Mrs. Lincoln would always wear widow's weeds in public and black in private. Her trunks of former finery lay fallow in storage. It slowly dawned on her how unlikely it was that anyone or anything was going to come to her financial rescue, so she fell back on her own reserves.

She first wrote about the possibility of selling off unwanted goods in a letter to her friend Sally Orne in August 1865, hoping Sally might know someone willing to purchase privately the elegant lace dress she wore "ceremonially" at the second inaugural ball. She also made reference to other rich and beautiful materials with which she would be willing to part. This strategy seemed simple to her: to dispense with items of interest to others but no longer of interest to her to obtain cash to pay off debts.

Mrs. Lincoln may have been prompted in part by reports of sales of Lincoln mementos. Items were being auctioned off and fetching high prices. She also knew Robert was feeling the weight of the family's financial constraints, and wanted to relieve him of this. She knew he was in love with Mary Harlan, of whom she heartily approved, and he wanted to get on with his own independent life. She sought to lift financial burdens to benefit both her sons.

Late in the fall, Mary accompanied Robert to Springfield—her first visit in more than five years. This was not a poignant homecoming, as the only thing that induced her to come back was the removal of Lincoln's body from a vault to a temporary tomb. Going there, she exclaimed, would open her bleeding wounds afresh, but she felt she could not avoid any widow's duty, and therefore accompanied her son. She did not mention contact with or kindness from any family member except her cousin John Todd Stuart, who assisted her in negotiations with the Springfield group funding Lincoln's grave and memorial. This was a dispiriting exertion, and she returned to Chicago complaining of a cold derived from a "chill" on the train. Her indisposition could have stemmed in part from the cold shoulder she received. That even Springfield would snub her was yet another sign of her declining status.

Mrs. Lincoln kept hectoring Judge Davis about her husband's estate. The very same month that she concocted a scheme to sell off parts of her wardrobe, large store bills were due. She told Davis that she would be writing to merchants to take care of things herself. One bill, for over $600, included the purchase of eighty-four pairs of gloves. It is hard to imagine what Mrs. Lincoln considered a suitable budget when she was snapping up seven dozen pairs of gloves while claiming "rigid economy." She might have had a memory of her days in the White House, when the crush of hundreds in a reception line might force her to run through a dozen pairs in one day, but those days were over.

While she schemed to sell old clothes for new, she ensnared her son's former tutor, Alexander Williamson, in her plots to raise funds from East Coast sources—through Congressional appropriations or private donations, or both. She coerced him into becoming her personal lobbyist, a career at which he failed miserably. At times she wrote daily bulletins with advice about whom to consult and whom to avoid, how to make an approach, whom they could best coax: her insider's guide to the D.C. establishment. She asked him to telegraph daily—which was enough of an imposition, but she also made her request during the holiday season, and not long after Williamson had lost his mother. Mary's zeal to satisfy creditors blinded her to all other considerations. She would not rest until her debts were retired and her name was cleared. She felt this was the first step to restoring her reputation.

Shortly before Christmas in 1865, she despaired over unwelcome news that Congress would not give up four years of Lincoln's pay but was willing to part with only one year's salary, which after deductions was a sum of a little more than $22,000. She felt this made only a small dent in the amount she owed.

Mrs. Lincoln spent her first holidays without her husband in a downmarket Chicago hotel writing mournful predictions—"The family of the Martyred President shall be homeless wanderers forever." (Certainly not forever, but at least until the estate could be settled and she might secure funds for a permanent residence.) She began her new year with renewed resolve to clear up outstanding obligations—for example, by writing a letter to her favorite Washington jeweler begging him to take back goods worth over $2,000 (pearl, onyx, a clock, bracelets, and silver spoons) *at the price they were purchased.*

Her winter would be laced with even more disappointment when Congress proposed to give President Johnson $75,000 to refurbish the White House. The press complained that the Executive Mansion had been emptied and that outstanding bills Mary had left behind plagued city merchants. Mary fumed to her friends over such criticisms splashed within the pages of the *New York World*—claiming these were partisan attacks and should not be directed at her personally. But she did take them personally.

Her self-pity crested in the weeks before the "terrible anniversary," as she called the date marking her husband's death. But the anniversary of Lincoln's assassination was but one day in a parade of remembrances when Mrs. Lincoln allowed herself to be engulfed by anguish. She wallowed in memories of loss on her wedding anniversary in the first week of November; on her birthday, on December 13; a few days before Christmas, on Willie's birthday; at the beginning of each new year without her husband; on the first day of February, the date of Eddie's death; and her husband's birthday on February 12. Tad's April 4 birthday—one bright spot—was sandwiched among dates that evoked deep sadness. Mary would twice annually relive her husband's violent death—on Good Friday, which fell on March 30 in 1866, and on April 15, the calendar date on which Abraham Lincoln died. Marking time only caused Mary Lincoln a heightened sensation of melancholy as her first year without her husband drew to a close.

She spent her anniversaries still without a home, feeling dismal about her prospects. Simon Cameron, Lincoln's former secretary of war whose rumored corruption had forced him to step down, returned from his diplomatic post (in Russia) and promised to provide the former First Family with assistance. He pledged to give Mrs. Lincoln $1,000 and to circulate a petition soliciting funds from other wealthy Republicans. Her spirits temporarily lifted with Cameron's promise, and she began to shop for a house. She settled on a handsome row house at 375 West Washington Street. Unfortunately, Cameron's fundraising, like Williamson's, did not bring satisfactory results.

Mary had her heart set on a new home, and she ended up paying out nearly all her cash on hand from

Lincoln's salary to buy the West Washington Street house, and went into debt to furnish it for herself and Tad (Robert had already moved into a bachelor apartment). Mary was happy to have Tad settled on the West Side in a home close to Union Park, where he might play with the offspring of other transplanted Kentuckians. Mrs. Lincoln felt, once again, as if she might be among her own element. She began to attend the Third Presbyterian Church, and Tad was enrolled in the Brown School. At the same time, economic woes persisted, as she spent much of the summer of 1866 trying to coax Cameron to make good on his promises so she might achieve fiscal solvency.

In August, the new president was planning a speaking tour, which would take him out to Illinois. Although Mary Lincoln wished she could see Mary Jane Welles, who would be in Johnson's party, she skipped town—to avoid the extreme unpleasantness of an encounter with Andrew Johnson. Her only legitimate excuse to be away from Chicago was a trip to Springfield, which she undertook reluctantly. She clouded matters by consenting to meet with William Herndon, Lincoln's former law partner. This was someone toward whom she had less antipathy than Johnson, but not by much. She considered both men uncouth, drunkards, and not worthy to polish the pedestal on which she placed her husband.

But Mary was willing to meet with Herndon because he was working on a book that she believed would exalt Lincoln. She wanted biographies to contribute to the growing recognition of her husband's status as savior of the nation. So as part of her duty as First Widow, she consented to an interview to contribute to this legacy.

Herndon's oral histories of Lincoln, collected over the months and years following the president's death, have become a valuable resource, as David Donald, Douglas Wilson, and other prizewinning scholars have noted. However, many of Herndon's findings have spawned the morass of clashing interpretations, ongoing scholarly debates, and popular misconceptions that such a source naturally elicits.

One of the most contentious topics stemming from his interview of Mary Lincoln on September 5, 1866, was her remark that her husband had never formally affiliated with a church, although he was a "true Christian gentleman." This interview and Herndon's use of her remarks would give Mary a wellspring of regret. In the years to come, she would repudiate his interpretation of her statements and confront the question of Lincoln's religious beliefs again and again. Insult was added to injury when his Springfield lectures on Lincoln in the fall of 1866 appeared.

Mary Lincoln claimed Herndon was at his worst when he offered up a sensational and—according to her—wholly fictitious historical account. Herndon suggested Lincoln had never loved his wife but pined for Ann Rutledge, who had died in New Salem in 1835. Herndon's lecture on November 16, 1866, about Lincoln and Ann Rutledge was not just a result of his friendship during nearly two decades as Lincoln's law partner, but was also based upon his fact-finding mission into Lincoln's former haunts, tracking down friends, relatives, and neighbors in his quest for authenticity.

This Rutledge bombshell hit Mary Lincoln very hard. Herndon's allegations about Ann Rutledge did not just upset her, they drove a stake into her heart. She knew that propriety forbade her from fighting back, despite what she felt was a direct attack upon her and her husband's memory. She knew she was the woman Lincoln loved, but here was Herndon claiming Mary was just a poor substitute for his first love. Such a blow struck at her self-esteem, at her very identity. William Herndon suggested that Mary was not the person she knew herself to be. He alleged that her whole life was some kind of charade. His characterization of Rutledge did not just supplant Mary, but *erased* her.

This may have been a psychological trip wire that contributed to Mary Lincoln's deteriorating mental condition. Historian Jason Emerson argues that Herndon's campaign was particularly traumatizing. In the months following Lincoln's death, Mary dwelled on the fact that she was her husband's "greatest" love, in fact that she had been, in his words to her, "the only one, he had ever thought of or cared for." This solo status provided her a comfort that was ripped away with this new discovery by someone who was supposed to be one of Lincoln's inner circle, indeed, someone to whom she had imparted confidences only months before.

Her son Robert was appalled, angered enough to put in writing that Herndon was making an ass of himself. Robert did everything he could to discredit and silence Herndon. He warned him that the subject of the Lincoln family was completely out of bounds for this manuscript under preparation. Robert included himself in this demand, placing his honor and reputation squarely on the line. He further argued that no one should be compelled to live in a glass house.

Mary must have felt that she *was* living in a glass house and everyone was taking aim at it. Herndon's accusations and, from her point of view, falsehoods in the fall of 1866 had an unhinging effect. By the winter of 1866–1867, Mrs. Lincoln was even more besieged. Her quest for congressional relief spearheaded by Williamson had become stalled, while her hopes for funds from Simon Cameron were dwindling. She was heartened by a letter to the *Chicago Tribune*, penned by James Smith, once from Springfield but now an American consul in Dundee, Scotland. Smith denounced Herndon in print—earning Mrs. Lincoln's eternal gratitude.

Yet despite this champion, she felt defeated:

There are hours of each day, that I cannot bring myself to believe, that it has not *all* been some hideous dream, and in my bewildered state, I sometimes feel that my darling husband, *must & will* return to his sorrowing, loved ones. . . . I am so miserable, that I receive but few strangers & consequently the friends of other days are dearer than ever to me.

This sense of isolation and deprivation was exacerbated by the fact that Mary was unsuccessful at settling in to her new home in Chicago. She felt her financial plight worsening with each passing month.

In the spring of 1867, she sought out a boarding school for Tad in Racine, Wisconsin, but vetoed it after a personal visit. Tad was managing well with his Chicago school and friends, while Robert was getting on with his legal career. But both sons were revisited by the details of their father's grisly death when called to Washington to testify in the John Surratt hearing in the summer of 1867. (Surratt had been abroad at the time of the assassination, but was extradited and returned to stand trial as coconspirator, a charge for which his convicted mother had died on the gallows two summers before.) Mary begged off this ordeal, pleading illness to avoid the summons. She felt she could not withstand such a spotlight, and wanted to maintain her low profile in Chicago.

By the time of the second anniversary of her husband's death, in April 1867, it became clear that her new house on West Washington Street was not going to work out—it was too costly to maintain on the money she had in hand. Money she hoped would be forthcoming had still not been realized. With regret, she was forced to rent out her new home to make ends meet. Mrs. Lincoln would have to move again and seek another way to manage economically.

She might have looked southward to Springfield, but several of her stalwarts there had disappeared. Her brother-in-law Dr. William Wallace died in May 1867, leaving his widow, Frances, in financial difficulties and thereby exacerbating Mary's sense of loss and alienation. She remained estranged from her sisters Ann and Elizabeth and had been spurned by her former Confederate family, even by her beloved Emilie, whom she had taken in when her husband died. She truly felt on her own, and she simply could not bear to return to Springfield, where she had met and married the man of her dreams, then followed those dreams to Washington, only to have them cut short by an assassin's bullet following four long years of war.

Thrown on her own resources, adrift again, Mary's life followed the bleak pattern of thousands of war widows—forced to handle matters beyond their own strength and ability to cope. But few of these women would endure the kinds of humiliations that lay ahead for Mrs. Lincoln.

16.

Widow of the Martyr President

One by one, Mary Todd Lincoln felt essential props pulled out from under her. First her beloved Willie's death, and then her husband murdered before her very eyes. Eighteen months into widowhood, her husband's former law partner publicly proclaimed that Lincoln had never loved her but spent his life pining over his first love, Ann Rutledge. During the fall of 1867, vulnerable and exhausted, Mary became entangled in an elaborate scheme that caused her pain and humiliation. After years of waiting for donors, with creditors and fears closing in on her, Mary decided to force the hand of benefactors who had reaped thousands in the Civil War windfall. This was a misguided series of calculations that went terribly wrong.

She decided to raise some cash by ridding herself
of unwanted clothing, perhaps with encouragement
from Elizabeth Keckly, who had patiently helped Mrs.
Lincoln pack up her more than threescore trunks
when she left the White House. Keckly had suggested
that she make donations—give away items she no longer needed to charity auctions. Mary had vowed to
spend the rest of her life swathed in black. This rendered scores of fashionable items from her Washington
years effectively useless. When Keckly visited Mrs.
Lincoln in 1865 during the Christmas season, she had
seen a robe that she had sewn for Mrs. Davis on display at a charity fair. Perhaps this triggered the notion
that Mrs. Lincoln should make some similar donations from her own wardrobe, as her gowns and personal ornaments had been well documented during
her tenure as First Lady.

Whatever the inspiration, shortly before moving
out of her new house in the spring of 1867, Mary
wrote to her dear friend, "Lizzie, I want to ask a favor
of you. It is imperative that I should do something for
my relief, and I want you to meet me in New York,
between 30th of August and the 5th of September
next, to assist me in disposing of a portion of my
wardrobe." Although the dates shifted, the plan remained in force. Mary promised to provide financial

assistance down the road. Mrs. Lincoln found a hotel and asked Keckly to book rooms for the two of them under the name of Mrs. Clarke, a pseudonym she had employed frequently during the Civil War.

Mrs. Lincoln placed Tad in the Chicago Academy for his schooling, then headed to New York, arriving on September 17. Upon arrival, Mary discovered there were no rooms at the St. Denis Hotel for "Mrs. Clarke." She panicked and sent off a hurried plea for Lizzie to come at once. From past experience, Keckly knew that Mrs. Lincoln often changed her schedule or her mind—leaving chaos in her wake. She knew as well that a woman of color would have trouble registering at the St. Denis.

Keckly sped to New York by train, and found Mrs. Lincoln eager to see her. Even though she registered as "Mrs. Clarke," Mrs. Lincoln was recognized for who she was—her name scrawled all over her trunks was a dead giveaway. The two women—one black and the other white—were told they could not have rooms on the same floor. Even for the former First Lady, the desk clerk refused to bend the rules. He assigned the pair to rooms in the servants' garret. In addition, they were refused seats together at the same table in the dining room. They went to bed without a proper supper, climbing steep steps to settle into bed.

When they took breakfast out the next day, Mrs. Lincoln confessed that she had visited a diamond broker the day before, seeking help. Her identity had been discovered, despite the use of a pseudonym, when a piece of engraved jewelry had betrayed her. Who was this particular merchant? Just a name pulled out of a directory, according to Mrs. Lincoln. But he happened to be a staunch Republican who volunteered assistance for the president's widow, which suited Mary. Especially when he moved the two women into rooms at the Union Place Hotel, and advanced Mrs. Lincoln $600 to cover her New York expenses.

Mrs. Lincoln entrusted her affairs to two businessmen—W. H. Brady and Samuel Keyes—who hoped to exploit the president's name for their own benefit. Brady believed that advertising the Lincoln name would ensure a bigger sale, claiming the people would not permit the widow of Abraham Lincoln to suffer. This was exactly what Mary longed to hear. When he dangled his anticipation of $100,000 in front of her, Mrs. Lincoln fell into the trap. Brady and his partner Keyes proposed a far-fetched strategy involving blackmail.

Brady wanted Mrs. Lincoln to suggest the names of prominent Manhattan politicians who owed their fortunes to her husband. He argued that these men would be willing to advance her money rather than having the world discover that she was compelled to sell her wardrobe. They wanted her to hint that she had held on to correspondence compromising many of those involved with wartime contracts. Brady wheedled and finally got Mrs. Lincoln to write him a series of letters suggesting worse to come. Mary backdated several letters to the period before she arrived in Manhattan, showing she was complicit in his plot. This whiff of blackmail, he promised, would coax wealthy Manhattan backers to open their wallets.

Meanwhile, Elizabeth and Mary invited secondhand-clothing dealers to their hotel to inspect merchandise for resale. Keckly and Lincoln prowled shops on Seventh Avenue, hoping to trade vintage clothes for new greenbacks. Between Brady's shenanigans and these hit-or-miss attempts to offload goods, a scandal was brewing. The two women had no luck with shopkeepers, and gossip began to circulate about this mysterious widow peddling her wardrobe. The whole project stalled, like Mrs. Lincoln's previous campaigns. Time was running out.

By the first week of October, with her (borrowed) funds nearly depleted, Mrs. Lincoln gave in to Brady's request to allow public exhibition of her wardrobe—presumably to stir up sales. He wanted to publish copies of her letters in the *New York World*, letters that he had coached her to write. He believed the Democratic paper would be eager to expose Republicans. Whether Mrs. Lincoln consented to this ploy, or whether Brady simply supplied these letters on his own, the first of the stories about her desperate circumstances appeared in print on October 7, 1867. Nothing before had ever elicited such a torrent of ridicule.

Her clothes, on display in Brady's showroom at 609 Broadway, provoked a cloudburst of sensational headlines and turned the situation into a perfect storm. Even more aggravating, publicity did not create commercial interest. Only scorn and rebuke followed, so Mary fled the scene.

It had been painful to be away from her sons, to manage outside her Chicago cocoon. After several weeks away from home, Mrs. Lincoln took the train back to the Midwest, leaving Keckly behind to represent her interests. Mary would write to her friend with sincere gratitude, "I consider you my best living friend, and I am struggling to be enabled some day to repay you."

The train ride home was more than a sad affair, a foretaste of horrors to come. Her Old Clothes Sale was discussed openly by strangers as they read the papers on the train. These fellow passengers presumably had no idea that Mrs. Lincoln was there, hidden under veils. Two men seated next to her even discussed whether the former First Lady might be able to raise enough cash to provide a decent burial for herself.

Mary found it even more mortifying to run into her old friend Charles Sumner, who turned as "pale as the tablecloth" when he recognized her in the dining car. Sumner at first seemed to duck his old friend, pained by her severely reduced circumstances. But taking pity on her, Sumner brought Mary a cup of tea before bowing out. She tossed the tea out the window, then sank into a fit of weeping over the indignities to which she had been reduced.

Arriving back in Chicago, Mary was heartened by her reunion with young Tad. She was unprepared for her encounter with Robert, however, who, after reading the papers, "came up last evening like a maniac. . . ." Robert was apoplectic for good reason. He had just weathered the Herndon revelations the year before. While he was trying to live quietly and climb the ladder at his law firm, his mother was splashing into the headlines. With Mary Lincoln's correspondence reprinted in the Chicago papers, Robert felt anger and defeat.

The public spectacle of Mrs. Lincoln's complaints in print was a terrible breach of Victorian conduct.

It was acceptable for men to fight her battles for her, but she was expected to refrain from making any protest, mute on the sidelines. Mary should never have given over copies of private correspondence to the press. The letters in her own hand were a breach of etiquette at best and evidence of criminal intent at worst.

With wide public discussion of the entire episode, naturally her conduct and her character came under intense scrutiny. And these new press attacks were more brutal than those in her bleakest days at the White House. Savage invective erupted in North, South, East, and West, raining down curses on Mary Lincoln.

The *Rochester Democrat* complained, "Mrs. Lincoln, the widow of the murdered President, has made an exhibition of herself. . . ." A Kentucky paper exhorted, "Will not somebody, for very shame sake, go and take away those drygoods the widow of the late Lamented persists in exposing for sale in a Broadway shop window?" A Springfield, Massachusetts, journal complained that Mary was a "dreadful woman" who forced "her repugnant individuality before the world."

Of the many mistakes made by Mrs. Lincoln during this period, perhaps none was so foolish as to cross Thurlow Weed. When the widow Lincoln sneaked into Manhattan, she was no match for the political kingpin nor his cronies. These powerful men, on home turf, ruled with an iron fist, clobbering any interference. The Republican guard was perfectly willing to swat down Mrs. Lincoln's feeble attempts, even if it meant knocking her down and holding her there. To the glee of the Democrats, Weed's *Commercial Advertiser* recycled reports of padded White House accounts and looted public rooms, smearing Mrs. Lincoln again. Weed's paper sanctimoniously pronounced: "Mrs. Lincoln invites and justifies animadversion upon herself."

Back in her rented rooms in Chicago, Mary was flabbergasted at the lava flow of venom pouring out of papers across the nation. Her scribblings to Mrs.

Mrs. Lincoln's Wardrobe Exhibition:
The Old Clothes Scandal

Keckly indicate that she understood the political nature of these attacks but could not believe their ferocity. On October 9, she wrote, "If I had committed murder in every city in this *blessed Union*, I could not be more traduced."

In the days that followed, she moved her household and hoped, despite all evidence to the contrary, that the sale would bring financial relief. She sent off daily bulletins to Elizabeth Keckly and promised she would not cry any more about all the "cruel falsehoods." While trying to backpedal out of this quagmire, she sprinkled around a few untruths herself. For example, she wrote to a friend that she had no idea how her letters ended up in print.

Robert Lincoln was mortified by this imbroglio and wrote to his fiancée, Mary Harlan: "The simple truth, which I cannot tell to anyone not personally interested is that my mother is on one subject not mentally responsible." His mother was impossible about *money*. Robert knew what his mother had been hiding for much of her adult life, that she suffered a disorder that might be characterized as financial bulimia: She felt compelled to hoard and spend in cycles of indulgence and regret. No matter how much she was told she was in fine shape, she continued to see herself as deprived— and to toss aside reason and judgment to decorate and bejewel herself or her homes. After she had slaked her thirst for finery, she would then become horrified by what she had done, yet the cycle never failed to repeat itself.

Robert knew that friends wondered why he did not just take charge of his mother's business matters. But he knew there was no simple financial solution, as these money issues were tied up with psychological complexities that he feared. Robert understood this situation would be a source of ongoing conflict, requiring great delicacy. He was afraid of what might develop in the future but felt hopeless to change the current situation. When his friend Edgar Welles offered him the chance to travel on a holiday to the Rocky Mountains, he fled.

Mary Lincoln also had a manic need to justify herself, which only increased with age. She complained to her former ally Alexander Williamson, "Those who swindled *our Government*, in war times, are the ones who most deprecate my movements." She had become a perfect paranoid, with many people truly out to get her. At times she even complained about her firstborn. She suggested that Robert was only pushing for the settlement of his father's affairs out of spite, "to cross my purposes."

During this period of escalating fears and public attacks, Mrs. Lincoln's personality began to disintegrate.

She had gamely confronted New York financiers, but found the game cost too much. She had left her faithful friend Keckly behind in New York to take care of business, but financial matters dissolved into chaos. She would end up sending Brady a check, as this misadventure not only failed to make a profit, but cost her money: salt on a wound.

Keckly tried to interest her friends in soliciting funds for the president's widow among the black community. Blacks might donate as a gesture of their abiding regard for Lincoln. Charity from African Americans was an offer that Mary initially spurned. By the time Mrs. Lincoln got around to responding with any enthusiasm to such offers—from Frederick Douglass, among others—their ardor had cooled. With the holiday season approaching, retail prospects for Mrs. Lincoln's wardrobe sale also cooled.

Mr. Brady proposed to stir up interest by taking the exhibit on the road and charging admission. The press had increased interest in Brady's showroom, with a parade of gawkers marching through. This publicity bonanza was something Brady had hoped to cash in on. The former First Lady's clothing was garnering attention, but no sales. Mary vetoed the plan for taking the show on the road, and instructed Keckly to send her costumes back without delay.

The spectacle of Mary's downfall—the "old-clothes scandal"—rippled outward from New York across the nation. She was humiliated and abandoned by the powerful men she had hoped to wheedle into providing monetary assistance. She felt betrayed by David Davis, who suggested to others that her clothes sale was proof of her diminished capacity. Part of Mary's justification for her desperate actions was that her husband's estate had not been settled. She maintained that her inheritance would reduce, though not eliminate, her financial woes.

Mary was relieved when the estate was finally settled—and grateful that Davis had increased the amount to over $110,000. She was entitled to one third of her husband's money, and needed the investment income for Tad's schooling and upkeep. Yet Mary was aggrieved it had taken so long. Plus, she continued to insist that she could not live comfortably on the $1,700 annual interest generated.

Shortly after collecting her inheritance, Mary wrote to her mother-in-law, Sarah Bush Lincoln, to mention that a few weeks before he died, her husband expressed a wish to erect a monument over his father's grave—a wish finally carried out not by Mary but by Robert Todd Lincoln, in 1880. Robert erected a stone in his grandfather's memory—a grandfather he had neither known nor even met. Mary also mentioned the blessed and deep comfort Tad afforded, as he was "named for your husband, Thomas Lincoln, this child, the idol of his father."

Mrs. Lincoln began to lean more and more heavily on her younger son, perhaps somewhat resentful that her elder son had moved out. Robert had a distinct preference for the finer things in life, and did not enjoy the less than luxurious quarters his mother maintained. She felt Robert pulling away from her—just as he had when he entered the army a few years before.

Mary's firstborn sought counsel from members of the male clique who once surrounded his father, and whom Mary perceived as indifferent to her. Mary's sense of abandonment deepened with Robert's struggle for autonomy. She did not accept this change gracefully, but knew she would have to defer to the circumstances. Robert would soon have another woman in his life, when he married Mary Harlan. Also, as he was launching his career, he would be well served to seek support from these men of influence, who were willing to play the role of godfathers.

Robert seemed to his mother to only resurface sporadically, and then only to level criticism at her. Feeling such displacement, Mary decided not to allow Tad to go away to school, but to keep him close. She would refer to him as her sunshine, and leaned on him even more as her bleak future unfurled.

One of the most painful breaks for Mary Lincoln developed in the spring of 1868. Mary had come to depend upon Elizabeth Keckly, who became Mrs. Lincoln's most reliable confidante. As a result, following Lincoln's death, Lizzie was given some of Mary's most precious relics. When she felt she had no one else to turn to, Mary asked for Lizzie's help during the sale in Manhattan. She had entrusted Lizzie with secrets and sorrows. When Keckly disappeared from Mrs. Lincoln's life, it left a terrible hole. Their falling out was a terrible blow—made even more tragic because she was banished by Mary herself.

Elizabeth Keckly had been a loyal friend to Mary Lincoln while others fell by the wayside. She remained devoted and pretended to believe promises made, even though she, of all people, knew it was unlikely that Mary could deliver them. Keckly dropped everything to travel with the former First Lady from the White House back to Chicago in the spring of 1865. She made a holiday visit later that year, and promised the Lincolns to return, if and when they could pay her fares. So when in the spring of 1867, her former employer summoned her to assist with her New York scheme, Keckly again abandoned everything to come to Manhattan. This move was at great financial and personal sacrifice. Then, after the fiasco of the old-clothes sale, she was left behind to clean up the mess. Stranded for months without support or compensation, Keckly gave up hope, forced to fend for herself. She became more and more desperate as her plight worsened.

First and foremost, Elizabeth Keckly wanted to help those of her race. She headed a fundraising drive for the Reverend Daniel Payne, one of the founders of Wilberforce College, a black campus in Ohio. The college had burned to the ground on the day of Lincoln's assassination, and was still seeking funds to rebuild three years later. Keckly promised to loan some of her Lincoln relics for a fundraiser, an exhibit traveling to Europe, where Lincoln's hat, cloak, comb, and brush, as well as Mary's gloves, might be put on display.

Understandably, Mary went wild at the thought of any more exhibitionism being attached to her name; she pleaded with Keckly to withdraw the items she had placed on loan. Keckly took back her promise of the gloves Mrs. Lincoln had worn the night of Lincoln's murder, but other souvenirs remained at Wilberforce's disposal, to Mrs. Lincoln's distinct displeasure.

Elizabeth Keckly believed Mrs. Lincoln had been unfairly maligned in the old-clothes scandal, and wanted to promote a more balanced view of her former employer. Keckly felt justified in trying to influence public opinion, as Mrs. Lincoln had even authorized Lizzie to release information to the newspapers on her behalf during the old-clothes debacle. Both women were keen to improve Mary's tarnished image.

Elizabeth Keckly had been contacted by antislavery journalist James Redpath, who also wanted to create a more sympathetic portrait of Mrs. Lincoln. Redpath urged Lizzie to share details. Keckly was encouraged by Redpath to put pen to paper, and spent weeks writing her side of the Lincoln saga. By March 1868, her efforts resulted in a finished manuscript: *Behind the Scenes; or, Thirty Years a Slave and Four Years in the White House.* The publisher, Carleton and Company, hawked the book as a sensational tell-all in advertisements in April. When the book finally appeared, the volume created another minor scandal, especially when letters from the First Lady appeared verbatim.

Once again, Robert Todd Lincoln exploded in rage. He had taken on the mantle of ferocious family guardian. He bought up as many copies of the Keckly book as he could, and had them destroyed. During *this* scandal, nearly all journalists took Mary Lincoln's side in their zeal to discredit Keckly.

Most infuriating to Robert and most offensive to the Victorian conventions of the era, several of Mrs. Lincoln's letters appeared in the appendix. This section of the book had been engineered by Redpath, apparently without Keckly's permission. Keckly sent word to Mrs. Lincoln that these private documents had been printed without her permission. Even if Mary did receive this news, she chose to ignore it. Mrs. Lincoln had vehemently professed her own innocence about letters in the *New York World* the previous year, yet both she and Lizzie knew better, so she may have discounted Keckly's claim.

In any case, Keckly hoped that her sympathetic portrait of Mary Lincoln would soften the harsh, ugly, and reckless headlines of the old-clothes scandal. But even journalists who had rabidly attacked the former First Lady the autumn before nevertheless rallied to her defense in the face of "back stairs gossip of negro servant girls."

Keckly tried to fight back, asking if she was being denounced only because "my skin is dark and that I was once a slave." But the press continued to have a field day. Realistically, it made no difference to Mary what Keckly's intentions were, as Lizzie had written her book. Keckly biographer Jennifer Fleischner observed that Mary Lincoln made only one subsequent direct reference to her former confidante, dismissing her as "the *colored* historian." Elizabeth was never to return. Mary was ill prepared for such a loss. It was one of many things that cast her adrift.

Mary Lincoln felt that she had been locked in the stocks, deserted by comrades and friends. Mrs. Lincoln was so shaken by articles that portrayed her as mentally unbalanced that she railed about them in private. At one point, Mary exaggerated a newspaper account that "says there is no doubt Mrs. L.—*is* deranged— has been for years past, and will end her life in a lunatic asylum. They would doubtless like me to begin it *now.*" Perhaps it awakened memories from childhood or simply smacked of her husband's threats to have her put in a mental hospital after Willie's death, but whatever it was, it struck a chord with Mary. If she could not learn to control her emotions, she might be committed.

Again and again, Mary tried to control circumstances, and when this failed, she gave in to hysteria, blaming others for her own miscalculations. She suffered from self-destructive patterns and maintained her dance of denial over mental problems throughout her long life. She might rightly feel abandoned, as the men and women in her life peeled away, one by one. But she was wrong to always see herself as the victim of a plot, or as someone deserted—when she was the one doing the exiling. Her precarious bid to maintain her grip on reality kept Mary in a constant state of exhaustion during this period. Severely weakened by her struggles, she decided she could no longer endure these campaigns.

During the summer of 1868, she claimed that Robert's upcoming wedding was the "only sunbeam in my sad future." But with her older son soon married and starting an independent life, Mary actually felt the four walls closing in on her. Her return to Chicago was not paying off. She had bought a house but couldn't afford the upkeep; she had to rent her property out and find her own rented rooms. In the fall of 1867, she sold her household furniture to John Alston of Hyde Park for a little over $2,000. Mary was able to keep this transaction out of the press.

Mary vacillated between the desire to forge a public role as First Widow—seeking attendant comfort and acknowledgment—and the need for anonymity, which gave her the peace she regularly required. Mary decided that it would be better to leave those places and people she associated with her husband's memory. She would undertake the long-awaited trip to European capitals where she and her husband had often planned to visit.

She and Tad would venture forth, heading abroad to find a place where her son's education and their upkeep would not come at such a costly price. It would be the first trip abroad for both mother and son. Mary had hopes that they might live comfortably in Europe until fortune might smile on her again.

Once she determined to head out of the country, no one could dissuade Mary from this course. She began her farewell tour with a brief trip to Springfield to visit Lincoln's grave site—where the memorial association was preparing a large monument in her husband's honor. Mary comforted herself that there would be a place for her as well—by her husband's side, where she longed to remain.

Next, she and Tad departed for a resort in Pennsylvania, taking a lengthy holiday. It was on this idyllic visit to the Alleghenies during the late summer of 1868 that Tad befriended another young boy, Daniel Slataper from Pittsburgh. Mary Lincoln and Eliza Slataper, Daniel's mother, bonded during this family vacation—and remained friends through their correspondence. The women kept in touch long past their first (and final) time together.

The two youngsters got into some hijinks, as when Tad was playing with other boys, courting danger by jumping onto passing trains, and nearly got himself killed. But not all boys have their failings advertised in the papers, with mention of being a president's son, as befell Tad that summer. Perhaps this contributed to Mary's steely resolve to get out of the country, to escape further harassment.

Robert had finally decided to marry his sweetheart, Mary Harlan. But Mrs. Lincoln was still fighting her fears and was not sure she could manage the Washington wedding planned. However, after a rest in the mountains, she was willing to detour to D.C. for Robert's marriage, before setting sail to Germany. She spent a matter of hours in Washington, swathed in her customary widow's weeds, to see Robert Todd Lincoln and Mary Harlan joined in marriage on September 24, 1868.

The Harlan-Lincoln nuptials went off without a hitch, with roughly three dozen in attendance. However, instead of being the golden moment of which Mrs. Lincoln had dreamed ever since Mary and Robert had glided round the dance floor at the second inaugural ball, Robert's wedding was a modest affair. His mother admired the two gold bracelets sent along as gifts by Robert's law partner, but perhaps wished there might have been more pomp to this circumstance. Still, she acknowledged that it was a lovely ceremony, although her role was as observer rather than as participant.

Mary felt dizzy and indisposed during her brief trip to Washington. The day after the ceremony, she begged off traveling with the wedding party on a special train to New York, but sent Tad along with the Harlans, who would return him after the celebrations in a few days. Mary felt the need to recover at a hotel in Baltimore, where she and Tad were scheduled to board their ship to Germany. While in Baltimore, she hoped her newest friend, Eliza Slataper, would visit before she crossed the Atlantic. She begged her to come— "If you love me . . . if you value my peace . . . come." But this new friend was too ill to travel, something Mary readily understood and forgave.

Mary confessed to Eliza that when she was in D.C., "the feeling of BEING THERE did not oppress me as much as I supposed it would." But she found the experience profoundly disturbing, and shut herself away in her hotel room. Despite its discomforts, the sea voyage may have looked more attractive as an escape from those places that fueled her grief.

Mrs. Lincoln's first grand tour could not have been very grand. She traveled as a private citizen, on a limited income. Yet as she put distance between herself and the United States, excitement built. She hoped she and Tad might make new lives for themselves.

Landing in Germany in October 1868, Tad and Mary Lincoln found the English-speaking community in Frankfurt large and welcoming. She was able to plop herself down in the lap of European luxury at the Hôtel d'Angleterre, and to place Tad in a proper academy, where he was tutored in both German and English. He had difficulty at first, but soon took private lessons to improve his speech. Mary was pleased at his progress.

At first, the newness of her European home engaged Mary, but soon this little slice of paradise began to fray at the edges. She was fluent in French, but knew no German, which kept her isolated. Money matters caused Mrs. Lincoln to relocate into a cheaper hotel, to economize.

While abroad—in a new atmosphere, granted the deference she felt she deserved—she thought she would be better able to control her emotions. Alone and on a fixed income, however, the widow let fears prey on her mind. Mary believed she was whispered about wherever she went—and clearly her elaborate widow's weeds did not diminish the possibility of comment.

Also, during this period, Mary was unable to find the right blend of medicine, diet, and exercise to lift her from her lethargy. She could not afford a nurse, and Tad abandoned his studies temporarily to tend to her. He would sit by her bedside when she needed him, where, she confided, "His dark loving eyes—watching over me, reminded me so much of his dearly beloved father's—so filled with *his* deep love."

Her German doctor advised an excursion to the baths in Italy for her health. Mary complained that this was beyond her limited means. She began to feel consumed with anxiety, as health and money worries engulfed her. And still she held out hope of congressional rescue, scanning the horizon for any gleam of federal funding to alleviate her burdens.

News from Washington relaunched her campaign for a pension. In December, she penned an open letter to the Senate (which was read into the official record on January 26, 1869), which echoed familiar themes: "Nor can I live in a style becoming to the widow of the Chief Magistrate of a great nation, although I live as economically as I can."

She wrote to friends about her new dressmaker ("Popp has made dresses for Queen Victoria's daughters"), trying to put her best foot forward. Tad was coming into his own—speaking with a German accent. He and his mother went on tours of the Black Forest and Heidelberg, spellbound by the sights, but still Tad remained homesick. Mary's letters were full of self-pity. She felt cheated, spending her fiftieth birthday swathed in black and alone in Frankfurt. All this loneliness fueled insecurities.

A few days after she passed the half-century mark, she wrote remorsefully to David Davis that although she had promised Robert some inherited bonds, she had since decided to retract her offer. She feared the risk of real estate speculation. This even caused her to fall ill. She asked Davis if he thought perhaps Congress would offer her a yearly stipend. She emphasized that she had medical needs, that she was in serious financial straits. Her doctor insisted that she make a pilgrimage to a spa, but she resisted because of high prices. Even as she crossed an ocean to escape crippling anxieties, she was drawn back into the web of worries from which she had just extracted herself. She persistently looked to external resources, rather than to any internal reform to resolve her dilemma.

When an Indiana senator introduced a joint resolution for Mrs. Lincoln's relief on January 14, 1869, this only renewed Mary's anxiety. She was desperate for news of its passage. Her old friend Charles Sumner had been lobbying behind closed doors, but the measure was voted down on March 3 in Congress's final session. The vote was twenty-three in favor, twenty-seven against, and sixteen absent. This last number gave Mrs. Lincoln hope—and Sumner as well. He immediately introduced a new bill during a special session of Congress, specifically requesting $5,000 a year for the president's widow. But Sumner was thwarted when the bill got sent back to committee and into legislative limbo.

This disappointing news came at a time when Mrs. Lincoln had escaped the dreary German winter for the south of France. She deluded herself that she was practicing economy because, she explained to a friend, going to Nice instead of Italy was so much cheaper. Plus she believed that because of her health, the spa was an absolute necessity.

Mrs. Lincoln was most likely plagued by problems associated with menopause, on top of other medical issues. In any case, her physical decline was precipitous during this period. Her body might have required the change of climate prescribed, but a change of scenery lifted her spirits as well. But upon her return to Frankfurt, she was disheartened by the news of Congress's failure to act on her behalf. She sent along her thanks to Sumner—and enclosed a handwritten note for presentation to Grant:

In memory of my great and good husband, will not President Grant exert his great influence with the Senate Committee & others, in order that an immediate pension maybe granted me?

Most respectfully, Mrs. A. Lincoln.

Mrs. Lincoln was shameless in her single-minded pursuit of her pecuniary interests, trying every angle. Poor Sumner played Mary's messenger boy, at a very low point in this campaign.

Again, to lift her spirits, Mrs. Lincoln decided to take Tad on an adventure for their summer holiday. They stopped off in Paris, which was a thrill for them both. The American consul in Dundee, Rev. James Smith, Mrs. Lincoln's dear friend from back home, met them in London. He then escorted them all the way back to the Scottish highlands. This was such an exciting journey for Mrs. Lincoln—a pilgrimage to the place whence her people came. This land of her ancestors provided both affinity and respite. The former First Lady was buoyant with Scottish pride as she made the circuit of historical sites.

She and Tad, traveling incognito, enjoyed castles as well as tours of mountains and lochs. They explored Glasgow, and tarried for almost a week in Edinburgh. They visited famous abbeys and Robert Burns's birthplace. Mary was particularly fascinated by royal sites—such as the deathbed of King Duncan and the stairs where Mary Queen of Scots fled Lochleven. She stopped in at Balmoral—telling a friend that she was sure she would have been welcomed by her "sister in grief," had the Queen been in residence. Mary enjoyed a rapturous five weeks, commenting, "It appears to me that we saw every place, yet I presume we might remain five months there, continually traveling round without doing so." By the end of her tour, she confessed, "Beautiful, glorious Scotland, has spoilt me for every other country!"

It certainly soured Mary when she returned to Germany in the fall of 1869. She was disenchanted with her new home, with its bad weather and costly climate. The only thing she had found at a bargain in Germany was the bombazine she required for her mourning wear. Mrs. Lincoln acquired bolts of this material, having paid ten dollars a yard in the United States but finding the same quantity was only fifty cents in Frankfurt.

By this time, Mary Lincoln had moved out of her luxury hotel and into the more economical Hotel de Holland. Concerned about Tad's health, especially his lack of appetite at boarding school, Mary took him out of halls as often as possible. He was able to eat out whenever Mary was there, and so she checked up on him regularly.

Mary was delighted when her son Robert became a father on October 15: a baby girl christened Mary, nicknamed Mamie. This new family member pulled her back from the orbit of her life abroad, reminding her of home.

Robert was a proud father and protective husband. He wrote less than a month after his daughter was born, telling his great friend Judge Davis that his wife seemed well, but she would have to be careful because it had been a long and difficult labor. He was also able to joke that his wife was happier about the sex of their first child than he was. At the same time, he wrote to John Hay, John Nicolay, and many of his father's friends about his contentment. His mother surmised that Robert had reached one of the happiest stages of his life, and she felt hopeful.

Mary Lincoln kept trying to lure her older son to Europe, begging Robert to bring his wife and child along. But instead, events conspired to lure Mrs. Lincoln back across the ocean. First and foremost, Mary was deeply troubled by Tad's longing for the States. He missed his Chicago friends and his brother Robert, and he was eager to enjoy his new status as uncle. Tad was keen to meet his niece, Mamie, and this became a priority for Mary as well. With each new year, Mrs. Lincoln increasingly felt her advancing age and her deep loneliness. Frailer and more disheartened by each passing season adrift, she wrote on December 12: "My husband was so richly blessed with all these noble attributes that each day makes me worship his memory—*more & more*—Tomorrow is the anniversary of my birth—I will be 46—and I feel 86."

This comment is doubly revealing, for it was written when Mrs. Lincoln was fifty-one. She had begun to lie about her age even before she went to the White House, but after she was widowed, she began claiming she was nearly fifteen years younger than her husband when they met. At the same time, she did not try to minimize her age to make her more attractive to suitors, as this was not part of her plan.

The idea of remarriage was anathema. Mary was deeply offended when a rumor began to circulate that she was going to marry a count. She seemed shocked by such a suggestion, even violated by the very notion that she would remarry. Her identity was wholly bound up with remaining Mrs. Lincoln.

With the approach of 1870, Mary wrote to her close friend Sally Orne, wishing they might have a reunion in the south of France, suggesting the French spa town of Pau. Mary felt desperate about her own situation and wrote to Orne in the winter of 1869–1870, "I am *alas!* Out in THE COLD . . . the demon—*want*—will still pursue me." When she read a report in a British journal that her pension bill was being tabled (both lack of precedent and lack of need were cited), she became despondent. Mary contacted Henry Seligman, one of her husband's most loyal admirers, who had helped her once before. Seligman joined with Sally Orne's brother, politician Charles O'Neill, to try to drum up support for Mrs. Lincoln.

It was cold comfort during this bleak period that First Lady Julia Dent Grant had been accused of improprieties during the gold scandal of 1869. Mary wrote to Orne, "As, in the midst of all my wickedness & transgressions I never indulged in gold speculations, perhaps a more dreadful woman than *Mrs. L.* may yet occupy the W.H." Julia Grant was also charged with "indiscretion," which had the same vagueness in the nineteenth century as it does in the twenty-first. Primarily, this smear was not about sexual impropriety, but about acts of disloyalty.

The Senate floor once again became a gladiatorial ring, with Senator Simon Cameron, former secretary of war, reminding his colleagues of all Mrs. Lincoln had suffered. He suggested that during Lincoln's presidency, gossipmongers focused on the First Lady to try to hurt the president. Cameron suggested, "The gossips of the town did all they could to make a bad reputation for Mrs. Lincoln." Another senator, Reuben Fenton of New York, conceded that Mrs. Lincoln might have been unwise while in the White House. Nevertheless, from his observations of Mrs. Lincoln's conduct, both at home and abroad, Fenton favored the bill to support Mrs. Lincoln.

Sumner rose in the Senate halls to eloquently plead her case: "Surely the honorable members of the Senate must be weary of casting mud on the garments of the wife of Lincoln. . . . She loved him. I speak of that which I know. He had all her love." These persuasive appeals finally resulted in her pension's approval—on Bastille Day (July 14) in 1870. She did not receive the full $5,000 she had hoped for, but a decent $3,000 annual grant. She sent lavish thanks to Charles Sumner, among others, for devotion to her cause.

But no sooner had she finally secured this bounty than the outbreak of the Franco-Prussian War convulsed the Continent. Events propelled her—with Tad in tow—back to England. She contacted many old friends when she landed in London. Mary found the sights and sounds—the people and politics of the capital city—totally absorbing. She felt lucky to find a rigorous British tutor to continue Tad's academic progress, and settled into the spa town of Leamington. As the Christmas holidays approached and her boy wanted to be back in his home country, she promised Tad that he could sail back to the United States on his own. But when faced with the prospect of giving up his company, she backed out of the plan, citing the dangerous seas as a reason for postponement. She wanted them to stay in Europe a while longer.

Although London gave Mary many opportunities for socializing, both she and Tad were plagued by feeble health, aggravated by constant colds. She escaped the damp by heading down to Italy, while Tad remained behind in a boarding school in Brixton. While Mary toured castles and sailed on Lake Como, she felt the spirit of her husband hovering nearby. She visited the Florentine studio of sculptor Larkin Goldsmith Mead, who had been commissioned to carve statuary for the Lincoln monument in Springfield.

Finally, Tad won out and convinced his mother their homecoming was overdue. Mother and son sailed back to Manhattan on May 11. The storms at sea did nothing to improve Tad's weak lungs, and he arrived stateside with a serious chest ailment. While Tad indulged in bed rest at their hotel to fortify himself for the journey out to Chicago, Mary Lincoln visited with her friend Rhoda White. Rhoda encouraged Mary to think about giving up on Illinois and relocating to New York. But Mary and Tad both seemed eager to get back to Chicago. Mary was thrilled at the prospect of seeing her first grandchild, her little Mamie. Yet she was also just as concerned that her younger son not overtax himself, as his health seemed particularly fragile.

Tad at eighteen,
shortly before he died

When they finally arrived, it was a grand homecoming. Mary and Tad settled into her older son's new home, in a fashionable district along Wabash Avenue. She was enchanted by her new granddaughter, *her* namesake. Mrs. Lincoln seemed to ignore, in her inimitable way, the fact that Robert's daughter bore her own mother's name as well. Tad's homecoming was interrupted when Mary Harlan Lincoln was called to Washington to look after her mother, who had been stricken by sudden illness and requested her daughter's company. Mary understood that her daughter-in-law was needed elsewhere, and perhaps the exit had benefits all round.

Mary had become so very proud of her younger son, who had shot up in height. She seemed especially pleased when some suggested he resembled his father. And like his father, he was incredibly solicitous of Mary. Tad officially had reached adulthood. When he begged his mother, she made a special gesture and took off her mourning weeds for a day: on April 4 for his eighteenth birthday. (This is the only time recorded since her husband's death that Mary failed to wear black.)

When they returned to America, she was happy to show off her son, newly polished from European study and travel. But in his frail and weakened condition, after his exhausting sea and rail journeys, the Lincolns were forced to move into Clifton House, where Mary was able to nurse Tad.

In letters to friends, she complained bitterly about her situation, whining to her New York acquaintances, "My husband, so fondly indulgent, would have shrunk back, in horror, if he could have imagined *his* loved family, *thus* domiciled." Her priorities shifted dramatically, when by June 8, Tad was not improving and indeed "dangerously ill." Due to his breathing problems, she had been up with him night after night, more than ten nights in a row. His recovery seemed slow and steep, with severe setbacks all along the way. Despite the moment of hope her return to Chicago represented, darkness threatened as Tad began to decline.

On the evening of July 14, 1871, exactly one year to the day after Mary had been granted her special pension, Tad Lincoln suffered a serious relapse—so serious that his brother Robert came to his hotel bedside to support him. Tad's physician, Dr. C. G. Smith, saw no hope for his patient's "dropsy of the chest." As Mary took her turn by her son's side, his breathing became impossibly labored. Tad could only sleep if he was strapped upright, and Robert and Mary kept vigil by his bedside.

On the morning of July 15, after battling for air all through the night, Tad's head slumped onto his chest and he died. Abraham Lincoln had died at 7:22 in the morning. His younger son's premature death at about the same hour just a few years later, following another long night's vigil, hit very, very hard—for both Mary and Robert. Tad's older brother took his death stoically, while his mother failed to cope with yet another devastating loss.

Mrs. Lincoln was able to attend the small funeral service held at Robert's home, but could not accompany the body back to its resting place in Springfield. Tad's death was mourned by the nation, for he had been well liked by the press during Lincoln's time in the White House. The photos of the two of them near war's end had become emblems of "Father Abraham" and a "favorite son." There was an outpouring of sympathy for Mrs. Lincoln, bowed down with all the sorrows her life had dealt her.

So many Americans had shared stories about Tad's central role in his father's life. He had been the one to curl up asleep on the floor near the desk of his "Papaday." He was a constant companion to the president during his final months—and the rock that gave Mary strength following her husband's death. Many other widows empathized with Mary's grief over the loss of one who sustained her during her prolonged mourning. They imagined how bereaved Lincoln's widow would feel, robbed of her much adored son. To have to *again* endure the bathing of the corpse and the dressing of the lifeless form of another of her boys seemed too much for any mother.

Many times, following her husband's death, Tad had given her a reason to go on living. But now, there were four in heaven and two left behind, to quote her son's words. The fifty-two-year-old Mrs. Lincoln bemoaned her fate: "There is no life for me, without my idolized Taddie."

17.
Rising from the Ashes

The Chicago to which Mary Lincoln returned in 1871 was not the town she had left in 1869. She had relocated there in 1865 because she needed a new home for her two fatherless boys. She chose Chicago because of its special association with her husband's career, because of its proximity to Springfield, and because she sought esteem and support during widowhood. Mary took her younger son abroad for his education when these dreams failed to be realized. Upon their return, Tad tragically succumbed to illness and died. However, Chicago, despite Mary Lincoln's many disappointments, remained full of promise.

The city grew exponentially after the Civil War, being at the crossroads of the expanding railways as well as at the intersection of the Mississippi and the

Great Lakes. Lumber, grain, pork, and beef poured into the city's sprawling warehouses as the town's factories and bustling streets filled up with immigrants.

Chicago city fathers had grand aspirations—especially Potter Palmer, who opened his six-story marble emporium in October 1868 and introduced his "Palmer system" of allowing customers to take home goods on approval, which lured women to his counters. His wedding gift in 1870 to the Kentucky heiress Bertha Honoré was a 225-room hotel, the Palmer House: an impressive, modern showplace featuring lavish French decor, but built to last—with alarms in every room and hoses on every floor, billed as the first "fireproof" hotel in the country. Over six hundred fires had plagued Chicago in 1870, and builders scrambled to protect their investments.

The town was getting taller, especially with the dedication of the Chicago Water Tower in 1869, the year of the city's centennial. This Windy City icon—154 feet high and quarried from Joliet limestone—symbolized the city's preparedness for firefighting.

The July that Tad died in 1871, no rain fell on Chicago. By summer's end, with less than an inch of rainfall in more than ninety days, the city was a virtual tinderbox. A fire broke out on a warm Sunday evening on October 8. The blaze rapidly spread from building to building as the wind rose, and smoke and sparks filled the air. Even brick and stone succumbed to the consuming inferno, and the ground shook periodically from the collapse of buildings. As if in some scene from Dante, Chicagoans fled their homes with whatever they could carry.

Thousands were herded to safety as the blaze burned brightly into the night, and all the next day, and into another night. Block after block of cityscape disappeared in a swath of flame. Nearly three and one half square miles were reduced to rubble.

Prominent lawyer and Lincoln collector Isaac Arnold's estate was a showplace for art and books. Arnold had assembled an impressive library of ten thousand volumes from his time in Congress, as well as from his association with Abraham Lincoln. He had published a book on Lincoln's years in the White House and was about to embark on a full-scale biography when the fire threatened his estate. He and his staff valiantly tried to fight off the flames, sprinkling the lawn until the pumps ran dry. Defeated, Arnold abandoned his collection and steered his flock to safety, using a rowboat to seek shelter at a nearby lighthouse, where they awaited rescue by a tugboat. The wall of flames proved a near impassable gauntlet, but Isaac Arnold lived to write his harrowing tale.

Robert Todd Lincoln was involved in a rescue operation of his own on the city's South Side, helping Mrs. Scammon, the wife of his law firm's senior partner. Scammon's family was forced to flee when fire threatened the family's home while he was out of town. Robert Lincoln stepped in to assist, presumably after his wife and child, as well as his mother, were out of harm's way. Robert volunteered to assist with the evacuation of his boss's library from his Terrace Row mansion, one of the last blocks of fashionable townhouses to fall victim to the flames.

As a drizzle began to fall, on Monday night, the fire was nearly extinguished—by firemen and dynamiters who had been working nonstop for nearly forty hours to contain the blaze. As the sun rose on Tuesday morning, October 10, the city faced grim statistics: three hundred people dead, nearly eighteen thousand buildings destroyed, and almost one hundred thousand homeless, nearly one third of the city's population.

Bodies were being stacked in a livery stable, makeshift hospitals treated the worn and wounded, while thousands wandered the burned-out streets in search of missing relatives. These scenes of horror renewed memories of Civil War loss—a hushed silence pierced only by the wails of those discovering lost loved ones. Weeping women and dumbstruck children wandered the cityscape.

As tent shelters sprang up in parks, the city fathers tried to put a brave face on this disaster. Nearly $2 million had gone up in smoke, but the city could now refashion itself in the Beaux Arts style while leaping ahead with steel framing and masonry construction. But what about the immediate tasks at hand? While architects and builders enthused over the opportunity to reinvent the city, townsfolk faced the herculean project of recovery.

Historian Taylor Patterson has observed that men tended to cast the fire's devastation in terms of financial losses, while women spoke in emotional terms about the significance of the personal and domestic items lost: "It is for these I grieve, not over the loss of money, but for 'my mother's Bible, a lock of my sister's hair, the clothes and toys of my dead children.'" The male instinct to confront and conquer the social and physical elements stood in direct contrast to women's sense of trauma and reflection.

We do not have any surviving account of Mary Lincoln's response to the fire, but having the city where she had staked her hopes reduced to rubble could not have been more dispiriting. She witnessed the town tumble into chaos and crisis. This was a fitting backdrop for her own personal season of grief.

Tad's death at age eighteen sent Mrs. Lincoln into a permanent malaise. Shortly after her boy's burial in Springfield, she divided Tad's inheritance with her only remaining son, offering him more than his legal portion. This was an act of supreme generosity, as Mary remained apoplectic about financial matters. But Robert had his own family now and was launching a promising legal career. Tad's illness and sad demise also caused his brother heartache and grief. Robert complained, "My own strength was then used up & I was compelled to leave my office for as long a period as possible." He headed out to Colorado to recover—while his wife and child went off to visit his in-laws in Iowa.

Left alone and behind, Mary felt even more forlorn. She was offered sympathy by a handful of Chicago intimates, including Harriet Farlin, who kept Mary company during the days immediately following Tad's death. Mary also spent time in the home of the Bradwells, a remarkable couple: Myra, a pioneering feminist and legal editor, and her husband, transplanted Englishman James Bradwell, who became a respected member of the bench in Cook County and also served in the Illinois state legislature. David and Elizabeth Swing, the minister of Mary's Presbyterian congregation in Chicago and his wife, had young children to whom Mrs. Lincoln became attached, especially at the time of her bereavement.

Mary had once felt extremely close to her daughter in law, but an unfortunate coolness developed. The former First Lady had written to Mary in 1870 that "I often tell Tad I can scarcely flatter myself he will ever marry to suit me quite as well as dear Bob." She had frequently instructed her son and daughter-in-law that any of the items in her trunks stored at their house were theirs for the taking, with specific instructions about what treasures they might borrow, what items they must be sure to use. But those honeymoon days were over.

The return to Chicago had brought about a dramatic sea change. Robert's mother was increasingly disenchanted with Robert's wife, and a fault-finding mother-in-law became an unwelcome presence. The two Mary Lincolns fell into intense dislike. They could no longer be contained under the same roof, causing Robert no end of headaches. The younger Mary Lincoln spent more and more time with her parents at their home in Iowa. Mary Lincoln had become a disruptive influence on her son's marriage, causing her to drift even further from her former haunts in Chicago.

The charred ruins in Chicago seemed to mirror Mary's own mood. The Great Fire injected yet another layer of dread into her life, as Robert reported that his mother insisted upon storing her trunks at the

Fidelity Safe Deposit Company building ever after. (She would later ship them out of the city in a panic when she had a premonition—a false alarm—about another fire.)

Over the course of the next few years, Mary became a nomad, peripatetically seeking comfort for real and imagined ills. Her neuroses multiplied, the longer she remained rootless.

Mentions of the former First Lady were sporadic in the national press. Mrs. Lincoln turned up at resorts in Wisconsin, in spiritualist communities in upstate New York and St. Charles, Illinois, and as far afield as Canada and northern Florida. She was plagued by nervous disorders throughout this period, and shamelessly played on others' sympathies to garner attention. Her letters are suffused with self-pity and complaint during this particularly challenging period.

After Tad's death, many of Mrs. Lincoln's activities remain obscure within existing accounts or correspondence. She was unhinged by what she felt were William Herndon's continuing persecutions. Stories about Ann Rutledge were revived, as well as rumors about Nancy Lincoln's (Abraham's mother's) illegitimacy. Mary was even more appalled by suggestions that Lincoln himself was born out of wedlock. Mrs. Lincoln felt embattled by such savage speculations.

In 1873, William Herndon delivered public lectures on Lincoln's lack of religiosity. Disputes followed, aired in such articles as Herndon's "Mrs. Lincoln's Denial and What She Says." Mary complained bitterly that books such as Ward Hill Lamon's *Life of Abraham Lincoln* (1872) contained "sensational falsehoods & false calumnies." She was especially disturbed by reports that Lincoln was "godless" and suggestions that she was not the great love of her husband's life. Her anguish over these attacks on her husband's memory and their marriage prodded her into deeper withdrawal.

During this period, Mary paid her infamous visit to William Mumler's Boston studio. In the 1860s, Mumler began dabbling in spiritualism. He reported that "a lady who was heavily veiled and wearing a black dress gave her name as 'Mrs. Tydall' when she called unannounced at the studio and asked to be photographed." Mumler ushered her to a seat, but she kept the veil over her face. He asked if she would remain hidden for the photo, and she replied, "When you are ready, I will remove it." One of the Mumler portraits reveals a plain, sad-looking woman with a mourning headpiece. But another image has provoked more comment.

The Mary Lincoln "spiritualist" portrait has received much play in popular culture. Mumler stated that it was only after the photo emerged that he realized his sitter was the former First Lady. He confessed surprise when he developed one of his prints, that "behind her stood the smiling image of Abraham Lincoln himself." This photograph has had wide distribution over the years, even after being dismissed as a clever hoax. The infamous image was included in a 2005 exhibit of spirit photography at the New York Metropolitan Museum of Art titled "The Perfect Medium: Photography and the Occult."

In early 1872, the mother of Eddie Foy (the popular vaudeville star) was hired as a paid companion for Mary—to serve as both nurse and guard. Mrs. Foy confided that due to Tad's death, Mary "suffered from periods of mild insanity." Mrs. Lincoln was a difficult charge, and Foy's mother quit more than once, but was induced by her family to return to the job. She remained—on and off—with Mrs. Lincoln until 1875.

Mary continued her frantic and unhappy wanderings. Her epistles from this period seem strange and random—clinging to the past, seeking a different kind of future, indicating personal limbo. She nursed a sense of indifference to whether she lived or died, yet fought valiantly against the encroaching ailments of aging, which steadily interfered with her mobility and progress. Reembracing spiritualism, Mary reached out to

Mary's "spirit" portrait (photographed
by William Mumler)

communicate with Tad, Willie, and Eddie, as well as
her beloved husband.

She exulted at the arrival of Robert's male heir, her
husband's namesake, Abraham Lincoln II, on August
14, 1873—sweetly nicknamed "Jack" by his family.

In August 1874, the *Illinois State Journal* reported,
"The following news item has been going the rounds of
the newspapers for several months, and inadvertently
appeared in the 'personals' of yesterday's [*Washington*]
Chronicle: 'Mrs. Lincoln, wife of the late President

Lincoln, is in Paris.' The distinguished lady referred to has not been abroad for several years. If anybody is in Paris, France, representing herself to be the widow of the late President Lincoln, she is an imposter." No one had seen Mrs. Lincoln for most of 1874, because she had been secluded in Chicago due to ill health.

Mary was deeply moved by the loss of Charles Sumner on March 11, for he had been her champion and defender. She wrote to a friend, "The sudden and unexpected news of the death of the good and illustrious Mr. Sumner, will cause much sorrow, over the whole world. . . . He, was the well beloved friend of my idolized husband, and personally, I have lost my dearest and best friend." Sumner would lie in state in the rotunda of the U.S. Capitol before being conveyed back to his home state of Massachusetts for burial in Mt. Auburn Cemetery. Sumner's death conjured up thoughts of her own mortality, and she suggested that she felt as if she had perhaps remained behind for too long.

Press reports suggested that Mrs. Lincoln would spend the winter of 1874–1875 in St. Charles, Illinois (a spiritualist haunt), and simultaneously rumors circulated that Mrs. Lincoln had arrived in Florida to spend the season there. (The latter rumors were true.) She was in a state of hyperanxiety and decline by the spring of 1875, showing the strains of her gypsy status. Also at this time, her movements and motivations came into sharper focus, in light of the events leading up to her institutionalization.

On March 12, exactly two weeks before Good Friday, which would mark the tenth anniversary of her husband's assassination, Mary Lincoln was overwhelmed by a foreboding of her son's endangered health back in Chicago—she was sure he was dying. Whether it was a particularly vivid dream or some kind of hallucination, she would not be soothed. Once seized by the idea of Robert's illness, she packed to leave Mrs. Stockton's boardinghouse amongst the orange groves of Florida. She would rush to the rescue, despite pleas for prudence from all around her.

Mrs. Lincoln insisted upon making the long journey back from Jacksonville to attend Robert's deathbed. She sent a wire to Edward Isham, Robert's law partner in Chicago, promising that she would return immediately. Mary also sent a telegram to Robert, begging him to hold on until she arrived. When her son began to receive frantic messages, he contacted the Western Union office in Florida to try to gauge what kind of trouble his mother was causing.

By March 16, she had made her way by rail to Indianapolis, and Robert was able to track her down.

She was met by her hale and hearty son at the Chicago depot, which amazed Mary. She was also full of fantastic complaints about the journey, far-fetched tales that set Robert's teeth on edge. He could see his mother was in full-blown delusional mode, and did not know how they would all weather this episode.

His wife was pregnant with their third child, and his mother was not welcome in his home because of a longstanding estrangement. Robert took Mary to the Grand Pacific Hotel, where he stayed with her to prevent further mishaps. During this period, Robert confided to his circle of intimates that his mother had the ability to seem perfectly coherent at one moment and then slip into irrationality the next, making her condition even more frustrating. Robert realized it was just a few days short of Easter weekend, a dreaded landmark of ten years since she lost her husband under such spectacularly traumatizing circumstances. He feared she was completely disintegrating in light of the upcoming anniversary. To play it safe, Robert ordered Pinkerton guards to shadow his mother whenever she left the hotel.

A few days after Easter, Mrs. Lincoln attempted to go to the hotel lobby half-clothed. Upon being restrained outside her room, she screamed at Robert, part of a long parade of public embarrassments. The staff had tried to humor Mary, to indulge her unpredictable behavior. Robert was at his wit's end—writing to relations, consulting with doctors, seeking advice from Justice Davis, among others. It was at this point that Mary predicted that there would soon be another devastating fire in Chicago—and insisted upon sending her trunks to Milwaukee for safekeeping.

She had been full of premonitions from the year before. She confided that her will was prepared so she would not die intestate. After consulting with a medium, she was sure of her impending death, that she would be "taken" sometime in September 1874 on a date named by her husband during a séance. The date passed and she survived, but anticipations and anxieties multiplied. All her preparations increased rather than relieved her fears. Spending too much time isolated and ill, Mary's previously engaged energies had gone haywire.

Dr. Willis Danforth had been treating Mary for migraines for years, serving as her personal physician in Chicago. Headaches were part of a larger constellation of ills. Migraines alone could and did cause hallucinations, and adding medication would cause even more possibilities for impairments.

On May 8, Dr. Danforth heard about Mary's incredible symptoms: Indians pulling wires from inside her head—which may have been a metaphorical reference to spiritualism. Danforth concluded that Mary had become a danger to herself, an ironclad justification for confinement.

Mary also returned to bad habits from the past—mixing medications, perhaps with alcohol. She resumed unhealthy spending patterns; evidence of her binges littered the room—carpetbags full of footstools and packages crammed with lace curtains. In fairness, she was *not* buying shawls and dresses for herself, as reported in the press, but rather household items—although she remained a woman without a house. She stockpiled handkerchiefs and could not resist jewelry. Her weakness for gloves—seventeen pairs purchased since her arrival in Chicago—attested to her full-blown manic state.

Robert became even more alarmed when his mother could not be dissuaded from carrying nearly $60,000 in bonds on her person, with wads of cash bulging from her pockets. Worried about thieves and shysters, Robert consulted with Leonard Swett, a legendary attorney and an old friend of the family, who helped Robert shape a plan of action. Robert had been a dutiful son, but his marriage could not withstand this latest assault, and no other family members were willing to help. Robert had shared the long night in the Petersen house and knew what a nightmare his father's death had been for his mother. But he had also seen her disintegrating over the past decade. And this new behavior and irrational talk were introducing new constraints: He had every reason to believe his mother was self-destructing.

Dr. Robert Patterson, a prominent local authority on medical insanity, was paid to consult, but had no personal contact with Mrs. Lincoln. He ran the Bellevue Place sanatorium in Batavia, Illinois, a facility that helped elite families deal discreetly with their "problem" females. Robert was worried both about the necessity of confining his mother and the mechanics of confinement. He was concerned about what public opinion might be. There was no hope that he could move his mother into an asylum without publicity. Indeed, Illinois was one of the few states in the nation requiring an open hearing with a jury for any commitment proceedings.

The Illinois personal-liberty law stemmed from an infamous case in 1860 in which a Calvinist minister named Packard was able to have his wife locked up at the Jacksonville State Hospital after a perfunctory medical examination that consisted of taking her pulse. On the sole basis of her husband's testimony, Mrs. Packard was confined for three years. Upon her release, Elizabeth Packard fought her husband's plot to have her put away again. Packard told not only her own sad tale, but shared stories of other women she had met while confined, women also subjected to unjust imprisonment. Myra Bradwell, in her paper, the *Chicago Legal News*, campaigned against such outrages and for women's rights. Packard's campaign and Bradwell's publicity ensured that any involuntary confinement must pass the test of an open courtroom.

Leonard Swett and Drs. Danforth and Patterson were all aware of these legal considerations and the pitfalls along the way. Swett counseled that he would help orchestrate the necessities. Robert worried most of all about the fallout from his actions, but was just as worried about the consequences of inaction. He conferred with a group of legal and medical experts on Sunday, May 16. He had hoped to persuade his mother to voluntarily enter a sanatorium, and submit to a financial conservator. John Todd Stuart, the mayor of Springfield and one of the few connections Mary maintained with her former hometown, advised him that this was something to which Mrs. Lincoln would never agree. Robert begged Stuart and another cousin, Elizabeth Grimsley Brown, to travel to Chicago to help with the situation—but neither one arrived in time. Finally, Robert felt forced to act.

On the morning of May 19, convinced that his mother might do herself harm and prodded by his team of advisers, Robert Lincoln filed an affidavit to have his mother brought to trial on charges of insanity. Once the papers were filed, Leonard Swett visited Mary's hotel room to escort her to the courtroom for a hearing that very day.

Imagine the fear and betrayal Mary felt when this old friend of her husband's appeared at one o'clock on a Wednesday afternoon, telling her he had come to collect her for a court appointment in one hour, a trial in which she was going to have to refute charges of insanity in front of a judge and jury. When she asked for her son, Mary was informed that Robert was already there, that this hearing was at his instigation. When she asked Swett to leave so she could change her clothes, he refused, suggesting that she might throw herself out the window if he took his eyes off her. She wept profusely and resisted stubbornly, to no avail. Swett threatened handcuffs if she did not come willingly.

The phalanx of men at the hearing would have unnerved even a well person, but by now Mary was fighting collapse. Isaac Arnold, another of her husband's dear friends, had been tapped to represent her. He tried to beg off, pointing out that he, too, thought she should be placed in custodial care. Shell-shocked,

Mary was seated at a table next to her son—who would join the parade of eighteen witnesses. Members of the hotel staff and medical community gave evidence of her unsound mind during the more than three hours of testimony. Robert broke down weeping on the stand while convincing the jury of his fear for his mother's safety. He argued that he had no alternative but to have his mother confined.

Mary Lincoln seemed struck mute by these proceedings. She was given no opportunity to mount any defense and offered no testimony at the hearing. Her lawyer was nearly as silent, offering no defense of any of her actions. The proceedings moved smoothly to produce the desired outcome: custodial care and incarceration.

Whatever the evidence against her, whatever the true nature or severity of her debility—and scholars continue to heatedly debate this issue—Mary Lincoln was not given a fair opportunity to oppose her own legal kidnapping. Rules were designed to keep opposition to a minimum. Leading mental health experts of the day, such as the editor of the *American Journal of Insanity*, suggested a public trial caused "unnecessary excitement" for the unfortunate defendant. Thus such a hearing was maintained only as a pro forma safeguard.

Historians Mark Neely Jr. and Robert McMurtry Jr. ask, "Can the word of the experts, all men, who testified to this famous woman's insanity be trusted?" They emphatically say yes, and historian Jason Emerson agrees. Emerson argues that all these men were her peers, and he believes there was more damning evidence withheld from the public proceedings. But historian Jean Baker's judicious concerns about the gendered nature of these proceedings still stand. Furthermore, Baker underscores the blatant disregard of procedure: Mary Lincoln was entitled to legal notice, which she was denied. In addition, this open hearing, with reporters scribbling down the details of Mrs. Lincoln's every false move, insured that Mrs. Lincoln would be destroyed by this sensational publicity.

From Mary's perspective, she was bushwhacked by her son and spirited into a rigged hearing by a cabal that included her own lawyer, who wanted to withdraw because he thought her guilty as charged. A jury of upstanding male citizens was assembled to weigh the evidence, but Mary was not given any time or opportunity to defend herself. When the jury adjourned, Mrs. Lincoln finally spoke: "O Robert, to think that my son would ever have done this." Within ten minutes, the jury returned a verdict of guilty: Mary Lincoln was ruled incompetent by reason of mental incapacity. Robert Lincoln was designated her financial conservator, and he determined to send her to Dr. Patterson's sanatorium in Batavia.

She was taken back to her hotel room to await transport the next day. She was forced to surrender her bonds, which she kept on her person. Mary was reduced to weeping and begging not to be robbed of what little she had left. But Swett insisted and remained until he could collect her bonds. Mrs. Lincoln was robbed of her dignity as well as having that which guaranteed her financial security for her old age snatched away.

Crushed by circumstances, as her plight went from bad to worse, Mary Lincoln managed to give her guards the slip and headed for a pharmacy, trying to obtain more sleeping powders, especially chloral hydrate. Her sentinels feared this behavior was prelude to suicide, panicked as she was about the Bellevue Place sanatorium and what might happen to her.

Even though this was a top facility with pleasant staff and grounds, several of the inmates were prone to violence and hysteria. Mary would be locked away with women so disturbed that they would stab themselves, threaten others, and create disturbances that might unbalance even a well woman, which Mary Lincoln clearly was not. The whole prospect of incarceration frightened her, but she had no one to turn to.

Meanwhile, the press was having a field day with this story. The newspapers splashed the case into the headlines, quoting Robert's attorney that his concerns were only for Mrs. Lincoln's *safety*. Robert preemptively denied that his motives involved financial gain. Contemporary critics leapt to Robert Todd Lincoln's defense. The majority opinion was best summed up by the *Chicago Times*, which suggested that Robert "discharged his delicate and unhappy task with filial thoughtfulness." Justice Davis applauded the fact that now Robert could go back to work, after his lengthy and debilitating ordeal. He also said that sending Mary to an asylum would remove "the unfavorable impressions created by her conduct since your father's death." These kinds of considerations give weight to the speculation that Robert was just getting his mother out of the way. His advisers' expressing such cold calculations puts a chill on any sympathy for Robert's plight.

Lincoln's only remaining son was cut to the quick by any suggestion that he was moved by self-interest, and he wrote long justifications of his conduct to John Hay and Sally Orne, among others who might question his decision. In an editorial for the *New York*

Tribune, John Hay suggested that Mary had never recovered from her husband's death, and would find sympathy "now that a court of justice has declared her bereft of reason."

Although contemporaries were quick to credit her son by measuring against the medical standard—would she hurt herself?—subsequent critics have focused on the bias involved in this judicial proceeding. Mrs. Lincoln was whisked to her hearing without due notice and with no one to offer a substantive defense on her behalf. She was condemned to incarceration with no appreciation of the consequences such a confinement might have for her future. Robert himself suggested that his role in the verdict, as well as the verdict itself, drove Mary to a suicide attempt on the evening of May 19. His promising career in politics was compromised by his mother's erratic behavior. Certainly such considerations must have had an impact on her son's decision. Yet many contemporaries believed it was for her own good as well as for Robert's peace of mind that his mother was put away.

Scholars disagree over whether or not Mrs. Lincoln made any genuine attempt to overdose on drugs on the night of May 19. Clearly, she was agitated as she moved from pharmacist to pharmacist, ordering powders on the night before she was to be confined to a sanatorium. There is no evidence that had she ever made any suicide attempt before or after, although she often made melodramatic references to not wanting to go on living—a constant refrain from April 15, 1865, onward. Surely even if she wanted to take a drug overdose, Mary's motivation might have been that she was wrongfully declared insane and not that she actually was. The murky events of those May days and nights in Mary Lincoln's life will never be fully understood.

Mrs. Lincoln later conceded that she might have appeared deranged during this period because she had used chloral hydrate too freely, to induce sleep while suffering from insomnia. Those scholars who dismiss chloral hydrate as nothing that could produce her symptoms or behavior are remiss. Chloral hydrate mixed with alcohol was the first Mickey Finn, the knockout drink often used for robbery, just as various sedatives combined with alcohol are used as date-rape drugs in the twenty-first century. Mary would not have been the first Civil War widow to drink alone or clandestinely. A chloral hydrate overdose, or mixing the drug with alcohol, can cause difficulty in swallowing, convulsions, severe drowsiness, troubled breathing, staggering, slow or irregular heartbeat—and confusion. Any or all of these symptoms might have contributed to Mrs. Lincoln's impaired condition, so it is not so far-fetched that Mrs. Lincoln's behavior might have been altered through her chemical intake.

Once she was publicly put on trial, stripped of all her earthy possessions, and humiliated, what must have gone through her mind? Mary was forced to board a specially appointed railway car assigned to take her to Batavia, along with Robert, Isaac Arnold, and the asylum's director, Dr. Robert Patterson. Accompanied by a new doctor she had never met and a son she felt she had never really known, she headed for a place where she would be locked in among strangers, cut off from family and friends.

The Bellevue Place sanatorium, roughly forty miles outside of Chicago, was a private facility with impeccable credentials. Established in 1867, Bellevue Place boasted a three-story main building, which housed "lady patients," plunked down in the midst of twenty acres of well-maintained gardens. The facility had the look of a resort. Patterson's sanatorium had gained a solid reputation for discretion among Chicago's elite families. Historical investigations may question some of the conditions that led to confinement, as women were routinely locked away to save their family embarrassment. Consorting with spiritualists seemed a disease requiring quarantine among some wealthy clans. For this and other reasons, rich fathers or husbands may have found it easier to exile an errant female to Bellevue than to subject the family name to scandal.

During her time at Bellevue, Mary suffered minimal physical restraint. There were no bars on her window, although netting was stretched across the outside of the window openings. It was called an "ornamental screen" by Bellevue's supervisor. She was locked in her room at night by a personal attendant, who slept in an adjoining bedroom in Mary's personal suite. Mrs. Lincoln was the most advantaged of Bellevue inmates, given privileges no other patient had. She took her meals either in her private suite or in the Pattersons' private dining room, with the doctor's wife and family. This would have prevented her from mixing with some of the more unpleasant patients. Her pampered treatment signified either the mildness of her symptoms or her elevated status, or both.

Mary could freely wander the more than twenty acres of grounds, well planted with roses. She always had a carriage at her disposal. Excursions outside the main building were shadowed by one of the more than a dozen nurses and attendants. Despite this bucolic setting, Mrs. Lincoln was far from content with her surroundings.

What must it have been like to be imprisoned in this manicured, lush setting, to have once been the

Bellevue Place sanatorium in Batavia, Illinois

First Lady of the land and now to be branded a lunatic and locked up in a rural hamlet, deprived of society, friends, and family? At Bellevue, Mary had everyone at her beck and call, with staff offering her the proper deference she required as First Widow. Yet she still felt terrorized by incarceration, and why shouldn't she? Her grandfather and brother, (both Levi Todds) had died in asylums, as did her niece Mattie Todd and her grandniece Georgie. She had long been fearful of having her name linked with "lunacy" and now, with all this sensational publicity, she saw no way to uncouple her reputation from the stain of insanity. The entire ordeal turned her hair white.

Mary's state of mind and her subsequent institutionalization have provoked at least two dramatic historical reexaminations in the past few decades. First, in 1975, James Hickey, Lincoln curator extraordinaire, found a cache of documents that had been squirreled away at Hildene, the Vermont family home where Robert Lincoln died. This remarkable discovery yielded forty pages of previously unpublished letters by Mary Lincoln. These documents formed the basis of Mark Neely Jr. and Gerald McMurtry Jr.'s *The Insanity File* (1987). Robert Todd Lincoln had tucked away fragments of letters, and some other documents, marking the entire cache "the Insanity File." Neely and McMurtry look at medical, legal, and other factors contributing to Robert's decision, providing a strong overview but one that contributed to a portrait of Mary Lincoln as a paranoid, diminished personality.

Then another cache of previously unpublished letters by Mrs. Lincoln came to light in 2006. Once again, it had been hidden away, this time in a trunk held by descendants of a Lincoln family lawyer. Prior to this recent discovery of eight new pieces of correspondence, published in Jason Emerson's *The Madness of Mary Lincoln* (2006), less than a dozen letters had been available from the period 1874–1875. These letters are crucial to any attempt to assess Mary's mental capacity and responses to these dramatic events. Surely she was scribbling away to anyone she thought would help her during these desperate times. But despite new documentation, clashing interpretations of events remain.

The question Was she crazy? remains unanswered, although many doctors seem to want to diagnose without reliable evidence. And no matter what evidence is unearthed, experts will always disagree and scholars will squabble, and Mary Lincoln's mental health will remain, like her husband's, a topic of endless interest and speculation.

It is evident that Mrs. Lincoln was able to write rational and convincing letters while incarcerated, letters pleading that she had been wrongly committed and deserved rescue from the Bellevue sanatorium. Within weeks of her insanity conviction, she mounted a campaign by enlisting friends and members of the press to engineer her release from Patterson's care. She finagled her way into the home of her sister Elizabeth Edwards, in Springfield. How and why she did this and what it says about her mental condition continue to provoke debate, but the significant cache of personal documents added to the previously public record—court documents, institutional records, and newspaper accounts—allows a sharper image to emerge.

Mary Lincoln was admitted to Bellevue on May 20, 1875, and was offered a rest cure of regular exercise, healthy diet, and regulated medication. Despite exhibiting evidence of depression, she gave her caretakers little trouble in the first few weeks. But by early June, she had become distraught, even distracted—postponing her carriage rides and unmoved by weekly visits from her son Robert, even when accompanied by his five-year-old daughter, Mamie. She generally kept to her rooms when visitors appeared on the grounds, deeply humbled and pained by her commitment.

Staff noticed by early July that Mrs. Lincoln was suffering from insomnia and seemed more agitated. She consented to an interview with a reporter, an unscheduled meeting arranged by Dr. Patterson. This journalist, from the *Chicago Post and Mail,* who had only come to see the grounds and was given the interview as a bonus, published an amazing account. The interviewer commented that the former First Lady wore a shabby dress and her hair had turned white. This alone would have been shocking to the Chicago readers who knew Mrs. Lincoln and her reputation.

The journalist reported that her conversation rambled and she spoke with imaginary companions if left alone in her room. Such an account demonstrated the wisdom of Robert's decision to send her away and supported the idea that institutionalization would benefit her. However, after the story appeared on July 13, 1875, things changed. Shortly thereafter, Mrs. Lincoln demanded that Dr. Patterson release her into her sister Elizabeth's care in Springfield, Illinois. This was a calculated risk for Mary.

Mary Lincoln had been estranged from her older sister Elizabeth on and off over the years. She had been quite close to her when she was younger, valuing her as an adviser and maternal figure. But during her long and difficult courtship with Lincoln, things shifted as Mary stubbornly stuck to her plan to marry this brooding backwoods lawyer over the objections of both Ninian and Elizabeth Edwards. There was a streak of rivalry between the two sisters in Springfield, but relations warmed up considerably when Lincoln's success turned the tables. The former son of the governor looked to the rising Republican star for loans and support. When Lincoln was elected president, Elizabeth and her daughter made more than one visit to her sister's home while she was First Lady of the land.

Lincoln had always valued Elizabeth's ability to soothe his wife, and especially appreciated her visit to them after Willie died. But unfortunately, while she was living in the White House, Mary opened a letter from her niece Julia (addressed to her mother

Elizabeth, a houseguest of Mary's at the time) with some less than complimentary remarks about her Aunt Mary. Thereafter, a permanent coolness developed between Julia and Mary. She repeated malicious gossip about her niece, who kept inappropriate male company and late hours, embarrassing her family with "loose" behavior. Indeed, Julia's increasingly serious indiscretions led to her family's conspiring to get her out of the country to prevent further fallout from her sexual escapades.

Julia exhibited some form of mental illness from the age of thirteen, which did not subside when she married in 1856 at age eighteen. She was apparently sexually overactive for a woman of her age as judged against the norms of the time, which led to severe family crises. Her mother suggested that she had exhibited signs of insanity, with symptoms exacerbated by postpartum depressions following the birth of each of her three children.

Julia became a serious family burden. Her father nearly went broke building her a new house and setting up his son-in-law in business during the years leading up to the Civil War. During wartime, Ninian Edwards sought advancement through Lincoln's patronage and protection, and was able to replenish his coffers through what was deemed greed and perhaps graft. Upon Lincoln's death, Mary might have expected support from her in-laws during her time of need, as the widow of one who had not turned his back on them. But offers and visits failed to materialize, and the sisters remained on bad terms. There is no record of correspondence or contact during the first ten years of Mary's widowhood, although clearly indirect contact was made through other family and mutual friends.

During these postwar years, Elizabeth endured the deterioration of her daughter's mental health and reputation. By 1872, Julia and her husband, under a cloud of scandal, departed for Argentina, where Edward Baker had secured a diplomatic post. Perhaps this close personal experience with mental illness within her immediate family contributed to Elizabeth's eventually sympathetic attitude toward her sister. Elizabeth seemed to mellow once Mary's plight was placed squarely before her. Elizabeth indicated to Robert that she chalked up Mary's situation to family illness—inherited insanity. She also took pity, as she recognized that her sister was someone who had been given much to bear, handled it poorly, and "made the world hate her."

Mrs. Lincoln told medical staff at Bellevue that Elizabeth had raised her and she regarded her as sort of a mother—even though they had fallen out. Robert had apparently consulted with Elizabeth when Mary fell ill in the spring, but Elizabeth had begged off taking any responsibility at the time, particularly as she was facing surgery. Following her medical recovery, Robert's aunt responded to the reports of Mary's institutionalization and the publicity about her plight. It was indeed a shame and dishonor that no family had stepped in to rescue her, especially as Mary's husband had always shown such charity to his Todd relatives. By late summer 1875, Elizabeth volunteered to take Mary in, directly from the sanatorium.

Beginning in July, Mary's insistence that she be allowed to go to Elizabeth's became part of a larger scheme. She wrote to the Bradwells on July 28, asking them to pay her a visit, and also requesting that they contact General William Farnsworth (a prominent Chicago attorney who had been present at Lincoln's deathbed) and William F. Storey, the editor of the *Chicago Times*, one of the most popular Chicago tabloids, the motto of which was "to print the news and raise hell." Mary was assembling a powerful group of advocates, and planned a major assault to change her situation.

After the Bradwells showed up in Batavia on July 30, Mary launched a campaign for early release.

Despite Mary's lip service to Patterson's staff, she was deeply aggrieved about Bellevue and told Bradwell, "It does not appear that God is good, to have placed me here. . . . I have worshipped my son and no unpleasant word ever passed between us, yet I cannot understand why I have been brought out here."

The Bradwells not only contacted Mary's sister Elizabeth, they began a networking campaign to involve others. They asked Elizabeth and Ninian Edwards to visit Mary at Batavia and to bring her back with them to Springfield. They employed some of the hyperbole common to Mrs. Lincoln's style, suggesting that Mary needed to get out from "behind the grates and bars"—even though there weren't any at Bellevue.

Mrs. Bradwell was, above all, a great publicist. She convinced *Chicago Times* writer Franc Wilkie to meet with Mary on August 7, bringing him to the facility while Dr. Patterson was away. Ironically, Patterson was in Chicago consulting with Robert Lincoln, concerned that the Bradwells were unraveling all the progress that his mother had made thus far. When Patterson came back and discovered that his patient had been meeting with a "stranger," he wrote Bradwell a stinging rebuke. The doctor informed her that she could no longer visit Mary or have any contact without Robert's explicit permission.

Then the battle began to heat up—and all the parties played a dramatic role in self-justification. Mrs. Edwards told Myra Bradwell that although she felt her sister was mentally unbalanced, Mary should never have been institutionalized. Robert cast aspersions on Myra in correspondence with his Aunt Elizabeth, saying she was "a high priestess in a gang of Spiritualists and from what I have heard it is to their interest that my mother should be at liberty to control herself and her property." (This was a slander he was forced to recant.)

Elizabeth was off on a vacation at Rye Beach in New Hampshire, and so talk about releasing Mary to her was, for a time, purely hypothetical. Shipping Mary to Springfield was inconvenient as well as medically ill advised, according to her current physician. Dr. Patterson became alarmed at the undermining of his authority. He got Robert to agree that Mary should not leave Bellevue and that the doctor should "cut off absolutely all communication with improper persons." (And presumably Patterson would determine who was "proper" and thus control all access to Mary.)

When Robert lowered the boom, the Bradwells went on the offensive with a carefully orchestrated campaign. First Myra leaked the news on August 21 to the *Chicago Post and Mail* that Mrs. Lincoln was better and this improvement meant that she would be going to her sister's in Springfield soon. This stood in direct contradiction to the express wishes of all other parties involved but Mary. On August 23, Judge Bradwell followed up with a story in the *Chicago Post and Mail* saying it was an outrage to hold Mrs. Lincoln against her will, and that if she were not released from Bellevue Place, "startling developments" would follow. The Bradwells targeted Patterson, not Robert Todd Lincoln. On August 24, in a coup de maître, Franc Wilkie's *Chicago Times* story of his August 7 interview with Mrs. Lincoln was published under the headline REASON RESTORED. Advocates suggested that even if Mrs. Lincoln had been impaired at the time of her trial, by August she deserved release.

In the wake of Wilkie's exposé, Robert's supporters declared the Bradwells' version of events fictitious, scandalous tittle-tattle. From the pages of the *Chicago Tribune* all the way to the *Boston Globe*, partisan opinions clashed. Finally, Dr. Patterson responded in anger with a letter to the *Chicago Tribune* on August 29: "It has been stated that she [Mary] has been 'kept in close confinement,' 'virtually imprisoned behind grates and bars,' 'locked by her jailer as a prisoner.' . . . These and other harsh terms are not used in the interest of truth." Patterson held his ground: "It has been publicly stated that I have 'certified' to the recovery or mental soundness of Mrs. Lincoln. This is not true. She is certainly much improved, both mentally and physically; but I have not at any time regarded her as a person of sound mind. . . . I believe her now to be insane."

Robert's letters indicate he was, once again, devastated about his mother's "making herself talked about by everybody." This is a difficult matter to dissect. Even if Robert were pleased to have his mother returned to reason, he was naturally concerned about how she would paint his role in this debacle. With all the publicity and gossip, a natural outcome of the public insanity hearing required by law, there would be public discussion of the Lincolns' private matters, which pained him deeply.

First and foremost, his mother's airing of the family feud offended him deeply. Robert took the time to pen defensive letters. In early August, he explained to Elizabeth Edwards, "I would be ashamed to put on paper an account of many of her insane acts—and I allowed to be introduced in evidence only so much as was necessary to establish the case." The spin campaigns were in full blossom by August 1875, at which time Robert's filial devotion seems to have been fraying. Clearly he had forgotten what a formidable foe his mother could be.

The campaign of agitation paid off on September 10, when Mrs. Lincoln was allowed a "visit" to her sister in Springfield. Patterson and Robert called it an "experiment," but Mary (prematurely) called it a triumph. She clearly made concessions: Mrs. Lincoln was accompanied by a nurse and allowed only three trunks as luggage. She was on her best behavior. Within a fortnight, her sister Elizabeth reported, "She has dined at Mrs. Smith's [her sister Ann], taken tea at our sister Frances's and received every visitor, with a manifestation of cheerfulness, and pleasure, as has surprised me. . . ." By September 20, Patterson pronounced the experiment a success, and Mrs. Lincoln would remain indefinitely in her sister's home. Robert paid his aunt $100 a month to care for his mother, from her considerable funds, over which he had total control.

By November 1875, Mary became more animated, demanding that the Bradwells supply her with a copy of her 1873 will and suggesting that her bonds should be returned to her. She volunteered to place the bonds in the hands of Springfield banker Jacob Bunn, where they could remain on deposit for the remainder of her life. But she wanted her finances out of Robert's control and remained obsessive about money matters. Robert resisted her entreaties and feared that his mother could ruin his careful financial planning. He believed she might bankrupt herself to spite him.

In the new year, Robert received disturbing news from his Uncle Ninian: Mary was consumed with rage over her son's machinations, and she had taken to carrying a pistol. She ranted and raved over the indignities to which she had been subjected. Elizabeth and Ninian did not know how seriously they should take such threats, but they felt it was necessary to tell Robert.

Scholars have speculated that Mary may have wanted genuine harm to come to her son, to avenge what he had done to her. Carrying a gun might or might not have led to violence. Far more likely, Mary Lincoln may have held onto a gun for its symbolic power. After being manhandled, deprived of her liberty, locked away, and deeply traumatized, Mary may have sustained a stress disorder. This may have led her to dig out her son's old pistol (which she had tucked away in a trunk as a memento) to hang on to as a kind of totem.

Possession of a handgun during this period was far more common, raising fewer eyebrows than it would today. Abraham Lincoln himself was carrying a pistol on the night he died. Mary Lincoln might well have armed herself the moment she escaped the long arm of Bellevue, and the fact may only have been discovered in January. Whatever the case, while her hosts fretted, Mary Lincoln fumed.

Ninian Edwards suggested to Robert that he remove some of the financial restraints imposed. Robert did send his mother cash, but not the bonds. Mary began to make unreasonable demands—that paintings, books, and specific items she listed be returned to her. She hired her own lawyer, and the negotiations continued until the clock ran out. The law allowed for restoration of legal rights after one year had passed, but nothing could be done by the court before that date. Robert could have given her control over her finances, but would have had to countersign any legal documents. Instead, he just dutifully managed her money for her. While she boarded with her sister, she was not privy to the details of her own accounts.

A hearing was scheduled in Chicago on June 15, 1876, to address Mary's legal status. This time Ninian Edwards represented his sister-in-law. When a jury again convened to determine her status, Mrs. Lincoln was conspicuously absent, but better represented. Her strong petition was read into the record. Finally twelve men on the jury determined "that the said Mary Lincoln is restored to reason and capable to manage and control her estate." Mary was thrilled with the outcome, considering it a real vindication.

A flurry of letters sought out former friends, especially those who had assisted with the effort to free her, such as the Bradwells. Mary also sought out others. She wrote to Myra Bradwell, requesting that she help her track down Ellen Johnson, a former washerwoman, to explain to her how "a *faithful, devoted* son failed in his attempt to render a deeply bereaved mother *insane*." Her need to put the episode behind her was balanced by her drive to expose the wrongs her son had committed when he legally confined her.

She confided her innermost feelings to Myra Bradwell in a letter that has only recently come to light, highlighting her fierce indignation. She wrote in July 1876, complaining, "None of my treasure in the way of rich and rare presentations that were made me have been returned to me."

Correspondence in the weeks following her legal emancipation demonstrates the towering rage she maintained toward Robert for his betrayal. Mary wrote an incendiary letter to her son, crowing about those who supported her and condemning him. She was iron-willed in her determination to get her own property back: "You have tried your game of robbery long enough." She went so far as to consult attorneys and again drew up a list of property she wanted returned, including clothing, books, and paintings. She called Robert the "monster of mankind."

In response, Leonard Swett made a swift legal rebuttal. He could prove that these items were gifts that Mary had freely given her family over the years, documented by letters that Robert had saved. Mary raged against Swett, as well, in newly discovered correspondence, blaming him for stirring up even more trouble for her. Mary suspected plots against her and asked Judge Bradwell to look into matters for her. She remained restless and suspicious, seeing chicanery at every turn.

Mary felt permanent estrangement from Robert but grew closer to other relatives, most especially her siblings in Springfield who had taken her in. She also became attached to Lewis Baker, the seventeen-year-old son of Julia and Edward Baker. His parents had gone to Argentina on a diplomatic mission and left him behind with his Grandmother Edwards in Springfield. In their mutual abandonment, the widow and her grandnephew formed a special relationship. Mary also gave $600 to her sister Frances Wallace, who had been widowed since 1867 and suffered strained financial circumstances. She might have stayed behind to continue to rebuild bridges in Springfield, but she could not salvage her wretched relationship with her only remaining son. Mary decided she would have to escape the country to be safe from Robert, who might try to lock her away again. This paranoia was not misplaced, as she heard from Ninian and others about the great plans Republicans had for Robert. No one wanted his crazy old mother standing in the way of Robert's brilliant political career. Mary wanted to remove herself from the possibility of reinstitutionalization, so she decided to go abroad once again.

To have had her child turn on her after what she must have felt was a life of sacrifices on behalf of her family must have done Mary irreparable damage. She was trying to keep afloat in a sea of despondency; she'd been cut loose into a permanent storm, far from any harbor. She would have to flee so as never to allow herself any temptation of reconciliation.

Mary swore her sister Elizabeth to secrecy about her travel plans, making her promise to keep the date and destination a secret from Robert. She took a roundabout route from Illinois to the East Coast, escorted by her new favorite, Lewis Baker. The two made stopovers in Lexington and New York, and took in the last days of the Centennial Exhibit in Philadelphia. When she stepped aboard her ship bound for France on October 1, Mary proclaimed her fate: "I go in exile, and alone!"

18.
Smoldering Embers

M ary Lincoln's last years were spent wandering on the Continent, until ill health forced her to return to the States. She spent her last months moving between New York and Illinois, seeking relief from infirmities and indignities until she reached her final resting place in Springfield. The years from autumn 1876 until spring 1881 witnessed Mrs. Lincoln moving in and out of her newly adopted European berth in Pau, a small provincial capital in the southwest corner of France. Little is known about this period of her life, but the notion that she went quietly back to her hometown of Springfield after breaking free of the asylum ignores this critical part of her life and what it reveals about her personality.

Why she settled in Pau remains a mystery. She later claimed she had friends there, and Mary had a wide network of friends at various times, a fact that makes tracking her correspondents so challenging. From her earlier sojourn in Europe with Tad and from her extensive network of contacts from her White House days, Mary had the opportunity to visit people all over the Continent. So there was nothing that dictated a French residence. But there is evidence that Pau recommended itself as a place to preserve her delicate health and to stretch her limited income.

Nestled not far from the Spanish border, Pau stood twenty miles from the Pyrenees. The town, built high above a river, sported expansive boulevards lined with lush vegetation, fanning out in half-circles from the town center. The houses in this French hill town, with their tile roofs and stucco walls, reflected Spanish influence. Pau boasted a royal château, which housed

The city of Pau, France

the private apartments of Louis Napoléon and his wife Eugénie. Perhaps Mary was attracted by the compactness of this eye-catching crossroads. Gamblers and invalids, saint-seekers and social butterflies kept alighting in this village full of lovely architecture and meandering streets. The entire town could be crossed in a brisk twenty-minute walk, although by 1876 Mary would not have done much walking, at least not briskly.

Although the architecture might have been pleasing to Mary Lincoln, she would have been even more pleased to confirm the attractive climate—specifically recommended for someone with her tendency to collapse. The stillness of atmosphere was a main feature. However, one British naval captain decamped with the complaint that he rarely could find even "a capful of wind."

Pau was the favorite of a competitive spa industry in the southwest of France during the middle decades of the nineteenth century. Unlike Italian spa towns plagued by seasonal winds, Pau remained temperate. During much of the winter when the Continent was chilled to the bone, Pau escaped extreme temperatures. When pestilence swept coastal Mediterranean towns, Pau dodged these threats. Claims for Pau's superiority were promoted by a British physician, Dr. Alexander

Taylor, whose 1856 volume, *The Climate of Pau* (to give a short version of its fashionably cumbersome title), put the town on the proverbial map. Taylor's book went into several editions and he kept up a formidable campaign of publicizing the town's major attractions. As a whimsical aside, Taylor noted, "Ladies find that their hair retains the curl much better than in England." A street named after Taylor in the town center today attests to his significant role in building the city's reputation as a resort.

Dr. Taylor not only charted the daily mean temperature of the city—with elaborate calculations over a period of years; his volume also provided detailed analysis of the most healthful aspects of Pau's location and weather. Dr. Taylor's own facility welcomed patients and guests seeking all kinds of improvements. He offered detailed comparisons with Montpellier, Nice, Rome, Pisa, Florence, Naples, Biarritz, and other spa towns. Mary Lincoln had already been to several of these places during her previous European tour, and deliberately selected Pau instead.

If Mary used Taylor's book as her bible, as did many seasonal residents, she would have learned that "the price of substantial articles of living are somewhat less than in England." She was grateful to find a place to live comfortably on her widow's pension.

Taking the waters was an increasingly popular therapy among the wealthy. The late nineteenth century was perhaps the peak era of this medical phenomenon, after Italians revived the ancient practice of natural-springs therapy. Savonarola published a treatise on the baths and thermal waters of Italy in 1498, and the European leisure classes embraced his prescription.

When Henry IV assumed the throne of France, he honored his childhood in the Pyrenees by establishing a system of royal superintendents for baths, mineral fountains, and other such facilities throughout his kingdom. Visitors flocked to Pau's nearby hot springs. Their water is warm but never boiling, their pools laced with healthful ingredients, especially limestone.

Mary Lincoln was familiar with the medical wisdom about the value of drinking and bathing in these waters. Consuming two to six tumblers a day was thought to improve "congestion of the brain, which in persons of sedentary pursuits, and of advancing years, threatens apoplexy." Mrs. Lincoln visited Saint Sauveur, in the valley of Luz, which was open seasonally from May to October and claimed to be one of the finest thermal establishments in the Pyrenees. Certainly Saint Sauveur's proximity to Pau recommended it, and it became Mary's top choice, perhaps because Taylor mentioned that its accommodations were "cheaper."

Mary Lincoln—plunked down in a new country and still fearful of running out of funds—sought both comfort and economy. She also sought a location where she might easily draw out her funds quarterly. As an American widow anchored in France, her concerns about money continued to be paramount.

Her first letter from abroad to her seventeen-year-old grandnephew Lewis was written on October 17, 1876, from an unnamed hotel ("very delightful") in the port where she landed, Le Havre. Stopping over to recover from her sea voyage, she delighted in being treated royally by local aristocrats. She took carriage rides daily, "with coachman & footman in livery," enjoyed a five-volume history of France (ordered from a Parisian bookseller), and instructed her nephew to "write at once & direct to Pau, France." Her destination fixed, she boarded the steamer *Columbia* to Bordeaux, where she took the next train bound for the spa town.

We have no idea of the exact date of her arrival. In Pau a biweekly paper known as the *Journal Étranger* oriented newcomers and provided expatriates with valuable information. This paper ran advertisements for the scores of hotels, many with large and small apartments available. Some places rented single rooms that could be had for as little as two francs per day. Establishments varied from deluxe to spare, with everything in-between. Rooms with three meals ranged from nine to twelve francs per day, on average.

Most crucial to the paper's success was the listing of new arrivals. Each tourist hotel in town sent the paper a roster of foreign visitors who had recently checked in. Relying on the handwriting of their guests, perhaps a clerk misread Mrs. Lincoln's signature: A "Mrs. Linham" was listed as a new arrival at the Hôtel de France, a fashionable residence for English-speaking guests, in the October 28, 1876, issue. This would have been a likely time for her to arrive in Pau, if she had landed in Le Havre the second week of October and then traveled on.

By November 1876, Mary Lincoln was writing from the Hôtel de la Paix, where she stayed for several months. The hotel was listed in a pocket guide to Pau (printed with side-by-side translations, in English and French) and advertised in an 1883 guide: "Hôtel de la Paix: Recommandé par sa bonne situation sur la Place Royale. Pensions de famille, table d'hôte restaurant à la carte. Prix très modérés. Chambres depuis 2 francs." The economy afforded by this well-located hotel no doubt attracted Mrs. Lincoln when she arrived well into the season—September through June.

English was a familiar tongue in Pau. Hundreds of British expatriates had established several churches as well as a literary society, founded in 1856. The most fashionable residents joined Le Cercle, a club modeled on London invitational societies that limited membership to one hundred. Applicants were kept on waiting lists for years. A lavish casino perched at the top of a bluff offered six concerts per week, charging monthly or seasonal rates. The casino also supported a private club that admitted *only* foreigners.

At this stage, Mrs. Lincoln sought a self-proclaimed exile—from her family, from her nation, from her past. She had always banished the reminders that brought her the most pain—mementos like Eddie's clothes, Willie's toys, and her husband's canes—as if she could send away her memories along with these relics. She had left behind many trunks, in hopes she might start afresh.

She had always wanted to live the life of a grande dame, and perhaps this European escape gave her some small opportunity for wish fulfillment, despite all the other dashed dreams along the way. Her last adventure, as an American expatriate, reflected great determination, for she was frail and alone, but she chose to make her own way despite the considerable obstacles.

Certainly one of the main attractions of Pau for Mary Lincoln was its large community of English-speaking visitors. The population of the town was only 25,000, but there was a British consulate. Pau offered a range of concert halls as well as the lavish casino. The only thing lacking, visitors suggested, was good theater. This was inconsequential if not preferred by Mrs. Lincoln; she had not set foot inside a theater since the night of her husband's death. Much like the widow's weeds she donned daily, absenting herself from this former indulgence was part of her permanent agenda, a reminder and ritual she would never forsake, even if someone were so unwise as to offer her a box at the theater.

Robert's return of control of her estate in June 1876 revealed that she was a widow with considerable property and a government pension of $3,000. Yet Mary clung to her lifelong worries over money. She sought the least expensive lifestyle she could manage, while requiring comfort and class. Regardless of the situation and without fail, Mary always claimed to be short of funds.

It was supremely ironic that she had spent so much of her adult life yearning for financial independence, and now that she had it, she was no longer surrounded by an intimate circle who might benefit from her financial largesse. She was not shy about funding personal comforts, and she hoped to live out the rest of her life with dignity and ease, avoiding the limelight she had once craved. Mrs. Lincoln did not avoid sharing the fact that she was the widow of the revered president of the United States, but she also did not imperiously make the demands she once had.

If she could live peacefully within this Pau community, its medical establishment and large expatriate population boded well for Mary's long-term prospects. Her fluency in French delighted the natives and eased her transition. She was eager to put down some roots.

However, no sooner had she settled in France than disturbing news arrived from Illinois. When a master engraver, Ben Boyd, had been imprisoned for counterfeiting, his gang found themselves running out of loot. They plotted to steal the body of Abraham Lincoln, planning to demand Boyd's release and $200,000 in gold in exchange for returning the corpse. This kind of stunt elicited a dramatic response. The body of New York merchant A. T. Stewart was also stolen, robbed from Manhattan's St. Mark's Church and held for ransom, amid a public furor.

Luckily, one of the gang plotting the Lincoln heist solicited the services of an undercover lawman for the conspiracy, and when the robbers headed for Oak Ridge Cemetery, they were interrupted by detectives lying in wait—but not before the gang had sawed a padlock off the door of Lincoln's tomb and pried a marble lid off his sarcophagus. They were unable to spirit the coffin away in a nearby wagon, as they had planned, but they did manage to escape.

Members of the gang were caught in Chicago over a week later, and Robert Todd Lincoln supervised their prosecution. Their trial resulted in convictions, with terms of one year in Joliet State Prison. Robert was so grateful that he presented a portrait of his father to Patrick Tyrell, one of the guards who helped foil the attempted larceny, with his sincerest gratitude.

As a result of this sad episode, security was stepped up at the cemetery, and the newly assembled Lincoln honor guard redoubled their efforts. The town took more seriously the significance of safeguarding and honoring the president's sacred resting place. On April 15, 1880, more than three hundred braved the rain for the first memorial ceremony at the grave, scheduled for 7:22 a.m. to honor the exact moment of President Lincoln's passing. Also, in response to public outcry, the state of Illinois amended its law to allow a sentence of up to ten years in prison for anyone convicted of grave-robbing.

Robert did not contact his mother about these events, knowing what kind of response he would get and perhaps waiting for time and distance to heal the hurt. Indeed, her letters to family back in Illinois demonstrate the grudge she continued to hold, as when she complained to Lewis Baker, "I was so cruelly persecuted, by a bad son, on whom I had bestowed, the greater part of my all," a fact that continued to rankle "deeply in my heart."

For his part, Robert seemed to have no interest in tracking his mother's condition or whereabouts. He could have easily located her if he had wanted. Although Mrs. Lincoln roamed during her European sojourn—to Nice to seek out friends, to Paris to buy lace and fine goods, she turned up in Pau like clockwork to collect her checks. She was particular about what she called her "rent" from Robert: His monthly mortgage payment flowed through her Springfield banker, Jacob Bunn.

Unfortunately, the quarterly disbursements of her United States government pension could only be processed following a complicated procedure of filling out forms that had to be sent first to Bunn, then forwarded to Mary in France. Mrs. Lincoln signed these forms and had her pension documents notarized before returning them to Bunn, who was authorized to act on her behalf back in the States. This was a very drawn-out process, and Mary constantly complained about delayed or lost mail.

The letters from Mary to Bunn—nearly a hundred over a four-year period—are sometimes little more than receipts. Yet most demonstrate Mary Lincoln's steel-trap grasp of the details of her situation. Her faculties appear razor sharp as she peppers the Springfield banker with requests, offering intricate details. Her incredible precision and her ferocious attention to minor matters belie any suggestion of failing mental faculties or encroaching dementia.

During 1875, Mary Lincoln's name had appeared on the "Lunatic Record" for Cook County, Illinois. A little over a year later, her rights had been restored, and she was moving among fashionable spas, calculating her profit on the dollar-to-franc exchange rate, and speculating whether or not the disputed U.S. election of 1876 would result in war. Tilden had won the popular vote, but Hayes would win the vote in the electoral college (still the only vote that counts) when several states with disputed returns struck a bargain to support the Republican candidate in exchange for the withdrawal of federal troops from the South, thus ending Reconstruction. She certainly hoped this imbroglio would not re-erupt into armed conflict, even though she had tried to leave such worries behind in the United States. Her money and her liberty were all that concerned her during this stage of her life.

Mary Lincoln's letters to family and friends during this period signify her deep appreciation of the kindness extended to her following her release from custodial care. She appeared particularly reconciled with her Springfield sisters. Mary wrote with enthusiasm about a trunk full of clothing she had assembled, making a special trip to Paris to send these fine woolen goods to her sister Frances in the fall of 1877. Frances was in considerably reduced circumstances and had to seek out a position as postmistress for economic reasons. Mrs. Lincoln remembered the support Frances's husband, William, had provided the Lincolns in earlier days, so she maintained contact and extended charity toward Frances Wallace during her hard times.

Mary Lincoln's letters displayed sorrow and eloquence following the death of Elizabeth Edwards's four-year-old granddaughter, Florence, and she maintained an avid interest in the family matters of her sister, a maternal figure who had charitably taken her in when no one else would. She expressed sad resignation to Bunn about a parade of Springfield deaths—including that of James Matheny, the best man at her wedding. Perhaps she felt the little provincial capital of Pau was the ideal place to weather the storms of the little capital in Illinois, half a continent and an ocean away.

Within a short while, Mary Lincoln regained her sense of entitlement, and her correspondence began to reflect her characteristic spirit and trademark spite. She railed against the widow of President John Tyler, who sought a pension in 1880, outraged that someone who had been so "bitter against our Cause" during the Civil War would seek government aid. As someone who had been branded a Confederate sympathizer during the war, someone who believed she had proven her loyalty above and beyond the call of duty, Mary Lincoln would not waste one drop of mercy for the widow of a man who had been so staunch a Rebel during the Union crisis. Mary had no sympathy for Julia Tyler's "impudent request," her plea that she was impoverished and needed a pension—a plea that echoed Mary's of years before. Even in exile, she was ferociously patriotic and irate at any dissent.

Surprisingly, Mrs. Lincoln perked up when she came across an account in a journal called the *American Register* that discussed the son of Stephen Douglas and her own son Robert in glowing terms. Any prediction of Robert Todd Lincoln's bright political future—perhaps seeing a presidential race in his future—excited Mary in spite of herself. The idea of Robert's flourishing prospects sent her into a swoon of excitement—imagining her darling granddaughter, Mamie, romping around the Cabinet room. She gave in to an old weakness and mentioned in a letter that, while First Lady, newspapers claimed she was "the power behind the throne." This kind of grandiose remembrance became a staple for Mrs. Lincoln with advancing years.

Mary Lincoln still held a grudge about her son's treatment of her, but she weakened. Even as she nursed her anger toward her son, she sent along gifts for her namesake, Mamie, via the Edwards family. And Mary loyally tracked Robert's career in the international papers.

Robert maintained public indifference to his mother's fate. When a friend asked about her in 1877, he said he did not even have her address, but knew she was "somewhere in Europe." Nevertheless, Robert hoped for a reconciliation and claimed he would write to her at once, if he thought it would do any good. Mary was perhaps chagrined that her son turned down President Rutherford B. Hayes's offer to join his Cabinet as assistant secretary of state in 1879, but their estrangement kept her from contributing her advice. Mother and son suspended direct contact for years after Mary's institutionalization. She soothed her frayed nerves with time in Pau and excursions to Vichy, Avignon, Biarritz, Marseilles, Naples, and Rome.

Imagine how irritated she must have felt when the past intruded in an unexpected way. On December 8, 1878, former President Ulysses S. Grant and his wife,

Julia, arrived in Pau, where they spent three days being toasted and hosted by the city fathers. A local guide-book described the festivities on December 11: "Le soir, un somptueux banquet fut organisé à l'Hôtel de France, en la présence des Américains et des Anglais les plus notables de Pau. . . . Le propriétaire de l'Hôtel de France, Gardière, m'avait finement baptisé sa bombe glacée du nom de la plus célèbre victoire de General Grant: Vicksburg."

In the days following Lincoln's assassination, Julia Grant sought out Mary Lincoln to express in person her condolences about Mary's loss—and to show her will-ingness to overlook the unhappy circumstances of their meeting the week before at the Virginia headquarters of her husband. Mrs. Lincoln rebuffed Julia Grant's overtures, and there is no evidence the two women ever spoke or corresponded. It seems certain they did not after Grant's election to the presidency. Having Julia Dent Grant ensconced in the White House was some-thing Mary Lincoln could not easily abide. Moreover, she seemed to revel in the scandals that plagued the Grants while he was in office.

When Mrs. Grant was accused of taking money to influence appointments, and other forms of corruption, Mary Lincoln seemed eager to credit such stories and to take some small pleasure that her former detractor was mired in scandals. This conflict between the two was well known within Washington circles, yet the friction contributed to Mary's reputation for uncharitability.

Abroad, foreign dignitaries would not have known about this estrangement. So it was disappointing that the government officials celebrating President Grant's visit to Pau did not bring Mrs. Lincoln into the fold as one of the American celebrities available. The Grants, perhaps completely sincere, declared that they had no idea that Mrs. Lincoln resided there; why would they, as her own son didn't know where she was? By contrast, Mary Lincoln was thrilled when her friend Myra Bradwell paid a visit so they could enjoy a fond reunion.

At first, Mrs. Lincoln seemed to relish being able to follow a completely independent, if not whimsical, path. But after four years as a gypsy, always seeking and never finding, Mary developed a serious case of homesickness and decided to return to the States in 1880.

This decision did not mean that she had mellowed, but rather that she had become exhausted by the challenges of expatriation. First of all, she was completely disenchanted with her adopted country, complaining about the French as the most *"unprincipled, heartless, avaricious"* people, adding, "with the exception of a very few, I detest them all." Second, she had become a semi-invalid after a series of mishaps, and required better and more constant medical care. A fall from a stepladder in December 1879 injured her back; then she suffered even more disability following another fall, down a set of stairs in June 1880.

When discontent was visited by disaster, Mary wanted to return to familiar surroundings. When she crossed the Atlantic to America in October 1880, she felt defeated by the elements and wrote poignantly to her grandnephew: "In ill health & sadness, quietude & loved faces, are far best."

While on shipboard, unsteady on her feet, Mary nearly fell overboard when she pitched down an exterior stairway. A woman passing her on the stairs, the actress Sarah Bernhardt, instinctively reached out to grab the older woman and thereby saved her. Bernhardt warned the woman draped in widow's weeds to be careful, saying, "You might have been killed." Mrs. Lincoln did not express gratitude with any grace, and when Madame Bernhardt discovered the identity of this figure wrapped in black, she suggested that perhaps she had not done the poor widow any favors. Even to a casual observer, Mary Lincoln appeared indifferent to her fate and ready to abandon the woes of her life.

Once again she sought Lewis Baker's assistance to ease her arrangements. She wanted him to meet her at the docks with American money in hand, and to help search out medical treatment in Manhattan. She wrote instructions for him with near military precision: "*Rely*, upon my passage, upon the *Amerique* October 16th—six days *later—without fail* start for me."

Lewis showed up when the ship docked, but they both were shoved aside as reporters flocked to cover the arrival of "The Divine Sarah," an international sensation. Mary received a brief mention in the press: "The widow of ex-President Lincoln has arrived in New York, from France, in good health. She is stopping at the Clarendon Hotel, and will in a few days go to Florida." But the paper was wrong on two counts, she did not feel well, and Florida, where her troubles all began five years before, would not be her destination when she left.

Mary let the press make their random speculations while she consulted with the eminent physician Dr. Louis Sayre—a pioneering orthopedic surgeon who she thought might be able to help her with her spinal problems. Rest was the main recipe he prescribed. After a few days, Mary decided the best course was to make her way back to Springfield, where her sister Elizabeth would take her back into her home.

Elizabeth Edwards supervised Mary's homecoming, assigning a granddaughter, Mary Edwards, to help her Great Aunt Mary around the house. The sixty-one-year-old Mrs. Lincoln was not easy to please at this stage, considerably crankier than during her previous visit to Springfield. By the time she settled back into her sister's home, her various ills left her barely able to walk for any length of time. Mary was also less and less able to see, having severely impaired eyesight due to cataracts. Being so disconsolate hardly made Mary an ideal houseguest.

Sixteen-year-old Mary Edwards found her grandmother's new addition to the household puzzling in other ways. Mrs. Lincoln brought with her more than forty trunks, which weighed over eight thousand pounds. She insisted upon having her things with her—an inconsiderate burden for the Edwards household. The pile of trunks caused one maid to flee because she didn't want to sleep under a ceiling weighed down by Mary's excess baggage. Mary claimed she was too feeble to go out anywhere, to make any effort to socialize or exercise. Yet she spent many of her waking hours pawing through the stored treasures and packed debris of her life. She kept sorting and sorting, but never seemed to make much progress to those who witnessed her repetitive and prolonged agitation.

When James Garfield was elected president shortly after Mary returned to the States, her hopes about her son's career began to inflate again. When he joined the government, appointed secretary of war, Robert Todd Lincoln seemed a man on the rise. Yet Mrs. Lincoln experienced one of her dramatic premonitions, fearing that Robert, like his father before him, might be shot. Mary had not made any move to reconcile, so Elizabeth Edwards intervened and engineered a reunion between Robert and his mother. Granddaughter Mamie came with Robert to visit Mary in Springfield in May 1881. Mary Harlan kept away from this first reunion, which resulted in a tentative reconciliation.

Two months later, on the morning of July 2, 1881, four months into the term, Robert was walking with Garfield through the Washington railway station when a disgruntled office seeker, Charles Guiteau, pulled a gun and fired two shots at the president. Garfield survived the initial attack, but deteriorated as infections set in.

President Garfield was wasting away in the White House, so in September he was taken to the Jersey shore to help him recover. His wife, recovering from malaria, had begun her summer at Long Branch, and she hoped the sea air might improve her husband's deteriorating condition. Despite doctors' constant care, on Monday,

September 19, the president died after suffering for eighty days.

Over the summer of 1881, Mrs. Lincoln relived the horror of her own husband's death, as the nation was transfixed by Garfield's slow decline. Following Garfield's burial, Mary found herself slipping in strength and energy. The president's shooting, and fears for her son, brought Mary to the brink of despair. She journeyed to St. Catharine's, Ontario, Canada, in search of radical medical care, hoping for better results. Then she returned to Dr. Sayre's establishment on West Twenty-sixth Street in New York, sampling some of the Turkish and electric baths. She spent a prolonged period in Sayre's care, undergoing treatment, and Robert and his wife Mary paid her visits during this medical ordeal.

In October 1881, the press reported that Mrs. Lincoln was "obliged to guard her health with great care. She has been distressed greatly by the sad death of President Garfield, and feels deep sympathy for his widow and children."

Mrs. Lincoln *was* sympathetic to Mrs. Garfield, but even in her feeble state, she became keen where money was involved. When she discovered there was a move afoot to offer government funds to provide for Garfield's widow and five fatherless children, Mary, not uncharacteristically, thought of herself. When Congress voted

Mrs. Garfield $5,000 a year, Mrs. Lincoln immediately sought parity. She begged Hannah Shearer's brother, the Reverend Noyes Miner, to lobby on her behalf in Washington, which he did.

In January 1882, Congress voted to increase Mrs. Lincoln's annual pension by $2,000. This meant, pending approval, that Mary Lincoln would also receive $5,000 a year. Congress generously voted her a $15,000 bonus as well as making the increase retroactive, so Mrs. Lincoln could have greeted 1882 with contented glee. She felt the country was finally forced to reward her with this increased compensation, and believed it was better late than never.

Although Mary was now without question a wealthy woman, she still suffered from her lifelong qualms about financial insecurity, even as this hefty rise in annual income provided her with an added cushion. But while her bank balance steadily grew, Mary's capacity for enjoyment of these funds slowly diminished. Physical infirmities plagued her. She returned to Springfield from her medical sojourn in New York with little relief of her symptoms and pain. She complained in March 1882 that she still felt partial paralysis. Increasing sensitivity to light rendered her eyes nearly useless. Were all these symptoms indicative of lupus? Scholars continue to debate her physical maladies: Diabetes? Arthritis? Scoliosis? Rheumatism? Small strokes? All of the above? Puzzling out her medical condition is as impossible for scholars today as it was for Mary's own physicians.

Seeking a cure posed a particular challenge for Mrs. Lincoln by the 1880s. In 1875, she believed doctors had conspired against her—and managed to lock her away against her will. Naturally, she was suspicious of physicians thereafter. But her precipitous physical decline was so severe that she made the long journey from Illinois to New York to seek out the best medical care, trying to find relief. Money apparently could not buy her a cure, and so she continued her downward spiral during the spring of 1882.

Anniversaries still shattered Mary, marking the passage of time, which weighed so heavily since her husband's death. This feeling had been compounded by her son Tad's tragic passing. By now, Mary had to face the numbing reality that nothing could alleviate her pain or restore her fading eyesight or mobility. There seemed no way to slow the steady march of mortality.

She was, in fact, practically bedridden, refusing any light but candlelight within her dim surroundings. She had lost a great deal of weight, especially compared to the "bloat" she had complained about in the late 1870s. Elizabeth Edwards took pity on the pathetic figure her sister had become—wasting away in her bedroom, refusing to take the air or enjoy company. Mary abandoned her bed on rare occasions to ransack her trunks, but refused entertainment or excursions. Whether or not sisters from a local convent took turns with visitations to this lonely figure remains a matter of local folklore.

When the gravity of all those who had gone ahead of her outweighed Mary's will to live, the end was near. Mrs. Lincoln had always felt the pull of the great hereafter, and her will to live was now like a dying ember. Many townsfolk were not sure if she was still in Springfield, but to those who personally recalled her husband, to those who revered his memory, she had become a legendary recluse. And in her seclusion, she drifted into a state that could have but one outcome.

Mary Lincoln died at the age of sixty-three on July 16, 1882. She died on the very day she had lost her father thirty-three years before, a day that always brought with it a special sadness. The day before had been the eleventh anniversary of Tad's death, yet another day that brought back memories of unbearable loss. She had survived her husband Abraham Lincoln by seventeen years and three months—and those days and nights had been filled with an unending regret that he had died, while she had lived.

The Edwardses' parlor, where Mary and
Abraham were wed

Mary's wedding ring had been removed during
the last days of her final illness, but it was restored to
her hand for burial. Mary would be laid to rest in a
new dress ordered by Elizabeth Edwards, sent down
from Chicago in one last, lavish gesture to reflect an
older sister's loving indulgence. Once her corpse was
settled into its coffin, Mary lay in the front parlor of
the Edwardses' home, the very room where she had
been joined in marriage to Mr. Lincoln on November
4, 1842, the most defining moment of her life.

After a packed crowd bade her farewell at the First
Presbyterian Church of Springfield, Mary Lincoln's
casket was conveyed to Oak Ridge Cemetery and

placed in a vault alongside her husband and three of her four boys. The minister's eulogy spoke of the couple by invoking a metaphor of their lives like trees twisted together, inextricably bound.

"Love is Eternal" was etched onto the wedding band she wore to her grave, and it was a creed to which she had clung. With her last breath, she had been determined to rejoin her husband, despite years of painful separation. Mrs. Lincoln believed she would at long last find the safe harbor of eternal rest next to the man whose love never wavered, to be with him where the wicked would cease from troubling and she, so very weary, might be at rest.

Her sole surviving son, Robert Todd Lincoln, inherited his mother's estate of more than $80,000 in cash, deeds, and bonds. Robert served under Garfield's successor, Chester Arthur, until 1885. He was mentioned as a possible Republican presidential candidate in several elections beginning in 1884, but he never sought higher office. He did accept an appointment as a minister to the Court of St. James's in 1889, an honor of which his mother would have heartily approved. She also would have been enthusiastic about his association with the Pullman Company, especially after he went on to become its president from 1901 to 1911. While he had been taunted as the "Prince of Rails" in college, he had become a captain of industry and a veritable King of the Railways by the twentieth century. Robert would amass a significant fortune and retire to his estate, Hildene, in Manchester, Vermont, a town where his mother first took him on a summer holiday during the 1860s. Robert predeceased his wife, Mary, in 1926. She died in 1937, leaving an estate worth over $1 million.

Robert and Mary lost their sixteen-year-old son, Jack (Abraham Lincoln II), to blood poisoning when he died in England in 1890. They returned his body to Springfield, where Robert expressed a wish to be buried. But after Robert passed away—a wealthy old man dying in his bed at eighty-two—his widow wanted him to have his own place in the sun. She told a Todd relation she did not want Robert overshadowed by his father in death as he had been in life. So she kept his body away from the grand Lincoln memorial vault in Springfield, which attracts more visitors than any other graveyard in the country, except for the one Robert's widow chose for her husband: Arlington National Cemetery.

Robert qualified for his Arlington burial plot courtesy of his few weeks of military service during the Civil War. Mary Harlan Lincoln was determined to honor her husband, and eventually she was buried alongside him. She brought the remains of their son, Jack, from the Springfield vault and buried him at Arlington as well. In 1997, a tombstone for the last of the Lincolns was added to the family plot there.

Robert, as Lincoln's only surviving son, was expected to carry on the family line and name. But first his son died, and the family name died out. Then Robert's female children and their children produced few heirs. Children of Robert's female offspring failed to reproduce, although as with so many aspects of Lincoln scholarship, there was a disputed end, and if the line truly did die out, it seemed as if it was by some inglorious curse. This was sad for the Lincolns, but positively shocking for the lineage-proud Todds.

Mary Lincoln had hoped to remove herself from wickedness when she died, but her critics have kept her name linked with scandals. Her legacy has been weighed down with controversy for well over a century since her death—and a recent exhibit about her life and times, mounted in 2007 at the Abraham Lincoln Presidential Library and Museum, was titled First Lady of Controversy.

Those who sought to defame her in life did not cease and desist with her death. Indeed, the attacks upon her during congressional debates over her pension seem to have dramatically escalated. Those congressmen who may have thought her foolish and even questioned her loyalties perhaps would be shocked to learn the levels of vituperation heaped on her by modern critics—that she was scheming, criminal, and even diseased. One recent volume suggested that Mary Lincoln's physical ailments at the end of her life stemmed from the effects of syphilis.

Abraham Lincoln's reputation steadily grew following his death, and his popularity rose exponentially after the centennial of his birth in 1909. The Lincoln Centennial initiated an outpouring of laudatory literature. Polls in the second decade of the twentieth century demonstrated Lincoln's eclipse of George Washington as the most popular president, and Lincoln has maintained that stature relatively unchallenged ever since. His international reputation has soared as well.

Lincoln's literary revival owes much to Ida Tarbell's heroic *Life of Abraham Lincoln* (1895), which outshone and outsold William Herndon's three-volume work published in 1889. Tarbell capitalized on Lincoln's rising reputation and her biographical prowess with her 1924 bestseller, *In the Footsteps of Lincoln*. Carl Sandburg's *The Prairie Years* appeared in 1926, and in many ways it is the literary equivalent of Eastman Johnson's superb 1868 painting, *The*

Boyhood of Lincoln. With the work of James G. Randall, among others, the next generation of Lincoln scholars launched a fleet of distinguished studies, ensuring that the field would become both scholarly and popular. A steady flow of scholarship and debate continues to flourish without any promise of slowing down. The approach of the 2009 Lincoln bicentennial has prompted another fresh flood of investigation into Lincoln's legacy.

During the fertile decade of the 1920s, while Lincoln biographers gathered momentum, Honoré W. Morrow claimed, "It began to look as if there were a conspiracy of silence about Lincoln's wife." When Morrow began research for a novel, she claimed to discover "Mary Todd Lincoln to be one of the most lied about women in the world." Morrow's fictionalized biography of this controversial First Lady appeared in 1928, a few years after *The True Story of Mary: Wife of Lincoln,* a study by Katherine Helm, Mary's niece. These volumes were intended to contradict the harsh, unsympathetic portrayals in earlier works, by drawing on private documents within the Todd family circle.

These studies took direct aim at negative images perpetuated by William Herndon. Certainly for the nineteenth century, this Lincoln biographer proved the most persistent and damning critic of Mary

Lincoln. The reverberations from Herndon's harsh charges are still with us. The personal antipathy between the two has been well documented. Mary Lincoln scholarship has been deeply affected by this breach. William Herndon's research material has gone through alternating cycles of use and abuse, recovery and reevaluation—with Herndon's stock seemingly on the rise in the opening decade of the twenty-first century.

The letters preserved between the Lincolns may be few, but that is probably because the Lincolns burned much of their private and personal correspondence before leaving Springfield for the White House. Even more to the point, Robert Todd Lincoln—so painfully reticent about family matters that he sealed his father's papers until twenty-five years after his own death—confessed that he sought out and collected his mother's letters. Sadly, there was never any release of Mary Lincoln manuscripts during or after Robert Lincoln's lifetime of buying up his mother's letters. Experts generally agree that he destroyed pages and pages of Mary Lincoln's personal documents to prevent them from making their way into the public eye—and especially into print. Keeping her letters away from posterity was not only Robert's crusade, but also a vigil maintained by his widow, Mary Harlan Lincoln.

Mary Lincoln has been subjected to a flurry of fictional treatments in the twenty-first century, from popular novels to literary fiction, from an off-Broadway musical to a twenty-first-century rock ballad. But sadly, she has become a punch line in popular culture—whether it's a lame joke about enjoying a play or as a poster girl for shopaholics. She just as frequently turns up in unflattering political cartoons.

Her legacy offers us insight into modern politics, for Mary Lincoln may have been the first White House spouse to lay claim to having been the fuel that fired her husband's presidential career. Also, Mrs. Lincoln may have been one of the White House's most recognizable victims—hounded by the press in and out of office. Only Thomas Jefferson before and Franklin Roosevelt and John F. Kennedy after have had more ink spilled about their private lives. But Mary Lincoln holds a special place in the popular imagination because she was the first First Lady to carve out a separate and distinctive role for herself. These distinctions became fascinating as the nation split in two and Mrs. Lincoln's dilemmas mirrored those of many during war. Only Eleanor Roosevelt, during her prolonged, controversial tenure as First Lady, and more recently Hillary Clinton, during her eight years in the White

House, have even approached the level of vituperation generated by Mary Lincoln as First Lady.

At war with herself as well as with social convention
and polite society, Mary Lincoln failed miserably in the
court of public opinion. She was flawed and brilliant all
at once, and never rose to the heights of humanitarianism that her husband so admirably achieved. Yet she
provided Abraham Lincoln with the space and support
he required to achieve his goals, and with the emotional
yeast he needed to become the wartime president he
became.

Her unconditional love sustained Lincoln's growth
to greatness. She was a woman of intense intellect and
passion who stepped outside the boundaries her times
prescribed, and suffered for it. She was someone who
endured more personal loss and public humiliation
than any other woman of her generation. Despite all
that she endured, Mary gracefully greeted death surrounded by memories of loves lost and found. Near the
end of her life, she would dwell on remembrance of the
stormy day in the parlor of that Springfield house when
she married her young lawyer, taking a leap of faith by
becoming Mrs. Lincoln.

Notes

INTRODUCTION

PAGE

2 *"She won't think anything about it"*: Lincoln's last words—*Bartlett's Familiar Quotations,* 17th ed. (Boston: Little, Brown, 2002)—were to his wife, about their relationship.

3 *White House*: Although "White House" did not come into popular usage until later in the century, the term will be used interchangeably with the phrase "Executive Mansion," which was the common parlance, or "the President's House."

3 *the largest national manhunt*: See James Swanson, *Manhunt: The 12-Day Chase for Lincoln's Killer* (New York: William Morrow, 2006).

3 *part of a larger conspiracy*: See Michael W. Kauffman, *American Brutus: John Wilkes Booth and the Lincoln Conspiracies* (New York: Random House, 2004).

5 *In Victorian America*: See Drew Gilpin Faust on the "good death," in *This Republic of Suffering: Death and the American Civil War* (New York: Knopf, 2008), p. 10.

5 *Edwin Stanton uttered his famous*: On whether Stanton said, "Now he belongs to the ages" or "Now he belongs to the angels," see Adam Gopnik's "Angels and Ages" (*New Yorker*, May 28, 2007). Gopnik's discussion of Lincoln scholarship highlights the marginalized role Mrs. Lincoln has been accorded, as he only mentions her *once* in this lengthy piece on Lincoln's dying words: "The doctors had to remember to cover up the blood with fresh towels when Mrs. Lincoln, fallen into a grief from which she never recovered, wandered in." Gopnik's treatment reflects the way in which Mary Lincoln continues to be characterized in Lincoln literature.

8 *Was she crazy*: In the twenty-first century, "Was Lincoln gay?" has become a part of the popular imagination as well, and a query that frequently gets posed to scholars studying Mary Lincoln.

11 *More than three million*: Faust, *This Republic of Suffering*, p. 39.

11 *The embalming business*: Ibid., p. 93.

ONE : KENTUCKY HOMES

14 *the Lincoln birthplace*: Dwight Pitcaithley, "Abraham Lincoln's Birthplace Cabin: The Making of an American Icon," in *Myth, Memory and the Making of the American Landscape*, ed. Paul Shackel (Gainesville: University Press of Florida, 2001).

14 *The Todd name*: The Todds struggled up from humble beginnings, in a pattern familiar throughout English-speaking colonial North America. They hailed from Scotland, but were banished from their birthplace for their refusal to convert to the Church of England. Two Todds, Robert and James, were arrested and sent into exile to North America. But they were not the first Todds to reach American shores. These hapless brothers disappeared following a shipwreck off the Orkney Islands. A third brother, John Todd, fled persecution to settle in Ireland in 1679. His Irish-born sons, Andrew and Robert, migrated to Pennsylvania in 1737. The relative prosperity of "Robert the Emigrant" (who was born in 1697) allowed his children, and especially his grandson Levi Todd, to aspire to a higher status. By the age of twenty, Levi had transplanted himself to the frontier state of Kentucky, where he studied law and became a surveyor, two dynamic professions on the emerging borderlands. With the outbreak of the American Revolution, Levi Todd entered the Continental Army and served under George Rogers Clark. He married his bride, Jane "Betsy" Briggs, in 1779. Their daughter Hannah, born in February 1781, was reputedly the first white child born in Kentucky. As a founder of Lexington, General Todd built a brick manse on the Boonesboro Road in December 1781, naming his estate Ellerslie, after his family's ancestral home in Scotland. Levi's seventh child and Mary Todd's father, Robert Smith Todd, was born at Ellerslie in 1791. Like his father before him, Robert grew up with substantial material comfort and enjoyed a superior education. After the death of his mother in

1800, when Robert was only nine, his father remarried. Mary's maternal line boasted a distinguished pedigree as well. Her great-grandfather, Andrew Porter, rose to the rank of general in the Continental Army and was a close friend of George Washington's. Porter's brothers scattered across the expanding nation: George B. Porter became governor of Michigan, David R. Porter served as governor of Pennsylvania, and James Madison Porter was appointed secretary of the navy. General Porter's daughter Elizabeth married Major Robert Parker (a distant cousin of Levi Todd's), and Parker became the first surveyor of Fayette County, Kentucky. Upon his death, he left his widow and daughter a small fortune. Thus when Mary's parents, Elizabeth Parker and Robert Smith Todd, were joined in marriage, they represented several branches of wealthy, patriotic, and politically connected clans grafted onto a formidable family tree.

16 *a family of middling wealth*: John Todd (Mary's granduncle), a Princeton graduate and distinguished Presbyterian theologian, ran the Virginia school where three of his nephews, John, Robert, and Levi (Mary's grandfather), matriculated. The donation of John Todd's library to the Transylvania Seminary contributed to the transformation of the institution into Transylvania College.

16 *"Eliza was a sprightly"*: Carl Sandburg, *Mary Lincoln: Wife and Widow* (New York: Harcourt Brace, 1932; repr., Bedford, MA: Applewood Books, 1995), p. 14.

16 *wed his teenage sweetheart*: Katherine Helm, *The True Story of Mary, Wife of Lincoln* (New York: Harper & Bros., 1923), p. 15.

17 *died in 1825*: "Yourself and family are invited to attend the funeral of Robert P. Todd infant son of Mr. R.S. Todd, from his residence on Short Street, this evening at 5 o'clock." Lexington, July 22, 1822. Clipping file at Filson Historical Society.

17 *taking a new wife*: As an engagement gift, Robert Todd presented Elizabeth Humphreys with a miniature portrait by Matthew Jouett.

18 *his "angel mother"*: Joshua F. Speed, *Reminiscences of Abraham Lincoln* (Louisville, KY: John P. Morton, 1884), p. 19.

19 *This friction stimulated*: When Betsy Humphreys was on a visit to New Orleans during their courtship, Robert warned her that "an occasion of this kind is seized hold of for the purpose of destraction [*sic*] and to gratify personal feelings of ill-will and indeed oftentimes how much mischief is done without any bad motive. May I be permitted to put you on your guard against any persons of this description." See also: Stephen Berry, *House of Abraham: Lincoln and the Todds, A Family Divided by War* (New York: Houghton Mifflin, 2007), pp. 10–11.

19 *forced to sell the family estate*: He had to shutter his dry-goods business during the panic of 1819. See Jennifer Fleischner, *Mrs. Lincoln and Mrs. Keckly: The Remarkable Story of the Friendship Between a First Lady and a Former Slave* (New York: Broadway Books, 2003), p. 20.

19 *A maternal uncle*: Kentucky Gazette, Feb. 12, 1802.

20 *She impressed her new granddaughter*: Helm, *Wife of Lincoln*, p. 34.

20 *Mrs. Parker deeply resented*: Ibid., p. 19.

21 *eight more Todds*: Robert Todd Smith (who died after a few days), Samuel, David, Alexander (all three sons who died in Confederate ranks), Margaret, Martha, Emilie, Elodie, and Katherine. Ibid., p. 17.

22 *appear "more human"*: David Herbert Donald, *Lincoln* (New York: Simon & Schuster, 1995), p. 28.

22 *Lincoln would refer*: Ibid., p. 28.

23 *"did not amount to one year"*: Ibid., p. 29.

23 *In stark contrast*: As Dr. William Augustus Evans has argued: "No better educated woman entered the White House in the first hundred years of the Presidency." Louis A. Warren, "The Woman in Lincoln's Life," reprinted from the *Filson Club Historical Quarterly*, Louisville, KY, July 1946. Filson Historical Society Collection, Louisville, KY.

24 *walk to school several city blocks*: Helm, *Wife of Lincoln*, p. 21.

24 *insulting him or laughing at him*: Ibid., p. 56.

24 *Mary's "retentive memory"*: William Augustus Evans, *Mrs. Abraham Lincoln: A Study of Her Personality and Her Influence on Lincoln* (New York: Knopf, 1932), pp. 102–3.

24 *finishing the task ahead of her cousin*: Letter of Mrs. Elizabeth Norris, in Helm, *Wife of Lincoln*, p. 21.

25 *polished her French*: Several Kentucky settlements, such as Versailles and Paris, demonstrated the affinities felt by frontier pioneers.

25 *a very renowned academy for young ladies*: C. Frank Dunn and William Townsend, "The Boarding School of Mary Todd Lincoln" (Privately published, Lexington,

1941). Filson Historical Society Collection, Louisville, KY.

25 *Lafayette paid a visit*: "A sumptuous dinner was served Henry Clay at Maysville, at which Lafayette was toasted." Evans, *Mrs. Abraham Lincoln*, p. 98.

25 *not far from Clay's estate*: Henry Clay's son would marry the Mentelles' daughter in 1836.

26 *The profits from his cotton manufacturing*: Interview with Gwen Thompson, curator at Mary Todd Lincoln Home, Lexington, KY, June 16, 2006.

26 *"My early home"*: Dunn and Townsend, "The Boarding School of Mary Todd Lincoln."

26 *remembered her as merry and smiling*: Sandburg, *Mary Lincoln*, p. 31.

26 *circle of girlhood friends*: William Townsend, *Lincoln and the Bluegrass: Slavery and Civil War in Kentucky* (Lexington: University of Kentucky Press, 1955), pp. 52–53.

27 *speaking French as fluently*: Helm, *Wife of Lincoln*, p. 53.

27 *"Mary's love for poetry"*: Ibid., p. 32.

27 *a flair for theatrics*: Sandburg, *Mary Lincoln*, p. 30.

27 *"She heard politics discussed"*: Helm, *Wife of Lincoln*, p. 40.

27 *One oft-repeated tale*: Ruth Painter Randall, *Mary Lincoln: Biography of a Marriage* (Boston: Little, Brown, 1953), p. 23.

27 *"If I am ever President"*: Helm, *Wife of Lincoln*, p. 3.

27 *"I wouldn't think"*: Ibid., pp. 42–43.

28 *"like an April day"*: Ibid., p. 32.

28 *"I always thought of tea roses"*: Ibid., pp. 19, 46.

TWO : MAKING HER OWN HOOPS TO JUMP THROUGH

29 *Mary sobbed over this episode*: Helm, *Wife of Lincoln*, pp. 26–27.

31 *Emilie's "uncommon beauty"*: Elizabeth L. Norris to Emily Todd Helm, September 28, 1895, in Emily Helm Collection, Kentucky Historical Center, Frankfort, KY.

31 *"I cannot think how"*: Ibid.

32 *This kind of cultural delusion*: See Catherine Clinton, *Tara Revisited: Women, War and the Plantation Legend* (New York: Abbeville Press, 1995).

32 *where slavedealers came to show their "stock"*: Ishbel Ross, *The President's Wife: Mary Lincoln, a Biography* (New York: Putnam's, 1973), p. 66.

32 *affixed to a black-locust log*: Sandburg, *Mary Lincoln*, pp. 26–27.

36 *When Mary was nineteen*: Jean Baker, *Mary Todd Lincoln: A Biography* (New York: Norton, 1987), p. 68.

36 *The Lexington of Mary Lincoln's youth*: Townsend, *Lincoln and the Bluegrass*, pp. 93–94. Townsend recounts the fate of Mrs. Turner, who was eventually murdered by one of her husband's slaves.

36 *In 1662, the Virginia assembly*: Joel Williamson, *New People: Miscegenation and Mulattoes in the United States* (New York: Free Press, 1980), p. 7.

37 *The color line was meant*: Paul Finkelman, "Crimes of Love, Misdemeanors of Passion: The Regulation of Race and Sex in the Colonial South," in *The Devil's Lane: Sex*

and Race in the Early South, ed. Catherine Clinton and Michele Gillespie (New York: Oxford University Press, 1997). See also: A. Leon Higginbotham Jr. and Barbara Kopytoff, "Racial Purity and Interracial Sex in the Law of Colonial and Antebellum Virginia," *Georgetown Law Journal* 77 (1989): 1989–2008.

37 *Elaborate systems of sexual and racial etiquette*: Peter Bardaglio, " 'Shamefull Matches:' Regulations of Interracial Sex and Marriage," in *Sex, Love, Race: Crossing Boundaries in North American History*, ed. Martha Hodes (New York: New York University Press, 1999).

38 *"a stranger would not suspect them"*: He also noted "such uncommon aptness in these two girls to take learning, and so much decent modest and unassuming conduct on their part, that my mind became much enlisted in their favor." Townsend, *Lincoln and the Bluegrass*, p. 76.

38 *"Evry thing that was"*: Ibid., p. 77.

38 *Lexington* Observer & Reporter: This was fortuitous, for when Johnson died, his brother went to court to declare that his brother had left no heirs. Leland Meyer, *The Life and Times of Richard M. Johnson of Kentucky* (New York: Columbia University Press, 1932), pp. 322–23.

39 *"a young Delilah"*: Ibid., pp. 341, 422.

39 *Johnson's liaisons*: When he was nominated at a Democratic convention in Baltimore, the Virginia delegation walked out. See Catherine Clinton, *The Plantation Mistress: Woman's World in the Old South* (New York: Pantheon, 1982), pp. 216–17.

39 *The author of the Declaration*: Townsend, *Lincoln in the Bluegrass*, p. 78.

40 *"If Col. Johnson"*: Ibid.

40 *Both these liaisons*: See Catherine Clinton, "Breaking the Silence: Sexual Hypocrisy from Thomas Jefferson to Strom Thurmond" (forthcoming).

41 *Mary Humphreys freed*: "Mary Brown Humphreys, Todd's second mother-in-law, was from a vigorously antislavery family whose members had sought to outlaw the institution in Kentucky's first constitution." Baker, *Mary Todd Lincoln*, p. 67.

41 *Humphreys had not only emancipated*: Helm, *Wife of Lincoln*, p. 51.

41 *Grandmother Parker*: Baker, *Mary Todd Lincoln*, p. 67.

41 *Illinois trumpeted that it was the first state*: Vermont and New Hampshire may beg to differ, but this claim is found in many Illinois history texts.

42 *Mary's hometown of Lexington*: "The terrible ravages of the cholera in 1833 will ever keep that fatal year memorable in the annals of Lexington." George Ranck, *History of Lexington Kentucky* (Cincinnati: Robert Clarke & Co., 1872), p. 325.

42 *Within ten days*: Ibid.

42 *One observer noted*: Ibid.

42 *"Father had all the trunks"*: Helm, *Wife of Lincoln*, pp. 49–52.

42 *One account described*: Ranck, *History of Lexington*, p. 325.

42 *"the whole city was in mourning"*: Ibid., p. 326.

43 *She was confirmed*: Rufus Rockwell Wilson, *Intimate Memories of Lincoln* (Elmira, NY: Primavera Press, 1945), p. 80.

43 *migrations to Crab Orchard Springs*: Helm, *Wife of Lincoln*, p. 54.

43 *One of her nieces claimed*: Ibid., p. 22.

43 *"The water has the pleasant flavor"*: Clinton, *Plantation Mistress*, p. 149.

43 *Visits to the springs*: The splendor and luxury of spas may have varied, but most innkeepers at fashionable southern resorts lured visitors with the promise of luxurious accommodations. As planter St. George Coalter confided to his wife, not only were 270 people in residence at White Sulfur Springs in Virginia, but also 100 slave attendants. Most facilities maintained a variety of cabins surrounding a complex of central buildings that included a dining hall, a ballroom, and a drawing room. The owners might sponsor boxing matches and gambling tables, and some even held fencing tournaments. Wealthy Southerners were building their own cabins at the springs, in rows that earned nicknames like Bachelors' Row and Paradise Row. One planter described a visit in July 1839 wherein his cabin was handsomely fitted up with "carpets, painting, sofas, lounges, Borking chairs and every comfort that man can devise." Clinton, *Plantation Mistress*, p. 149. See also: Percival Reniers, *The Springs of Virginia: Life, Love and Death at the Waters, 1775–1900* (Chapel Hill: University of North Carolina Press, 1941).

44 *Thomas Lincoln might "slash"*: Donald, *Lincoln*, p. 32.

44 *Also, when Abraham was weighed*: Ibid., p. 33.

46 *four oxen to "broke the prairie"*: Ibid., p. 36.

46 *with "argument so pithy"*: Ibid., p. 41.

47 *Even though Lincoln would lose*: Ibid., p. 46.

48 *Sometime during the spring of 1835*: A recently discovered entry in Sangamon County Record Book H reveals that sisters Frances Todd and Mary Todd witnessed a sales transaction between their brother-in-law, Ninian Wirt Edwards, and Samuel Wiggins on May 16, 1835. The entry clearly places both Todd sisters in Springfield much earlier than previous biographers had thought. The Sangamon County Record Books are at the Illinois Regional Archives Depository (IRAD) at the University of Illinois at Springfield. The specific entry is on pages 310–11 of Book H. See Thomas F. Schwartz, "Mary Todd's 1835 Visit to Springfield, Illinois," in the *Journal of the Abraham Lincoln Association*, vol. 26, no. 1, Winter 2005.

48 *Frances and Mary*: Frances may have enjoyed the town more than her younger sister, as she was still in Springfield on June 15, when she witnessed another document for Edwards. Ibid.

49 *Mary Mentelle, the daughter of her favorite teacher* Evans, *Mrs. Abraham Lincoln*, p. 101.

50 *"Virgin's milk"*: Daniel Sutherland, *The Expansion of Everyday Life, 1860–1876* (New York: Perennial Library, 1990), p. 59.

50 *"cut deeper than she intended"*: Helm, *Wife of Lincoln*, p. 55.

52 *six foot ten*: Donald, *Lincoln*, p. 60.

53 *a public meeting to defend Clay*: Townsend, *Lincoln and the Bluegrass*, p. 73.

54 *The median age of marriage*: Clinton, *Plantation Mistress*, p. 60.

54 *"to abandon his house"*: William Townsend, *Lincoln and His Wife's Home Town* (Indianapolis: Bobbs-Merrill, 1929), p. 67. Mary made a rueful comment in a letter in 1848 alluding to this situation as well.

THREE : ATHENS OF THE WEST

55 *"She could make a bishop"*: Helm, *Wife of Lincoln*, p. 81.

56 *"eyebrows cropped out"*: Donald, *Lincoln*, p. 115.

56 *"I love to dig up"*: Ibid., p. 99.

56 *When Lincoln arrived to serve*: Ibid., p. 53.

56 *Lincoln grew on the job*: For an interesting discussion of Lincoln's role in this move and the "logrolling" of party politics at the time, see Douglas L. Wilson, *Honor's Voice: The Transformation of Abraham Lincoln* (New York: Knopf, 1998), pp. 159–64.

57 *he treated the legislators*: Donald, *Lincoln*, p. 64. Perhaps this celebration began Lincoln's love affair with oysters. The bill totaled more than $200, which was much more money than Lincoln ever saw in a year during his early days.

57 *"Springfield is a small village"*: B. F. Harris, quoted in Evans, *Mrs. Abraham Lincoln*, p. 122. William Cullen Bryant had visited the place and described it as "having an appearance of dirt and discomfort." Albert Beveridge,

Abraham Lincoln, 1809–1858 (New York: Houghton Mifflin, 1928), vol. 1, p. 207.

59 *Yet the town opened up*: When the cornerstone of the new statehouse was laid on July 4, 1837, and Edward D. Baker gave his oration, it must have been one of the most memorable Independence Days of Lincoln's life.

59 *who offered Lincoln a roof over his head*: And infamously a bed they would share for four years.

60 *"Our town will soon rival"*: In 1835: Fleischner, *Mrs. Lincoln and Mrs. Keckly*, p. 91.

60 *any woman who arrived "would do well"*: Helm, *Wife of Lincoln*, p. 69.

60 *she "never at any time showed"*: Ibid., p. 55.

62 *"shut-mouthed man"*: Paul M. Angle, ed., *Herndon's Life of Lincoln* (New York: A. & C. Boni, 1930), p. xxxiv.

63 *shortly after their engagement*: Wilson, *Honor's Voice*, pp. 114–17.

64 *living on a farm outside the village*: On land that her fiancé actually owned but allowed the Rutledges to occupy. C. A. Tripp, *The Intimate World of Abraham Lincoln* (New York: Free Press, 2005), p. 70.

64 *"seemed to be so much affected"*: Wilson, *Honor's Voice*, p. 116.

65 *controversy for more than a century*: For a good summary of the scholarship and his current assessment, please consult David Herbert Donald, *"We Are Lincoln Men"*: *Abraham Lincoln and His Friends* (New York: Simon & Schuster, 2003). For dissenting views, see also: C. A. Tripp, *The Intimate World of Abraham Lincoln*,

and Joshua Shenk, *Lincoln's Melancholy: How Depression Challenged a President and Fueled His Greatness* (New York: Houghton Mifflin, 2005).

65 *outbreaks of malaria followed*: Tripp, *Intimate World*, p. 74.

65 *many feared for his sanity if not his safety*: Scholars have debated both the depth and causes of this particular period of depression. There seems to be little dispute that Lincoln did fall into a depression in the late summer of 1835, but its meaning continues to be the source of conflict among Lincoln interpreters.

66 *"I do know he was staying"*: Wilson, *Honor's Voice*, p. 120.

66 *his ongoing battle with melancholy*: On the extent of Lincoln's mental illness at this time, please consult the work of Joshua Wolf Shenk, who has made some very compelling arguments about Lincoln's state of mind in *Lincoln's Melancholy*. At the same time, I am not convinced that Lincoln's illness was linked as closely to the weather as Shenk seems to suggest; clearly weather plays a large role in many depressives' mood disorders, which often get worse with a lack of sun or are triggered by an overabundance of rain. However, to blame the weather disproportionately for such depressions is an overstatement, and, like many such interpretations, undermines rather than strengthens Shenk's central claims.

67 *"to older, married and hence unavailable, women"*: Donald, *"We Are Lincoln Men,"* p. 24.

67 *"a cultivated woman"*: "Register of Herndon's Informants," in Douglas Wilson and Rodney O. David, eds.,

Herndon's Informants: Letters, Interviews, and Statements about Lincoln (Urbana: University of Illinois Press, 1998), p. 738.

68 *When she failed to match him*: Donald, *"We Are Lincoln Men,"* p. 23.

68 *"The longer I can avoid"*: AL (Abraham Lincoln) to Mary S. Owens, May 7, 1837, *The Collected Works of Abraham Lincoln* [hereafter abbreviated *CWL*], ed. Roy P. Basler (New Brunswick, NJ: Rutgers University Press, 1953), vol. 1, p. 55.

68 *"There is a great deal of flourishing"*: AL to Mary S. Owens, May 7, 1837, ibid., vol. 1, p. 79.

69 *"our further acquaintance"*: Ibid., vol. 1, p. 95.

69 *"saw no good objection"*: AL to Mrs. Orville H. Browning, Apr. 1, 1838, ibid., vol. 1, p. 117.

70 *"deficient in those little links"*: Mary Owens qualified this criticism by adding, "at least it was so in my case, not that I believed it proceeded from a lack of goodness of heart, but his training had been different from mine, hence there was not that congeniality which would have otherwise existed." See Wilson, *Honor's Voice*, p. 138.

70 *"I was really a little in love"*: Ibid., p. 138.

71 *"I can never be satisfied"*: *CWL*, vol. 1, pp. 118–19.

71 *dance with her*: Helm, *Wife of Lincoln*, p. 74.

72 *Frances when she first arrived*: "He took me out once or twice, but he was not much for society." See Douglas Wilson, *Honor's Voice*, p. 180.

72 *the Coterie*: Fleischner, *Mrs. Lincoln and Mrs. Keckly*, p. 93.

73 *a poem handed down through the years*: "The Story of the Early Days in Springfield—and a Poem," *Illinois State Historical Society*, vol. 16, nos. 1–2, April–July 1923, p. 1245.

73 *was she noticed*: Herndon was definitely *not* part of the smart set that Mary Todd favored. Rather he was the son of a Springfield hotel keeper, Archer Herndon, who owned the Indian Queen. "Billy" Herndon, as he was known, had gone to local school and then attended college in Jacksonville, Illinois, for a year before dropping out. He returned to clerk in Joshua Speed's store, and became a part of the Whig gang who distinguished themselves from the more aristocratic, Eastern-born branch of the party. Herndon earned Mary Todd's enmity early in their acquaintance, allegedly for a compliment that misfired. When she first came to town, Herndon danced with Mary at a party and then told her that "she seemed to glide through the waltz with the ease of a serpent." Mary fired back, "Mr. Herndon, comparison to a serpent is rather severe irony, especially to a newcomer." Fleischner, *Mrs. Lincoln and Mrs. Keckly*, p. 106.

73 *"dashing, handsome—witty"*: Donald, *Lincoln*, p. 84.

74 *"and gaze on her as if drawn"*: Donald, *"We Are Lincoln Men,"* p. 41.

75 *"liked nothing better"*: Helm, *Wife of Lincoln*, p. 32.

75 *In late February*: Fleischner, *Mrs. Lincoln and Mrs. Keckly*, p. 107.

75 *Judge Todd had left Kentucky*: Helm, *Wife of Lincoln*, p. 78.

76 *He was sending off legislative proposals*: Earl Schenck Miers, ed., *Lincoln Day by Day: A Chronology, 1809–1865* (Washington, DC: Lincoln Sesquicentennial Commission, 1960), pp. 126–28.

76 *"When I mention"*: Justin G. Turner and Linda Levitt Turner, eds., *Mary Todd Lincoln: Her Life and Letters* (New York: Knopf, 1972), p. 16.

77 *"two sweet little objections"*: Ibid., p. 26.

77 *Ann was expected to find a beau*: Deposition, April 1852, Robert Todd Papers, Special Collections, University of Kentucky.

77 *allow her to go to the theater*: Helm, *Wife of Lincoln*, p. 119.

77 *"My hand will never be given"*: Turner and Turner, *Mary Todd Lincoln*, p. 18.

77 *obsessed over the topic*: David Donald has suggested, "They were in love with the idea of being in love." See Donald, *"We Are All Lincoln Men,"* p. 39.

77 *"Mr. Speed's ever changing heart"*: Turner and Turner, *Mary Todd Lincoln*, p. 20.

78 *Speed confided to his sister*: Donald, *"We Are All Lincoln Men,"* p. 41. This is not to be mistaken for Lincoln's malady: to want what he couldn't have, as opposed to Speed's desire for the hunt.

78 *an unfathomable aspect of Mary's experience*: One of Lincoln's earliest biographers and a friend who spent years collecting interviews, William Herndon, drew some outlandish conclusions that have plagued scholarship, causing wrangles and debates ever since. Whatever his claims, it is difficult to untangle the venom from some of his contradictory accounts, especially concerning

Lincoln's courtship and marriage. More recently, scholars who have suggested Lincoln was gay and authors who have depicted Lincoln as suffering from clinical depression have presented contrary, clashing views of this critical period as well.

FOUR : "CRIMES OF MATRIMONY"

81 *Lincoln failed to appear*: Helm, *Wife of Lincoln*, p. 89. Again in this area of the debated issues, Douglas Wilson has posited that there was some evidence to suggest independently that Lincoln had "stood up" Mary at a social event, which led to conflict. See Wilson, *Honor's Voice*, p. 252.

82 *"the fatal first"*: Both Joshua Shenk (*Lincoln's Melancholy*) and Douglas Wilson (*Honor's Voice*) have discussed alternative scenarios for "the fatal first" in their fine studies—and provide interesting insight—but neither has convinced me that the crisis Lincoln suffered in January 1841 did not involve his feelings about his relationship with Mary Todd.

82 *If Mary was the belle of Springfield*: Helm credits the notion of Mary Todd's involvement with Stephen Douglas in *Wife of Lincoln*, pp. 87–88.

83 *"to polish a stone"*: Helm, *Wife of Lincoln*, p. 81.

83 *"my mother did what she could"*: Randall, *Mary Lincoln*, p. 40. Elizabeth and Ninian failed to volunteer their disdain for the match in their later recollections to Herndon. Whether this was suppressed memory or not, it was certainly shrewd to mask it during conversations on the record, which they did.

83 *"Lincoln's people had not"*: Helm, *Wife of Lincoln*, p. 86.

84 *"Speed's* grey suit *has gone"*: December 15, 1840, Mary Todd to Mercy Ann Levering, Springfield, Illinois, Turner and Turner, *Mary Todd Lincoln*, pp. 20–21.

84 *"Harriet Campbell appears"*: Ibid. Mary Todd and James Conkling, Mercy's flame, both attended Harriet Campbell's wedding.

85 *derailing plans to wed Mary*: Wilson, *Honor's Voice*, p. 235.

85 *stories about the crisis*: For a very passionate advocate of Matilda Edwards's prominent role in the breakup, see Douglas Wilson, *Honor's Voice*, pp. 221–27. It seems surprising that none of the biographers attempting to unravel these tangled emotions have suggested that Matilda became the flame to which Lincoln was drawn principally because she was the object of Speed's obsession—which is a quite common syndrome among homosocial male intimates.

86 *subject of gossip*: "The word spread that Mary Todd had jilted Lincoln." Sandburg, *Mary Lincoln*, p. 44.

86 *"I am now the most miserable man"*: Ibid., p. 46.

87 *What was the root cause*: For recent analysis, see Shenk, *Lincoln's Melancholy*, pp. 155–65.

87 *considerably "unmanning"*: See Joshua Shenk's very perceptive analysis of manhood and melancholy in chapter 3 of *Lincoln's Melancholy*. I am indebted to Shenk for his deft exploration of Lincoln's relationship with Speed, and his creative and nuanced observations. Although I do not agree with all of his interpretations,

especially several of his insights concerning Lincoln's relationship with Mary Todd, his work has been instrumental to improving the tone of many debates within Lincoln studies. However, he has taken a fairly radical stand on Joshua Speed, suggesting that he was Lincoln's most intimate contact. Others have suggested that Speed and Lincoln may have had an erotic component to their relationship, but on this point, I would argue that recent scholarship has debunked this theory; see for example, Doris Kearns Goodwin, *Team of Rivals: The Political Genius of Abraham Lincoln* (New York: Simon & Schuster, 2005), pp. 58–59.

88 *"Poor Lincoln"*: Sandburg, *Mary Lincoln*, pp. 45–46.

88 *Lincoln's sad fate*: Conkling recalled that Lincoln at this time "used to remind me sometimes of the pictures I formerly saw of old Father Jupiter, bending down from the clouds, to see what was going on below." He called him a "poor hapless swain who loved most true but was not loved again—I suppose he will now endeavor to drown his cares among the intricacies and perplexities of the law." Sandburg, *Mary Lincoln*, p. 47.

88 *"reduced and emaciated"*: Shenk, *Lincoln's Melancholy*, p. 57.

88 *"He came within an inch"*: Bell also credited the story that Lincoln was lovesick over Matilda Edwards, not Mary. This story was promoted by Orville H. Browning, who was boarding at the same place with Lincoln and who later suggested Lincoln's spell developed because "he was engaged to Miss Todd, and in love with Miss Edwards, and his conscience troubled him dreadfully." See Wilson, *Honor's Voice*, pp. 235, 237.

89 *"I fear I shall be unable"*: Sandburg, *Mary Lincoln*, p. 46.

89 *"neither dead, nor quite crazy yet"*: Miers, *Lincoln Day by Day*, February 3, 1841.

89 *moody and silent*: Shenk, *Lincoln's Melancholy*, p. 62.

89 *"There is more pleasure in pursuit"*: Wilson, *Honor's Voice*, p. 246.

90 *At some point*: Shenk, *Lincoln's Melancholy*, p. 56. The story about Rickard seems much less well substantiated than the one about Ann Rutledge, but it seems to have been transformed into reality by repetition.

90 *Sarah's role*: The most convincing argument concerning Sarah Rickard would seem to be one explored by Gary Lee Williams in his doctoral thesis, "James and Joshua Speed: Lincoln's Kentucky Friends" (PhD diss., Duke University, 1971), in which he suggests that Lincoln's mentions of Sarah in letters to Speed are to reassure Speed that Sarah is fine—perhaps after Speed's romantic involvement, of an unknown quality or extent—with this young girl. Williams's reading of the letters seems plausible and, indeed, much in character with Speed's personality. It seems a good deal more satisfying than suggesting that Lincoln made three proposals of marriage within roughly a six-month period to three different women. See also Richard N. Current, *The Lincoln Nobody Knows* (New York: Hill and Wang, 1963).

90 *"Mary had tears in her eyes"*: Helm, *Wife of Lincoln*, p. 90.

91 *kept him tied to her by invoking honor*: Fleischner, *Mrs. Lincoln and Mrs. Keckly*, p. 111.

92 *"Hart, the little drayman"*: *CWL*, vol. 1, p. 255.

92 *"I'm glad to hear Lincoln"*: Shenk, *Lincoln's Melancholy*, p. 62.

93 *"made a deep impression"*: Wilson and David, *Herndon's Informants*, p. 342.

93 *"regularly when I return home"*: *CWL*, vol. 1, p. 261.

93 *"there is but one thing about her"*: Ibid.

94 *"Something of the same feeling"*: Wilson and David, *Herndon's Informants*, p. 430.

94 *"You will feel very badly"*: *CWL*, vol. 1, p. 265.

94 *"in two or three months"*: Ibid., p. 266.

95 *"I have hardly yet"*: Shenk, *Lincoln's Melancholy*, p. 94.

95 *"I feel somewhat jealous of you both"*: *CWL*, vol. 1, p. 281.

95 *"There is one still unhappy"*: Wilson, *Honor's Voice*, p. 257.

95 *"I believe now"*: *CWL*, vol. 1, p. 289.

96 *"articles of the most personal nature"*: Wilson, *Honor's Voice*, p. 275.

97 *allowed the duel to be called off*: By one account, "Shields's note to Lincoln was withdrawn and Lincoln's written apology was accepted, and the duel called off." Beveridge, *Abraham Lincoln*, vol. 1, p. 352. In another account, Lincoln had originally offered to fight with "cowpies," and the return from Sunflower Island, also known as Bloody Island, was both comic and fearful when a log covered with a red shirt was mistaken for a corpse. See Ralph Gray, *Following in Lincoln's Footsteps: A Complete Annotated Reference to Hundreds of Historical Sites Visited by*

Abraham Lincoln (Illinois) (New York: Carroll & Graf, 2001), p. 9.

97 *protecting Mary's honor*: Lincoln reportedly was ashamed of his role in this entire episode, and while he was in the White House, when someone questioned him about "a duel & all for the sake of a lady by your side," Lincoln replied, "I do not deny it, but if you desire my friendship, you will never mention it again." Randall, *Mary Lincoln*, p. 60.

97 *"You have heard of my duel"*: CWL, vol. 1, pp. 302–3.

98 *"that was the beginning"*: Wilson, *Honor's Voice*, pp. 284–85. Another tale involving the Hardins suggests that Abraham and Mary first encountered one another again when Lincoln appeared at their house when Mary Todd was at their home alone. Wilson, *Honor's Voice*, pp. 284–86.

98 *"Are you now in feeling"*: CWL, vol. 1, p. 303.

99 *The couple rekindled*: Novelists and some others have suggested that the Lincolns may have engaged in premarital sex at this point, although there is no evidence from which to make this claim.

99 *"all eyes and ears"*: Carl Sandburg, *Abraham Lincoln*, vol. 1, *The Prairie Years* (New York: Harcourt Brace, 1926), p. 280.

99 his *crisis* and her *fidelity*: And over time, Mary hoped that even his crisis might be forgotten. She wrote to Josiah Holland, author of a biography of her husband in 1865, "My beloved husband had so entirely devoted himself to me, for two years before my marriage"— clearly contrary to the facts. See Mary Lincoln to Josiah

Holland, December 4, 1865, Turner and Turner, *Mary Todd Lincoln*, p. 293.

99 *Mary had jilted Lincoln*: Sandburg, *Mary Lincoln*, p. 44. Whether it was Dr. Anson Henry who prodded this letter from Mary or whether it was written at the urging of Ninian and Elizabeth Edwards remains debatable, and the letter has not surfaced. See Randall, *Mary Lincoln*, pp. 45–46.

99 *"a great many beaux"*: Wilson, *Honor's Voice*, p. 237.

100 *"deems me unworthy of notice"*: Miers, *Lincoln Day by Day*, June 17, 1841. She also alludes to Lincoln in this letter as akin to Shakespeare's Richard III.

102 *"to hell"*: Douglas Wilson rather imaginatively suggests that this was an allusion to Byron's *Don Juan*, one of Lincoln's favorite poems. Wilson, *Honor's Voice*, pp. 291–92.

103 *a noose around Abraham's neck*: In recent years, both Michael Burlingame and Douglas Wilson have been credited with the suggestion that Mary Todd and Abraham Lincoln had sex sometime in the fall of 1842 and that this act forced Lincoln to marry against his will. Within this framework, of course, Mary is painted as the temptress who lures Lincoln into her web and then holds his honor hostage to blackmail him to enter a loveless marriage. This scenario, Burlingame says, is the only one that "fits the facts." The idea that Mary lured her former beau into having sex with her and then demanded that he marry her at once following this indiscretion is not impossible, but it reflects character assassination more than reasonable speculation. The fact that

Robert Lincoln was born in August to a November bride is hardly evidence of premarital sex. Why would it not be that after nearly eighteen months of Lincoln's indecisiveness and her family's campaign to separate them, once the couple reconciled and decided to be together, they would wed immediately to prevent further delays? Is this not just as likely as to cover up a possible pregnancy? Mary might have been the one demanding haste, but sexual blackmail is a shaky explanation. This "smoking shotgun" of an interpretation has been popularized in a recent novel: Barbara Hambly, *The Emancipator's Wife* (New York: Bantam Books, 2005).

103 *"cheerful as he had ever been"*: Tarbell Papers, Allegheny College Library, Meadville, PA.

103 *Elizabeth Edwards had only hours*: Wilson and David, *Herndon's Informants*, p. 444.

103 *"Ginger bread"*: Wilson, *Honor's Voice*, p. 291.

103 *"it was a Friday."* Helm, *Wife of Lincoln*, p. 76.

FIVE : "PROFOUND WONDERS"

106 *"Nothing new here"*: AL to Samuel D. Marshall, November 11, 1842, *CWL*, vol. 1, p. 305.

106 *"They don't live"*: Sutherland, *Expansion of Everyday Life*, p. 49.

109 *"talking the wildest"*: Randall, *Mary Lincoln*, p. 67.

109 *"sombre and gloomy"*: Ibid.

109 *could not make the visit to Kentucky*: Evans, *Mrs. Abraham Lincoln*, p. 128.

111 *"Mary is very well"*: Randall, *Mary Lincoln*, p. 69.

111 *"About the prospects:"* Ibid. And William Butler apparently assured Speed in May 1843 that the Lincolns were expecting a child, to which Lincoln responded: "I have so much confidence in the judgment of Butler on such a subject that I incline to think there may be some reality in it. What *day* does Butler appoint?"

113 *"that the happiness of the conjugal":* Clinton, *Plantation Mistress,* pp. 151–52.

113 *"We are but two":* Randall, *Mary Lincoln,* p. 71.

114 *"washed and dressed the baby":* Wilson, *Intimate Memories of Lincoln,* p. 61.

115 *"I was very fond":* Ibid.

115 *"One misses a record":* Randall, *Mary Lincoln,* p. 73.

115 *he presented Mary with $25 in gold:* Fleischner, *Mrs. Lincoln and Mrs. Keckly,* p. 151.

116 *When he and Mary were at the Globe:* Eugenia Jones Hunt, *My Personal Recollections of Abraham and Mary Lincoln* (privately published, 1966), p. 9.

116 *They paid $1,200:* Randall, *Mary Lincoln,* p. 75.

116 *The cottage boasted:* Edwin C. Bearss, *Historic Structure Report: Lincoln Home National Historic Site* (Denver: Denver Service Center, Department of Interior, 1973), p. 11.

117 *Most typical wood-framed homes:* Sutherland, *Expansion of Everyday Life,* p. 32.

118 *He wrote to his new wife:* Fleischner, *Mrs. Lincoln and Mrs. Keckly,* p. 152.

118 *She set to work:* Miers, *Lincoln Day by Day,* May 3, 1844.

118 *"My mother does not recall"*: Philip Ayers, "Lincoln as a Neighbor," *American Review of Reviews*, Feb. 1918, p. 184.

118 *The ledgers of shopkeepers*: For a useful analysis of the Lincolns' patterns of consumption, see Harry E. Pratt, "The Lincolns Go Shopping," *Journal of the Illinois State Historical Society*, vol. 48.

119 *he was not good-looking*: Wilson and David, *Herndon's Informants*, p. 646.

120 *"the mistress and the maid"*: Sutherland, *Expansion of Everyday Life*, p. 55.

121 *she was plagued by gossip*: See, for example, the testimony of James Matheny, in Wilson and David, *Herndon's Informants*, p. 713.

121 *"impossible to get servants"*: Randall, *Mary Lincoln*, p. 81.

122 *a woman's workshop*: Sutherland, *Expansion of Everyday Life*, p. 62.

123 *Mary Lincoln's cake:* See F. L. Gillette, *White House Cook Book* (Chicago: R. S. Peale, 1887).

123 *the metal eggbeater*: Susan Strasser, *Never Done: A History of American Housework* (New York: Pantheon, 1982), p. 33.

123 *as a matron's pantry*: Sutherland, *Expansion of Everyday Life*, p. 64.

124 *In Pennsylvania in the 1840s*: Molly Harrison, *The Kitchen in History* (New York: Scribner's, 1972), p. 114.

124 *One woman complained*: Sutherland, *Expansion of Everyday Life*, p. 66.

124 *welcomed the cast-iron stove*: Strasser, *Never Done*, p. 33.

124 *requiring at least an hour a day*: Ibid., p. 41.

124 *Once, when the fire went out*: Randall, *Mary Lincoln*, p. 77.

125 *the scheme backfired*: On the subject of Harriet Hanks's time with the Lincolns, as with almost every aspect of Lincoln lore there is disagreement. Two biographers of Mary Lincoln have opposing interpretations, bolstered by selective quotations. Ruth Painter Randall suggests that Harriet was more sympathetic toward Mary than Jennifer Fleischner indicates. Randall cites Harriet as confiding, "She had an ungovernable temper, but after the outburst she was invariably regretful and penitent." Randall, *Mary Lincoln*, p. 82. Fleischner pointedly quotes Harriet as saying that she "would rather Say *nothing* about [Mary], as I could Say but little in her *favor* I conclude it best to Say nothing." Fleischner, *Mrs. Lincoln and Mrs. Keckly*, p. 153.

126 *more "like a servant"*: Hanks's reliability as a witness has been challenged by Douglas Wilson. For example, she told Jesse Weik in an interview that Mary Lincoln was engaged to Stephen Douglas, but became ill, and Douglas did not want to release her, but Mary's brother-in-law, William Wallace, induced Douglas to give Mary up. Wilson, *Honor's Voice*, pp. 239–40.

126 *Ruth Stanton*: Wayne Temple, "Ruth Stanton Recalls the Lincolns," *Lincoln Herald*, vol. 92, no. 3 (Fall 1990), p. 91.

127 *Jumping to such conclusions*: See Catherine Clinton, "Wife Versus Widow: Clashing Perspectives on Mary Lincoln's Legacy," *Journal of the Abraham Lincoln Association*, vol. 28, no. 1 (Winter 2007).

128 *"She had the strange"*: Randall, *Mary Lincoln*, p. 79.

129 *Clearly, Mr. Lincoln was tired*: Donald, *Lincoln*, pp. 100–102.

130 *his father packing his saddlebags*: Ibid., p. 109.

130 *"best stump speaker in the state"*: Ibid., pp. 112–13.

131 *Mary kicked up a terrible fuss*: Ruth Painter Randall, *Lincoln's Sons* (Boston: Little, Brown, 1955), p. 15.

132 *"found him, and had whipped him"*: Sandburg, *Prairie Years*, p. 318.

132 *"It is my pleasure that my children"*: Randall, *Mary Lincoln*, p. 101.

132 *"a little rare-ripe"*: Randall, *Lincoln's Sons*, p. 15.

SIX : PLAYING FOR KEEPS

134 *Mary's accompanying her husband was*: Indeed, one traveling companion on the journey, John Bradford, recalled that the Lincolns didn't even have a nurse. Mr. Lincoln would take care of Little Eddie while Mary and Robert ate at the dining table, and Bradford would try to help the couple with their arrangements. See Hunt, *Personal Recollections*, p. 12.

135 *A Humphreys relative*: Helm, *Wife of Lincoln*, pp. 101–2.

137 *They would renew*: Webster invited Lincoln to his Saturday breakfasts. New England cronies were charmed by this Western politician's brand of banter. Richard Nelson Current, *Speaking of Abraham Lincoln* (Urbana: University of Illinois Press, 1983), pp. 3–4.

138 *Douglas quoted Frederick the Great*: Sandburg, *Prairie Years*, p. 346.

139 *"The negroes have no idea"*: William Dusinberre, *Slavemaster President: The Double Career of James Polk* (New York: Oxford University Press, 2007), p. 15.

139 *"the spot of soil"*: Donald, *Lincoln*, pp. 123–25.

139 *"Benedict Arnold of our district"*: Ibid., p. 125.

139 *Herndon reported*: Lincoln's position especially created a dilemma for local supporters back in Illinois begging for military appointments from the government. Sandburg, *Prairie Years*, pp. 380–81.

140 *Lincoln would lay down his knife and fork*: Charles O. Paulin, "Abraham Lincoln in Congress, 1847–49," *Journal of the Illinois State Historical Society*, vol. 14, no. 1, p. 86.

141 *Keeping her sons out of trouble*: AL to MTL (Mary Todd Lincoln), April 16, 1848, *CWL*, vol. 1, p. 465.

142 *when two slavecatchers arrived*: Sandburg, *Prairie Years*, p. 381.

143 *"All the house"*: Fleischner, *Mrs. Lincoln and Mrs. Keckly*, p. 165.

143 *"When you were here"*: AL to MTL, April 16, 1848, *CWL*, vol. 1, p. 466.

144 *the couple cared for one another*: Although some scholars have suggested the Lincolns had a loveless marriage, this was not the case, as I have argued elsewhere. See Clinton, "Wife Versus Widow."

144 *"You are entirely free from headache?"*: AL to MTL, April 16, 1848, *CWL*, vol. 1, p. 466.

145 *"able arguments"*: Miers, *Lincoln Day by Day*, September 12, 1848.

145 *Lincoln was impressed*: Ibid., September 23, 1848.

146 *"It will now mortify me deeply"*: Current, *Speaking of Abraham Lincoln*, p. 4.

146 *besieged by requests*: His own father-in-law, Robert Todd, asked Lincoln's help to secure a position for his brother David Todd's son-in-law, Thomas Campbell, who needed a clerk's position. See Townsend, *Lincoln and the Bluegrass*, p. 166.

147 *George had moved*: Fleischner, *Mrs. Lincoln and Mrs. Keckly*, p. 64.

148 *When his father's will*: See Berry, *House of Abraham*, pp. 39–42.

149 *His death certificate*: Donald, *Lincoln*, p. 153.

149 *We miss him*: Townsend, *Lincoln and the Bluegrass*, p. 193.

149 *Dr. James Smith*: Sandburg, *Prairie Years*, p. 413.

150 *"Bright is the home"*: Fleischner, *Mrs. Lincoln and Mrs. Keckly*, p. 169.

150 *Mary officially joined*: Minutes of the Session of the First Presbyterian Church, 1828–1862, p. 82, Illinois State Historical Library, Springfield.

151 *Mary passed along his clothes*: See *Looking for Lincoln*, Downtown Springfield Exhibit Texts, Illinois Historic Preservation Agency, Springfield.

151 *a live-in Irish girl*: Bearss, *Historic Structure Report*, p. 21.

152 *one in ten children died*: Thomas J. Schlereth, *Victorian America: Transformations in Everyday Life, 1876–1915* (New York: HarperCollins, 1991), p. 274.

153 *she did "not feel sufficiently"*: Donald, *Lincoln*, p. 154.

154 *Abraham Lincoln was viewed as "tender"*: See Michael Burlingame, *The Inner World of Abraham Lincoln*

(Urbana: University of Illinois Press, 1994). See especially chapter 3: "'Unrestrained by Parental Tyranny:' Lincoln and His Sons."

154 *"Bob Lincoln was an awful tease"*: Burlingame, *Inner World*, p. 61.

156 *Lincoln's feelings about his father's death*: See, for example, Charles B. Strozier, *Lincoln's Quest for Union* (New York: Basic Books, 1982), pp. 53–55.

156 *family disharmony during these years*: For a laundry list of these claims, see Michael Burlingame, "The Lincolns' Marriage: 'A Fountain of Misery of a Quality Absolutely Infernal,'" chapter 9 in *The Inner World of Abraham Lincoln*. Certainly Herndon and Weik solicited and recorded several vivid examples of Mrs. Lincoln's alleged bad behavior. See, for example, the interview with Margaret Ryan (no. 484, p. 596) and the one with Milton Hay (no. 626, p. 729) in Wilson and Davis, *Herndon's Informants*.

156 *"I heard him say"*: Helm, *Wife of Lincoln*, p. 108.

157 *remembered Mrs. Lincoln serving*: See Philip Ayers, "Lincoln as a Neighbor," *American Review of Reviews*, Feb. 1918, p. 184.

158 *"Lincoln & his wife"*: Wilson and Davis, *Herndon's Informants*, no. 343, pp. 451–53.

158 *Lincoln indulged his wife*: Elizabeth Keckley, *Behind the Scenes; or, Thirty Years a Slave, and Four Years in the White House* (New York: G. W. Carleton, 1868; repr. New York: Oxford University Press, 1988), pp. 124–26.

158 *"Lincoln yielded to his wife"*: See Jason Emerson, *The Madness of Mary Lincoln* (Carbondale: Southern Illinois

University Press, 2007), p. 10. Weik reported this in his book, based on Herndon's interview with Gourley, but Wilson and Davis do not include this in *Herndon's Informants* (see interview with Gourley, pp. 451–53). Moreover, competing interpretations of this quote cause a good deal of controversy, which is why it is referred to as "ambiguous" in the text. I have found women respond to Gourley's statement as a positive assessment, even a compliment to Mrs. Lincoln, while men tend to characterize it as evidence of her "shrewishness"—thus offering insight into how these clashing interpretations develop.

162 *Mary used Julia's daughter as*: "Been to Springfield Lately?" *Lincoln Lore*, no. 1719, May 1981.

164 *She adored her sister Mary*: Helm, *Wife of Lincoln*, pp. 106–7.

164 *"said he was glad he had a wife"*: Ibid., p. 107.

165 *"kicking and squalling in the gutter"*: Ibid., p. 113.

167 *developments drew him out of retirement*: For Lincoln's reawakening, see Donald, *Lincoln*, chapter 7.

168 *Frederick Douglass pleaded*: Fleischner, *Mrs. Lincoln and Mrs. Keckly*, p. 174.

168 *Lincoln assailed slavery*: William Lee Miller, *Lincoln's Virtues: An Ethical Biography* (New York: Knopf, 2002), p. 285.

170 *The new Illinois constitution*: And indeed, his own Sangamon County had not been any bastion of antislavery, for when the new state constitution included a clause forbidding blacks to enter the state, it passed in 1853 by a two-to-one margin statewide, but in Sangamon the

measure carried three to one. So Lincoln's views on race and slavery were out of step with his own Illinois community. Ibid., p. 307.

171 *He went into the contest*: Donald, *Lincoln*, p. 183.

173 *"Matteson's double game"*: Miller, *Lincoln's Virtues*, p. 312.

173 *her sister kept her disappointment strictly to herself*: Helm, *Wife of Lincoln*, p. 108.

175 *Lincoln's moral and ethical principles*: On Lincoln's ethical choices and character, see Miller, *Lincoln's Virtues*, pp. 313–16.

SEVEN : ENLARGING OUR BORDERS

177 *"to pick up my lost crumbs"*: Donald, *Lincoln*, p. 185.

177 *"Fifteen dollars"*: Ibid., p. 148.

177 *he returned the check*: His spirit had been battered when he was summoned to Cincinnati as a local representative on a patent case, only to find that the senior counsel his client had hired, an established legal star from Pennsylvania named Edwin Stanton, ignored him entirely during the whole of the trial. He had prepared earnestly, but found himself outclassed by Stanton's courtroom skills. His pride wounded, he returned the check, but cashed it when it was sent the second time. See Donald, *Lincoln*, pp. 186–87; see also: Goodwin, *Team of Rivals*, pp. 173–75.

178 *Mary had gotten*: Harry Pratt, *The Personal Finances of Abraham Lincoln* (Springfield, IL: Abraham Lincoln Association, 1943; repr. Ann Arbor: University of Michigan Library, 2006), p. 88.

179 *the upstairs had eleven-foot ceilings*: Bearss, *Historic Structure Report*, p. 17.

179 *two front bedrooms*: Ibid.

180 *Mary took the opportunity*: Donald, *Lincoln*, p. 198.

181 *It was definitely familiar*: I am indebted to Louise Taper for reminding me of this, as well as sharing so many insights about the Lincolns over the years.

181 *"we have enlarged our borders"*: Turner and Turner, *Mary Todd Lincoln*, pp. 48–49.

181 *Emilie did not get to see*: The Helms were married on March 20, 1856, and Benjamin Hardin Helm spent a week with the Lincolns in 1857. Helm, *Wife of Lincoln*, pp. 127–30.

181 *"I think they will have"*: Donald, *Lincoln*, p. 197.

183 *the railway company that owed him money*: Lincoln had no luck with the company and would have to place a lien on railway property to collect his fees, but he seemed to enjoy the trip nevertheless. See Donald, *Lincoln*, pp. 196–97.

184 *"my next husband shall be rich"*: Turner and Turner, *Mary Todd Lincoln*, p. 50.

185 *"how different the daily routine"*: Ibid., pp. 50–51.

185 *Lincoln himself was struck*: Goodwin, *Team of Rivals*, p. 179.

186 *"a very little, little giant"*: Helm, *Wife of Lincoln*, p. 140.

187 *The New York Times declared*: Donald, *Lincoln*, p. 214.

187 *David Donald has suggested*: Ibid., pp. 214–15.

188 *here he was engaged in a debilitating marathon*: Goodwin, *Team of Rivals*, p. 208.

189 *"sink out of view"*: Donald, *Lincoln*, p. 229.

189 *Lincoln's Cooper Union speech*: For an excellent discussion of Lincoln during this period, see Harold Holzer, *Lincoln at Cooper Union: The Speech That Made Abraham Lincoln President* (New York: Simon & Schuster, 2006).

189 *"From Friday, May 18"*: Helm, *Wife of Lincoln*, p. 151.

190 *the doors were wide open*: Ibid.

194 *few of which pleased his critical wife*: See MTL to John Meredith Read, August 25, 1860, in Turner and Turner, *Mary Todd Lincoln*, p. 65.

194 *"solid substance"*: Bearss, *Historic Structure Report*, p. 22.

194 *Another visitor observed*: Ibid., p. 23.

195 *"who is really an amiable"*: New York Herald, June 26, 1860.

195 *"Quaker tint of light brown"*: New York Herald, August 13, 1860, New York Daily Tribune, August 23, 1860.

196 *"wild to see him"*: Turner and Turner, *Mary Todd Lincoln*, p. 64.

196 *It had been a great disappointment*: Robert later characterized his rejection thus: "On being examined I had the honor to receive a fabulous number of conditions which precluded my admission." See John Goff, "The Education of Robert Todd Lincoln," *Journal of the Illinois State Historical Society*, vol. 53, no. 4 (Winter 1960), pp. 343–44. See also Henry B. Rankin, *Personal Recollections of Abraham Lincoln* (New York: Putnam, 1916), p. 192.

196 *"In your temporary failure"*: Ibid., p. 345.

197 *Robert himself made fun of his status*: Ibid., p. 355.

198 *Mary especially was weary*: Turner and Turner, *Mary Todd Lincoln*, p. 66.

EIGHT : HOPE THAT ALL WILL YET BE WELL

202 *"Could he, with any honor"*: Elizabeth Todd Grimsley, "Six Months in the White House," *Journal of the Illinois State Historical Society*, vol. 19, no. 3, p. 43.

203 *Mary was chided*: "Among other interesting speeches of Mrs. L reported here is that she said her husband had to give Mr. Seward a place the pressure was so great." Kreismann to Charles Ray, Jan. 16, 1861. Manuscript Collection, Huntington Library, San Marino, CA.

203 *"The idea of the President's wife"*: Randall, *Mary Lincoln*, p. 193.

205 *"when honesty in high places"*: Turner and Turner, *Mary Todd Lincoln*, pp. 71–72.

205 *"Sewing Machine"*: *Chicago Tribune*, January 24, 1861.

207 *"a scandalous painting"*: Henry Villard, *Lincoln on the Eve of '61: A Journalist's Story*, ed. Harold G. Villard and Oswald Garrison Villard (New York: Knopf, 1941), pp. 52–53. In another case, an actual noose was reported to have been delivered to the president-elect.

208 *"My husband is my country"*: Joan Cashin, *First Lady of the Confederacy: Varina Davis's Civil War* (Cambridge, MA: Harvard University Press, 2006), p. 99.

208 *leave behind their dog, Fido*: Randall, *Lincoln's Sons*, p. 85.

208 *to bid farewell*: Visitors and townspeople paraded through the Lincoln home on the evening of February 6 from seven until midnight, and Mrs. Lincoln "was dressed plainly, but richly," an observer reported. Her hometown paper crowed: "She is a lady of fine figure and accomplished address, and is well calculated to grace and do honors at the White House." *Illinois State Journal*, Feb. 9, 1861.

208 *Lincoln tied up his own*: Jesse W. Weik, *The Real Lincoln: A Portrait*, ed. Michael Burlingame (Lincoln: University of Nebraska Press, 2002), p. 307.

209 *transportation arrangements*: Soon-to-be-infamous William Wood was put in charge of this operation, and it was with his entrée into the good graces of Mrs. Lincoln that he organized the arrangements for the Lincolns and their party.

209 *"be surrounded by his family"*: Randall, *Mary Lincoln*, p. 195.

209 *his own luxurious train car*: Ibid., p. 202.

210 *"Mrs. Lincoln was piqued"*: There is no reason to disbelieve that this snub of Kate Chase's absence during the Lincoln triumphal visit to Columbus may have initiated the freeze Mary Lincoln maintained toward her younger social rival. Goodwin, *Team of Rivals*, p. 309.

211 *remained uncharacteristically composed*: *Cincinnati Commercial*, Feb. 16, 1861.

211 *"he had found it very difficult"*: *Cleveland Plain Dealer*, Feb. 18, 1861.

211 *At their next stop, in Albany*: *New York Herald*, Feb. 19, 1861.

211 *by becoming "public property"*: Ward Hill Lamon, *Recollection of Abraham Lincoln, 1847–1885*, ed. Dorothy Lamon (Chicago: A. C. McClurg, 1895), pp. 34–35.

212 *significant comings and goings*: New York Times, Feb. 20, 1861, p. 1, col. 6; p. 8, col. 2. See also: *New York Tribune*, Feb. 21, 1861, p. 8, col. 1.

212 *he presented Mary*: Donna McCreary, *The Victorian Wardrobe of Mary Lincoln* (Charlestown, IN: Lincoln Presentations, 2007), pp. 30–31. In 1960, John Fitzgerald Kennedy presented his wife with an emerald and pearl set (necklace, bracelet, and earrings), which became her signature and was much copied during and after Jacqueline Kennedy's era as First Lady.

212 *the New Jersey State House*: New York Daily Tribune, Feb. 22, 1861, p. 5.

212 *stopover in Philadelphia*: New York Tribune, Feb. 21, 1861, p. 5, col. 4.

213 *"the entire female population"*: New York Tribune, Feb. 23, 1861; David H. Donald and Harold Holzer, eds., *Lincoln in the Times: The Life of Abraham Lincoln, as Originally Reported in the* New York Times (New York: St. Martin's Press, 2005), p. 72.

213 *threats to the president-elect*: Not all of Lincoln's enemies were as specific as the anonymous "Vindex" who sent the president-elect in Springfield this letter from Washington: "Dear Sir: Caesar had his Brutus, Charles, the First his Cromwell And the President may profit from their example From one of a sworn band of 10 who have resolved to shoot you in the inaugural procession

on the 4th of March 1861." Ernest G. Ferguson, *Freedom Rising: Washington in the Civil War* (New York: Knopf, 2004), p. 59.

213 *coldness and snobbery*: Sutherland, *Expansion of Everyday Life*, p. 77.

214 *cave dwellers, particularly women*: Evans, *Mrs. Abraham Lincoln*, p. 176.

214 *"canals of liquid mud"*: Noah Brooks, *Washington D.C. in Lincoln's Time* (Chicago: Quadrangle Books, 1971), p. 46.

214 *"It would be difficult"*: Michael Burlingame, *Lincoln's Journalist: John Hay's Anonymous Writing for the Press, 1860–1864* (Carbondale: Southern Illinois University Press, 1998), p. 49. As a diplomat's wife described it, "The city, notwithstanding some splendid public buildings, most of them still in construction, like the Capitol, resembled a very big village . . . and Pennsylvania Avenue, the principal street . . . was still in possession of pigs and cattle, which during the night slept on the sidewalks, even near Lafayette Square, opposite the White House." Agnes Elisabeth Winona Leclerq Joy, Princess zu Salm-Salm, *Ten Years of My Life* (Detroit: Bedford Brothers, 1878), p. 21.

215 *areas of the city that boasted palatial homes*: Margaret Leech, *Reveille in Washington, 1860–65* (New York: Harper Brothers, 1941), pp. 17–18.

215 *President James Buchanan supplemented*: Ibid., p. 17.

215 *Harriet Lane, James Buchanan's niece*: William Seale, *The President's House* (Washington, DC: White House Historical Association, 1986), vol. 1, p. 342. In 1860, an

"auction" of over fifty shabby and unwanted items were sold off from the White House.

216 *"Mrs. Lincoln is awfully western"*: Ibid., p. 363.

216 *"very coarse jokes about Abe Lincoln"*: William Howard Russell, *My Diary, North and South*, ed. Fletcher Pratt (New York: Harper & Bros., 1954), p. 53.

218 *commented on Mary Lincoln's "exquisite toilet"*: Ross, *President's Wife*, p. 102.

218 *"She is more self-possessed"*: Baker, *Mary Todd Lincoln*, p. 179.

218 *She wore the pearls*: McCreary, *The Victorian Wardrobe of Mary Lincoln*, pp. 30–31. Again, this was not unlike the craze for a Kenneth Jay Lane three-strand pearl necklace—much photographed during Jackie Kennedy's days on the campaign trail in 1960 and her years in the White House (1961–1963). Knockoffs of this necklace began to appear in stores across the country in imitation of the First Lady's style.

218 *turning their backs on Mary Lincoln*: Only two of Mary's three sisters (Elizabeth Edwards and Frances Wallace of Springfield) and two of her five half sisters (Margaret Kellogg of Cincinnati and Martha White of Selma, Alabama) attended the festivities. The inauguration was marked by an absence of the most powerful doyennes of D.C. society. The day before the inauguration, Horatio Taft visited the Lincolns, to pay a call with his wife, and noted the paucity of ladies present, and that the Lincolns were "not welcome." Miers, *Lincoln Day by Day*, Mar. 3, 1861.

219 *strain on his family, his staff, and him*: Seale, *The President's House*, vol. 1, p. 366.

220 *"recalled to the cares of state by the messenger"*: Grimsley, "Six Months in the White House," p. 54.

220 *"The Lincoln boys have dined"*: Julia Taft Bayne, *Tad Lincoln's Father* (Lincoln: University of Nebraska Press, 2001), p. 42.

220 *warmed to Julia*: The Tafts became so crucial to the Lincolns that when Mr. Taft lost his job and was offered one in New York, Abraham Lincoln intervened to secure him work in Washington so that his children might stay in close proximity to the White House.

221 *"see more varieties in one evening"*: Michael Burlingame, ed., *Dispatches from Lincoln's White House: The Anonymous Civil War Journalism of Presidential Secretary William O. Stoddard* (Lincoln: University of Nebraska Press, 2002), p. 154.

221 *"monster gathering"*: Esther Singleton, *The Story of the White House* (New York: McClure, 1907), vol. 1, p. 70.

221 *"We are rewarded"*: Ibid., pp. 75–76.

222 *"We must confess to a sigh"*: Grimsley, "Six Months in the White House," p. 49.

223 *Mrs. Charles Eames's Sunday salon*: Grimsley, "Six Months in the White House," p. 68. See also: Michael Burlingame, ed., *An Oral History of Abraham Lincoln: John G. Nicolay's Interviews and Essays* (Carbondale: Southern Illinois University Press, 1996), pp. 45, 205.

223 *This bounty of flowers*: "I found a magnificent bouquet of flowers, with a card attached to them, with Mrs. Lincoln's compliments, and another card attached to them, and another card announcing that she had a 'reception' at 3 o'clock." Russell, *North and South*, p. 32.

223 *protected from most of the hate mail*: And having her outgoing letters read in advance was a public-relations precaution. That way, no one could accuse her of carrying on clandestine correspondence with Confederate relatives—including half brothers serving in the Confederate Army and half sisters married to Confederate loyalists.

223 *Mary's frequently aroused fears*: Burlingame, *Lincoln's Journalist*, p. 61.

223 *hailed by the couple as Cousin Mary*: Helm, *Wife of Lincoln*, p. 188.

224 *"gave great offense to many Republicans"*: Ibid., p. 175. See also: *New York Times*, Feb. 26, 1861.

224 *Slights and insults abounded*: Helm, *Wife of Lincoln*, p. 188.

225 *"there were only two or three ladies"*: Russell, *North and South*, p. 32.

225 *"The Great Black Republican 'Wigwam,'"*: Apr. 6, 1861, in Herbert Mitgang, ed., *Abraham Lincoln: A Press Portrait* (1956; repr. Athens: University of Georgia Press, 1989), pp. 255–56.

225 *"both the President and his wife"*: Helm, *Wife of Lincoln*, p. 173.

226 *only ten matching place settings*: Jerrold Packard, *The Lincolns in the White House: Four Years That Shattered a Family* (New York: St. Martin's, 2005), p. 10.

226 *"The Washington ladies"*: Russell, *North and South*, p. 32. The shift was dramatic, as one commentator observed: "Among the strange faces may be detected something of all the isms and kinks and crotchets of our

Northern reformers—spiritualism, free speech, free soil, free men, free love, free farms, free rents, free offices, free negroes, woman's rights, bran bread and patent medicine . . . very few, indeed, were the familiar Southern faces of distinguished men and beautiful and accomplished women we have been accustomed to meet on such occasions." See also: Singleton, *Story of the White House*, vol. 1, pp. 75, 77.

227 *"the crowd will be gradually leaving the city"*: Mary Lincoln to Hannah Shearer, Mar. 28, 1861, Turner and Turner, *Mary Todd Lincoln*, p. 81.

227 *"under the gorgeous gas chandeliers"*: Ferguson, *Freedom Rising*, p. 7.

228 who *"was the only one who had the skill"*: Rankin, *Personal Recollections*, p. 174.

228 *"beginning with an effort to keep Sunday"*: Michael Burlingame, ed., *Inside the White House in War Times: Memoirs and Reports of Lincoln's Secretary, William O. Stoddard* (Lincoln: University of Nebraska Press, 2000), p. 9.

229 *"War has no Sunday"*: Ibid., p. 176.

230 *"Mr. Caesar Augustus must have felt"*: Leech, *Reveille*, p. 66.

230 *their cheering embrace*: Lucius Chittenden, a clerk at the Treasury Department, witnessed the Seventh's march into the District: "There were many wet faces as this noble regiment marched up Pennsylvania Avenue, and I do not deny that mine was one." In David Detzer, *Dissonance: The Turbulent Days Between Fort Sumter and Bull Run* (New York: Harcourt, 2006).

230 *Mary Lincoln reassured a Springfield friend*: MTL to Mrs. Samuel Melvin, Apr. 27, 1861, Turner and Turner, *Mary Todd Lincoln*, p. 85.

231 *the Lincolns bade him farewell*: He returned to Kentucky, joined the Confederate army, and ended up dying in combat, leaving Emilie a widow with three children.

231 *"he must have Kentucky"*: "Rev. M. D. Conway on His Late Visit to Washington," *The Crisis*, Feb. 12, 1862, vol. 2, no. 3, p. 24.

232 *"a magnetic, brilliant young fellow"*: Grimsley, "Six Months in the White House."

232 *to find him the "best position"*: Tripp, *Intimate World*, p. 117.

233 *"Are we to have pestilence"*: Leech, *Reveille*, p. 77.

233 *"If she but drives down Pennsylvania Avenue"*: Russell, *North and South*, p. 27.

233 *newsmen hounded Mrs. Lincoln*: The partisan campaign against Mary Lincoln was unrivaled until the twentieth-century attacks on the character of Eleanor Roosevelt and the barrage of bad publicity accorded Hillary Clinton during her tenure as First Lady.

234 *became the target of celebrity gawking*: "A great sensation was created amongst the members of Henry Ward Beecher's congregation yesterday by the entrance of Mrs. Lincoln and party during the morning service. The attention of the congregation was about equally divided between listening to the sermon and gazing at the unexpected distinguished visit, and at the close a tremendous rush for the door took place to catch a glimpse of Mrs. Lincoln's pleasant face as she left the church." *New York Herald*,

May 13, 1861, p. 5, col. 5. Another article noted that "The Band Played the 'Star Spangled Banner,' 'Hail to the Chief,' 'The Red, White and Blue,' 'Hail, Columbia,' 'Yankee Doodle,' 'The Last Rose of Summer,' and other popular selections. Mrs. Lincoln appeared at the window, dropped a bouquet to the band and bowed her thanks to the crowd, which cheered heartily in return." *New York Daily Tribune*, May 15, 1861, p. 5, col. 1; *New York Herald*, May 17, 1861, p. 4, col. 6.

234 *"Mrs. Lincoln . . . busily engaged in 'shopping'*: *New York Daily Tribune*, May 14, 1861, p. 8, col. 5. The next day, the same paper reported: "Mrs. Lincoln, who is still sojourning at the Metropolitan Hotel, was engaged nearly all day yesterday [May 14] in making extensive purchases at Lord & Taylor's and various other places about the city. A great many distinguished citizens called upon her in the afternoon, but few of them were fortunate enough to see her. She found time, however, to call upon Col. Anderson and his wife. . . ." *New York Daily Tribune*, May 15, 1861, p. 5, col. 1.

235 *Any move a Lincoln made*: *New York Daily Tribune*, May 12, 1861, p. 1, col. 2. And, for example: "At 9 o'clock this morning Mr. Lincoln attended Trinity church, in consideration of his friendship for Gen. Thomas, whose son was at that time married to Miss Maynard." *New York Tribune*, May 17, 1861, p. 5, col. 2.

236 *a pair of horses*: Michael Burlingame, ed., *At Lincoln's Side: John Hay's Civil War Correspondence and Selected Writings* (Carbondale: Southern Illinois University Press, 2000), p. 190; p. 273, n. 27.

236 *Gossips intimated*: Burlingame, *Inner World*, p. 292.

236 *"She is expending"*: Turner and Turner, *Mary Todd Lincoln*, p. 88.

237 *the First Lady refurbishing*: The issue of *whose* money was being spent was not in question during these early complaints, but it soon crept into caustic responses.

237 *"did not indulge"*: Grimsley, "Six Months in the White House," p. 59.

237 *"signifying the Union of North and South"*: Baker, *Mary Todd Lincoln*, p. 52.

237 *Grimsley insisted that*: Grimsley, "Six Months in the White House," p. 59.

237 *Although a payment*: Perhaps they did not pay the bill directly, *but* Lincoln's account was debited for the amount—to reimburse the government and to straighten out accounts. Turner and Turner, *Mary Todd Lincoln*, p. 87.

238 *Mr. Lincoln was not*: Grimsley, "Six Months in the White House," p. 59.

238 *participated in a criminal conspiracy*: As has been suggested by Michael Burlingame, Apr. 25, 2006, in a scholarly workshop at the Ritz-Carlton Hotel, New York City, and also in an interview Nov. 3, 2006, with Gerald Prokopowicz, host of *Civil War Talk Radio*, on a program with the title "Michael Burlingame: King of Lincoln Researchers."

238 *Chase apparently regaled his daughter*: This was particularly ironic, as Kate Chase married for money to finance her father's political ambitions, and ended up divorced and indigent. See Goodwin, *Team of Rivals*, pp. 752–53.

See also Peg A. Lamphier, *Kate Chase and William Sprague: Politics and Gender in a Civil War Marriage* (Lincoln: University of Nebraska Press, 2003).

238 *"scandalous reports" about Mrs. Lincoln*: Anson Henry suggested that Chase encouraged "the circulation of lying scandals" against the Lincolns to promote his own career. See Anson G. Henry to Isaac Newton, Apr. 21, 1864, Abraham Lincoln Papers, Library of Congress.

239 *Mrs. Lincoln's china*: Turner and Turner, *Mary Todd Lincoln*, p. 88.

239 *A diplomat's wife noted*: Salm-Salm, *Ten Years*, p. 45. This observation was echoed by William Russell, who noted at the state dinner on March 28 that there was "no parade or display, no announcement—no gilded staircase, with its liveried heralds." And again, he wrote, the dinner "was not remarkable for ostentation. No liveried servants . . . " Russell, *North and South*, pp. 23–24.

239 *Prince Napoleon was taken aback*: His diary recorded, "One goes right in [the White House] as if entering a café." John Whitcomb and Claire Whitcomb, *Real Life at the White House* (New York: Routledge, 2000), p. 136.

240 *These libels were published*: Turner and Turner, *Mary Todd Lincoln*, p. 87.

NINE : DASHED HOPES

244 *"pledge not for thirty days"*: Burlingame, *Lincoln's Journalist*, p. 61.

245 *The Zouaves won even more plaudits*: Ibid., p. 64.

247 *the hotel owner, James T. Jackson*: Ross, *President's Wife*, p. 115.

247 *The president found himself*: New York Herald, May 25, 1861.

248 *The Ellsworth funeral*: Tripp, *Intimate World*, pp. 120–21.

248 *"When I played [it] for Mrs. Lincoln"*: Bayne, *Tad Lincoln's Father*, p. 15.

249 *"first blood shed on secession soil"*: Ross, *President's Wife*, p. 116.

250 *to christen as "Camp Mary"*: New York Times, June 28, 1861.

250 *"Please accept, sir, these weapons"*: MTL to John Fry, Turner and Turner, *Mary Todd Lincoln*, p. 91.

251 *"Mrs. L. clung"*: New York Times, June 22, 1861. Mrs. Grimsley, who was also on the ride, gave a different version of events in a letter home to Springfield: "I am confident had it not been for Gen Wallbridge who was with us, Taddy would have been crushed by the wheel. You will certainly think Mary and I have changed characters as the papers represent her as acting with great coolness while I had to be assisted from the carriage. So much for reporters."

251 *"the women kind are giving Mrs. Lincoln"*: Elizabeth Blair Lee, July 14 [15], 1861. See Virginia Jeans Laas, *Wartime Washington: The Civil War Letters of Elizabeth Blair Lee* (Urbana: University of Illinois Press, 1991), p. 61.

252 *"If you love me, give me a favorable answer"*: MTL to Hannah Shearer, July 11, 1861, Turner and Turner, *Mary Todd Lincoln*, p. 95.

252 *Rose Greenhow, the widow:* For a full account of Greenhow's career, see Ann Blackman, *Wild Rose: The True Story of a Rebel Spy* (New York: Random House, 2005).

253 *the first major military loss:* This was a Confederate victory with which Rose Greenhow was credited, by allegedly passing on crucial information to Confederate commanders.

254 *"I have passed through":* MTL to Hannah Shearer, Aug. 1, 1861, Turner and Turner, *Mary Todd Lincoln,* p. 96.

254 *"that the Union will go on":* Seale, *President's House,* vol. 1, p. 385.

254 *On March 6, Mary Lincoln:* New York Times, Mar. 7, 1861, p. 1, col. 2.

255 *"the head surmounted by a gilded half coronet":* Seale, *President's House,* vol. 1, p. 386. See also: "The White House," *Architectural Digest,* Mar. 2008.

256 *This grand occasion boasted:* For Grimsley's description, see "Six Months in the White House," pp. 69–70, and for John Hay's, see *Inside Lincoln's White House: The Complete Civil War Diary of John Hay,* ed. Michael Burlingame and John R. Turner Ettlinger (Carbondale: Southern Illinois University Press, 1997), pp. 87–90.

256 *The press was equally:* New York Herald, Aug. 4, 1861, p. 1, col. 1.

256 *the complete works of Victor Hugo:* Miers, *Lincoln Day by Day,* Aug. 5, 1861.

257 *homesickness and money worries:* Elizabeth Grimsley was forced to write to her cousin John Stuart to request a loan. See Harry Pratt, *The Personal Finances of*

Abraham Lincoln (Springfield, IL.: Abraham Lincoln Association, 1944), p. 86.

257 *took their toll on the First Family*: "Yesterday in many places, the thermometer was at 96 degrees in the shade. Several of the soldiers were prostrated by the heat during the day." *National Republican*, Aug. 8, 1861. Several days later, "with the thermometer ranging from 100 to 120 degrees, the ice dealers announce their stock of ice is exhausted, and that there is no more to be had here this summer." *National Republican*, Aug. 13, 1861.

258 *Mary Lincoln and her sons stubbornly resisted*: See Randall, *Lincoln's Sons*, pp. 105–7.

258 *Uniformed sentries ringed the grounds*: Seale, *President's House*, vol. 1, p. 376.

258 *egress through the basement*: This allowed Lincoln a secret exit from the White House. Ibid., p. 377.

258 *turned the Oval Room into a family parlor and library*: Ibid., p. 379.

259 *Buchanan had installed running water*: Ibid., pp. 377, 379–80. Experts debate whether the poor quality of this water may have played a role in the illness that led to the death of Willie Lincoln.

259 *"Mrs. Lincoln instituted the daily drive"*: Grimsley, "Six Months in the White House," p. 55.

260 *Lincoln, without the slightest show of displeasure*: Rankin, *Personal Recollections*, pp. 178–79.

261 *"Mr. Lincoln would come in looking so sad"*: Grimsley, "Six Months in the White House," p. 55.

261 *"If left to himself, when"*: Rankin, *Personal Recollections*, p. 176.

262 *"disorder in Lincoln's apparel"*: Ibid., p. 164.

262 *"was dressed in gray woolen clothing"*: Benjamin Brown French, *Witness to the Young Republic: A Yankee's Journal, 1828–1870—the Diary of Benjamin Brown French*, ed. Donald B. Cole and John J. McDonough (Hanover, NH: University Press of New England, 1989).

262 *"they understood each others peculiarities"*: Rankin, *Personal Recollections*, p. 180.

262 *the White House was being invaded*: "During the absence of Mrs. Lincoln and her family, the White House is to be overhauled and refurbished . . . the outside of the building is now being repainted." *New York Times*, Aug. 9, 1861.

263 *stopped off in Manhattan for a courtesy visit*: *Washington Star*, Aug. 17, 1861.

263 *"Mrs. Lincoln's presence here"*: Burlingame, *At Lincoln's Side*, p. 97–98.

263 *She was very solicitous of her pregnant friend*: Ibid., p. 99.

264 *"I impressed her"*: John Hay to Mrs. Fanny Campbell Eames, Aug. 21, 1861. However, ten days later would find Hay in New York City, planning a prolonged visit to Illinois. See John Hay to J. G. Nicolay, Sept. 2, 1861, Michael Burlingame, ed., *With Lincoln in the White House: Letters, Memoranda, and Other Writings of John G. Nicolay, 1860–1865* (Carbondale: Southern Illinois University Press, 2000), p. 54.

264 *to fend for herself, among the swarm*: The phrase "first lady of the land" appeared in John Hay's anonymous press reports, so he may be credited with, if not coining

the term, popularizing it. See Burlingame, *Lincoln's Journalist*, p. 98.

264 *". . . if Mrs. L. were a prizefighter"*: Headline: PRESS HOUNDING MRS. LINCOLN, *Chicago Tribune*, Aug. 31, 1861.

265 *"Mr. Lincoln came this morning"*: Miers, *Lincoln Day by Day*, Sept. 8, 1861.

266 *"Let the children have a good time"*: Randall, *Lincoln's Sons*, p. 101.

266 *Williamson was detailed*: Bayne, *Tad Lincoln's Father*, p. 65.

266 *male White House staffers*: Seale, *President's House*, vol. 1, p. 394. See Burlingame, *Inside the White House*, p. 151.

267 *he bade farewell, and gave young Willie*: Goodwin, *Team of Rivals*, p. 381.

268 *more than just a tiff over dress code*: The diary of Fanny Seward (Sept. 9, 1861) records that the Seward women came to pay a call on Mrs. Lincoln at the White House. The group were seated, then told Mrs. L. was "very much engaged," so they filed out, but "the truth of Mrs. L.'s engagement was probably that she did not want to see Mother—else why not give general direction to the doorkeeper to let no one in? It was certainly very rude to have us all seated first. . . ." Fanny was extremely annoyed but goes on to compliment Mrs. Lincoln for begging for the life of a soldier. Perhaps Mrs. Lincoln was engaged in trying to deal with all the rumors of scandal and corruption that were raging around her renovations of the White House, which she discovered

upon her arrival back in Washington on September 5. It is also true that she repeatedly blamed Seward for circulating stories against her. In addition, missing callers was a common hazard, as Mrs. Seward reported the same week: Sept. 8, 1861, "Mrs. Bates Called last week—I did not see her."

268 *Mrs. Lincoln began to rely more heavily:* Seale, *President's House*, vol. 1, p. 395.

268 *She warned that Wood:* MTL to Simon Cameron, Sept. 12, 1861, Turner and Turner, *Mary Todd Lincoln*, p. 103.

269 *"He is either deranged or drinking":* MTL to Caleb Smith, Sept. 8, 1861, Turner and Turner, *Mary Todd Lincoln*, pp. 101–2.

269 *The Lincolns paid French a visit on September 13:* French, *Witness*, p. 375. By now Wood was also defaming William Stackpole.

269 *"I know him to be a Union man":* Mary Lincoln to Mr. John F. Potter, 13 September 1861, Turner and Turner, *Mary Todd Lincoln*, p. 103.

270 *The chevalier's campaign:* Seale, *President's House*, vol. 1, p. 397.

270 *"the poor lady is loyal":* Russell, *North and South*, p. 567.

271 *"ground was of pale green":* Seale, *President's House*, vol. 1, p. 386.

271 *hoping that after her redecoration:* Turner and Turner, *Mary Todd Lincoln*, p. 106.

271 *"The ladies in Washington delight":* Russell, *North and South*, p. 566.

272 *"It would stink in the land":* French, *Witness*, p. 382.

273 *"She is an admirable woman"*: Ibid., Dec. 22, 1861. He also wrote to his sister, "She is a lady and an accomplished one too, but she does love money—aye, better than I do." Seale, *President's House*, vol. 1, p. 390.

273 *the editor of the* New York Herald: See *New York Herald*, Oct. 21, 1861; and MTL to James Bennett, Oct. 25, 1861, Turner and Turner, *Mary Todd Lincoln*, p. 111. Although Bennett took this public stand of defending Mrs. Lincoln, in private he was less complimentary. He wrote to Mrs. Clement Clay, "The *ensemble* of the personnel of the White House has sadly changed, more befitting a restaurant than the House of the President." Seale, *President's House*, vol. 1, p. 391.

273 *"he assured Mrs. L"*: Burlingame, *At Lincoln's Side*, pp. 194–95.

274 *fans continued to champion her*: See Burlingame, *Inside the White House*, p. 182.

TEN : GRAND DESIGNS GONE AWRY

276 *She hired two other*: Fleischner, *Mrs. Lincoln and Mrs. Keckly*, p. 207.

276 *only four staff members*: Seale, *President's House*, vol. 1, p. 395.

277 *"the poor neglected contrabands"*: *Iowa State Register*, Jan. 17, 1862 (dateline Jan. 4).

278 *"The colored people who worked in the White House"*: John E. Washington, *They Knew Lincoln* (New York: Dutton, 1942), p. 107.

278 *"Tyler was a very cross man"*: William Still, *Underground Railroad* (Philadelphia: Porter & Coates, 1872), p. 55.

279 *he was born his master's son*: Ibid.

279 *"a Virginian of distinguished southern ancestry"*: Washington, *They Knew Lincoln*, p. 108.

280 *Slade would spend*: Ibid., p. 111.

280 *Mr. Lincoln sometimes took Tad*: Ibid., p. 109.

281 *"had her ways, but nobody"*: Ibid., p. 77.

281 *"Mrs. Lincoln could put"*: Ibid., p. 119.

282 *Lincoln collected William's pay*: Ibid., pp. 133–34.

282 *a frequent visitor to the White House*: Ibid., p. 212.

283 *Mary asked her to appear*: Fleischner, *Mrs. Lincoln and Mrs. Keckly*, p. 202.

283 *"If you do not charge too much"*: Keckley, *Behind the Scenes*, p. 85.

284 *endeared her to the president*: Fleischner, *Mrs. Lincoln and Mrs. Keckly*, p. 207.

284 *"I know you will be sorry to hear"*: Turner and Turner, *Mary Todd Lincoln*, p. 106.

284 *"more than she did on her own kinfolks"*: Washington, *They Knew Lincoln*, p. 101.

285 *she would need to rely on the woman*: Keckley, *Behind the Scenes*, 95.

285 *observed that "the carriages are rolling"*: Goodwin, *Team of Rivals*, p. 409.

286 *Her requests for appointments*: Ibid., p. 414.

286 *"in keeping with the institutions"*: Fleischner, *Mrs. Lincoln and Mrs. Keckly*, p. 228.

287 *"La Reine has determined"*: Burlingame, *With Lincoln in the White House*, p. 67. See also: Keckley, *Behind the Scenes*, pp. 96–97.

287 *"Levees will be held"*: *National Republican*, Jan. 7, 1862.

287 *"They are both novel and pleasant"*: Burlingame, *With Lincoln in the White House*, p. 66.

287 *"gaunt and careworn"*: "Miriam," *Iowa State Register*, Jan. 29, 1862 (dateline Jan. 16). Miriam was Cara Kasson, wife of the assistant postmaster general; she wrote regularly for the *Register*.

287 *Herrmann the Prestidigitator*: *National Republican*, Jan. 22, 1862.

287 *their White House entertaining*: Turner and Turner, *Mary Todd Lincoln*, p. 218.

288 *"Half the city"*: Fleischner, *Mrs. Lincoln and Mrs. Keckly*, p. 229.

288 *had to disabuse several dignitaries*: See Burlingame, *Inside the White House*, pp. xiv–xv.

289 *"Are the President and Mrs. Lincoln"*: Fleischner, *Mrs. Lincoln and Mrs. Keckly*, p. 230.

290 *assumed he was his father's favorite child*: Keckley, *Behind the Scenes*, pp. 106–7.

290 *Mary came to depend on him*: Later, she would reflect, "How much comfort he always was to me, and how fearfully, I always found my hopes concentrating on so good a boy as he was." See Turner and Turner, *Mary Todd Lincoln*, p. 128.

290 *"in no immediate danger"*: Keckley, *Behind the Scenes*, p. 100.

291 *remarked, "Whew! Our cat has a long tail"*: Keckley, *Behind the Scenes*, pp. 100–102.

292 *the president was "nearly worn out"*: *The Diary of Edward Bates, 1859–1866*, ed. Howard K. Beale (Washington, DC: U.S. Government Printing Office, 1933), p. 233.

292 *to leave Willie's side*: Bayne, *Tad Lincoln's Father*, pp. 199–200.

292 *Lincoln's official correspondence*: Ida Tarbell, *The Life of Abraham Lincoln* (New York: Doubleday, Page, 1895), vol. 2, p. 89.

292 *dragged into the headlines by midmonth*: Lincoln was reported to have gone to Capitol Hill to meet with Republican members of the committee to ask that they spare him any disgrace. Randall, *Mary Lincoln*, p. 304.

293 *"My poor boy"*: Keckley, *Behind the Scenes*, p. 103. See also: Burlingame, *With Lincoln in the White House*, p. 71.

293 *Mrs. Lincoln's grief*: Keckley, *Behind the Scenes*, pp. 103–4.

294 *Stoddard suggested*: See Milton Shutes, "Mortality of the Four Lincoln Boys," *Lincoln Herald*, vol. 57 (Spring–Summer 1955), p. 6. See also: *New York Herald*, Feb. 21, 1862: "His disease was pneumonia."

294 *Attorney General Bates sadly remembered*: Randall, *Lincoln's Sons*, p. 137.

295 *she was touched by the president's relief*: Anna Boyden, *Echoes from Hospital and White House* (Boston: D. Lothrop, 1884), pp. 51, 53–55.

295 *to invite Bud to have one last visit*: Bayne, *Tad Lincoln's Father*, p. 82.

297 *the trappings of mourning seemed to rouse her*: She sent specific directives to New York to order new wardrobe items, such as a "mourning bonnet—which must be exceedingly plain & genteel." Mary Lincoln to Ruth Harris, May 17, 1862, Turner and Turner, *Mary Todd Lincoln*, p. 125.

297 *Victorian mores dictated*: Faust, *Republic of Suffering*, p. 148.

297 *pointing out St. Elizabeth's, a mental hospital*: Keckley, *Behind the Scenes*, p. 104.

297 *"He retained his prairie habits"*: Keckley, *Behind the Scenes*, p. 107.

298 *"my presence here"*: Elizabeth Edwards to Julia Baker, Mar. 2, 1862, Illinois Historical Society. She ended up sleeping in the same room in which Willie died and pronounced it the Death Chamber.

299 *trying to weather the clouds*: "Mrs. Lincoln secludes herself from all society . . ." Boyden, *Echoes from Hospital*, p. 51. See also: Elizabeth Edwards to Julia Baker, Mar. 12, 1862, Illinois Historical Society.

299 *scandalmongers suggested that Mary was the one*: *The Diary of Orville Hickman Browning*, 2 vols., ed. Theodore Calvin Pease and James G. Randall (Springfield, IL: Illinois State Historical Library, 1925–1933), Mar. 3, 1862; and Burlingame, *At Lincoln's Side*, pp. 186–87.

299 *debates over the abolition of slavery*: "The President's Emancipation Message," *National Republican*, Mar. 10, 1862.

300 *"making too much of the Negro"*: Ishbel Ross, *Proud Kate* (New York: Harper & Brothers, 1953), p. 90.

301 *Cooke received contracts*: Goodwin, *Team of Rivals*, pp. 402–3.

301 *Some even hinted that Mary Lincoln's being laid low*: Baker, *Mary Todd Lincoln*, p. 215.

301 *Mary Lincoln, dubbed the Republican Queen*: Even the usually kind Rebecca Pomroy wrote to a friend, "Oh,

for a Martha Washington to set the women of America an example of economy and prudence, patriotism and nobleness. Mrs. Lincoln is not a woman of sufficient dignity of character to fill the post at the White House at any time, but now how much more do we expect to see her filling the place of at least a patriotic wife." Pomroy to unknown recipient, Feb. 21, 1862, Schlesinger Library, Radcliffe, Harvard University, Cambridge, MA.

302 *"The 'enemy' is still planning"*: Burlingame, *At Lincoln's Side*, Mar. 31, 1862, p. 18.

302 *"Madame has mounted me to pay her"*: Ibid., Apr. 4, 1862, p. 19.

302 *On April 9 he snarled, "The Hellcat"*: Ibid., Apr. 9, p. 19.

302 *Lincoln certainly wanted*: "Mr. Lincoln said to me—Mrs. Edwards—do stay with me—you have Such a power & control Such an influence over Mary." Wilson and Davis, *Herndon's Informants*, p. 444.

302 *"Your Aunt Mary wonders"*: Elizabeth Edwards to Julia Baker, Apr. 1, 1862, and Elizabeth Edwards to Edward Baker, Apr. 14, 1862, Illinois Historical Society.

303 *A banner in the* National Republican *trumpeted*: May 1, 1862.

303 *that included compensation*: "The *Baltimore American* of yesterday, howls dismally over the abolition of slavery in the District. . . . The proslavery people are circulating the story that he will veto the measure. This is an entire mistake, as will soon be found." *National Republican*, Apr. 15, 1862. The *National Republican* of Apr. 17, 1862, reported that Lincoln had approved the bill the previous day.

303 *sister-in-law observed "Mr. Lincoln looks"*: Elizabeth Edwards to Edward Baker, Apr. 14, 1862, Illinois Historical Society.

303 *the press reported*: National Republican, May 22, 1862.

303 *trained on the presidential box*: Burlingame, *Inside the White House*, pp. 105–6.

304 *a fondness for Shakespeare*: Ibid., p. 189.

304 *invited actors to the White House*: See William Kelley's recollections of taking John McDonough to the White House to meet with Lincoln, and their four-hour visit, after which Lincoln confessed, "I have not enjoyed such a season of literary recreation since I entered the White House, and I feel that a long and pleasant interval has passed since I closed my routine work this afternoon." Allen T. Rice, ed., *Reminiscences of Abraham Lincoln by Distinguished Men of Our Time* (New York: North American Pub. Co., 1886), pp. 264–70.

304 *Lincoln was reading aloud*: Sandburg, *Abraham Lincoln*, vol. 2, *The War Years* (New York: Harcourt Brace, 1939), p. 314.

304 *"Did you ever dream of some lost friend"*: Goodwin, *Team of Rivals*, p. 423.

304 *among other outings*: See for example, *New York Herald*, May 4, 1862, p. 5, col. 2: "The President and Mrs. Lincoln visited the Navy Yard this afternoon, to witness some interesting trials of a breech-loading cannon, invented and patented by parties in Cincinnati. . . . The first discharge of the piece interrupted one of the President's stories, after which the experiments proceeded in a satisfactory manner." See also: Miers, *Lincoln Day by Day*, May 3 and May 7, 1862.

305 *sending out bouquets*: Mrs. Lincoln sent Mrs. Gustavus Fox a bouquet. Miers, *Lincoln Day by Day*, Apr. 18, 1862.

305 *She wrote solicitously*: Turner and Turner, *Mary Todd Lincoln*, p. 126.

305 *She began supplying*: Boyden, *Echoes from Hospital*, p. 61.

305 *"swallow-tailed coat and kids"*: Burlingame, *Inside the White House*, p. 38.

305 *"bears up and teaches us a lesson"*: Turner and Turner, *Mary Todd Lincoln*, p. 128.

305 *"our home is very beautiful"*: Ibid.

ELEVEN : STRUGGLING AGAINST SORROWS

306 *The First Lady, heartened by her younger son's recovery*: Elizabeth Edwards complained that Tad locked her and the president into the park at Lafayette Square. See Wilson and Davis, *Herndon's Informants*, p. 445.

306 *"He has grown & improved*: Turner and Turner, *Mary Todd Lincoln*, p. 128.

307 *She glowed with pride over the fact that her son*: Packard, *Lincolns in the White House*, p. 149.

307 *"I don't think he distinguished himself as a scholar"*: Goff, "The Education of Robert Lincoln," p. 354. Lowell was commenting on Robert's service as ambassador to the Court of St. James's, a post that he had once held himself.

308 *She would not let him smoke in the house*: Hunt, *Personal Recollections*, p. 21.

308 *"a son of Secretary Seward"*: *National Republican*, July 24, 1862.

309 *some of Willie's favorite toys*: Elizabeth Todd Edwards to Julia Edwards Baker, Apr. 9, 1862, Illinois Historical Society.

309 *Lincoln first inspected the site*: Matthew Pinsker, *Lincoln's Sanctuary: Abraham Lincoln and the Soldiers' Home* (New York: Oxford University Press, 2005), p. 193.

310 *"go out to the 'Soldiers' Home'"*: Turner and Turner, *Mary Todd Lincoln*, p. 128. Indeed she suspended the practice of Marine Band concerts on the grounds, and even refused to allow them to play in nearby Lafayette Square. Mary Lincoln to John Hay, May 23, 1862, in Thomas Schwartz and Kim Bauer, eds., "Unpublished Mary Todd Lincoln," *Journal of the Abraham Lincoln Association*, vol. 17 (Summer 1996), pp. 4–5.

310 *The papers reported*: *National Republican*, May 30, 1862.

310 *"I have seen nothing"*: July 2, 1862, "Correspondence from Washington," *Iowa State Register*.

310 *fashionable ladies in the District*: Salm-Salm, *Ten Years of My Life*, p. 75.

311 *"The air is swarming"*: Burlingame, *With Lincoln in the White House*, p. 86.

311 *"If Congress should remain"*: A report from a Baltimore paper is reprinted as "The Health of Washington," *National Republican*, May 29, 1862.

311 *One woman complained*: 11 June 1862, "Miriam" [Cara Kasson], *Iowa State Register*, June 11, 1862 (dateline Washington, May 26, 1862).

311 *French reported that her removal*: French, *Witness*, p. 400. See also: Burlingame, *With Lincoln in the White House*, p. 81.

311 *The largest of these smaller dwellings*: This is where the acting governor of the Soldiers' Home lived until the Lincolns moved in and he moved out. See Pinsker, *Lincoln's Sanctuary*, p. 4.

312 *zealous office seekers*: Ibid., pp. 23–25.

313 *rank of "third lieutenant"*: Ibid., p. 78.

315 *Mary had not yet returned*: Ibid., p. 43.

315 *her only recorded purchases*: Turner and Turner, *Mary Todd Lincoln*, pp. 129–30.

315 *"fully appreciated her kindness"*: New York Herald, July 17, 1862.

315 *"more pleasant than I"*: Turner and Turner, *Mary Todd Lincoln*, p. 130.

316 *Mrs. Lincoln threw herself headlong*: "Donation for the Sick," New York Tribune, Aug. 13, 1862.

316 *a rigorous campaign of regular visitation*: "Mrs. Lincoln Among the Sick," Washington Star, Aug. 29, 1862.

316 *"A noble example was set by Mrs. Lincoln"*: National Republican, June 7 and June 13, 1862; New York Herald, June 12, 1862.

316 *"walking for hours through the wards"*: Mrs. E. F. Ellet, *The Court Circles of the Republic* (Hartford, CT: Hartford Pub. Co., 1869), p. 526.

316 *Mrs. Lincoln wrote letters and performed mundane tasks*: On August 16, she requested $200 worth of lemons and $100 worth of oranges to distribute to hospitals. On October 4, she sent one thousand pounds of grapes to hospitals. Miers, *Lincoln Day by Day*.

317 *"almost regal in her deep black and expansive crino-line"*: Helm, *Wife of Lincoln*, p. 199.

317 *"burst into a passion of tears"*: Pinsker, *Lincoln's Sanctuary*, p. 30.

318 *Mrs. Lincoln threw herself more ferociously into the cause of dying soldiers*: "Mrs. Lincoln has been quietly engaged, for some weeks, in a systematic visitation of the hospitals of this city and vicinity." *National Republican*, Aug. 27, 1862. See Miers, *Lincoln Day by Day*, Aug. 16, 1862, and "Mrs. Lincoln Among the Sick," *Washington Evening Star*, Aug. 29, 1862.

318 *"Surely the women of America"*: *National Republican*, Aug. 22, 1862.

319 *Margaret Fuller featured in her review of Wesley's life*: Barbara Weisberg, *Talking to the Dead: Kate and Maggie Fox and the Rise of Spiritualism* (San Francisco: HarperSanFrancisco, 2004), p. 25.

321 *Lincoln was curious*: Wilson, *Intimate Memories of Lincoln*, pp. 61–62.

321 *a tight-knit circle of supporters*: Amy and her husband, pharmacist Isaac Post, were among the most prominent of the Quakers in the Rochester area. They were abolitionists and entertained many reformers in the parlor of their Sophia Street home, which was rumored to be a stop on the Underground Railroad in later years. Members of the Posts' immediate group of Waterloo Congregational Friends were among the first converts to spiritualism. Bret Carroll, ed., *Spiritualism in Antebellum America* (Bloomington: Indiana University Press, 1997), pp. 29, 91–92. Nothing was detected by this first

investigative panel, and a certificate of authenticity was issued. Yet dissenters continued their campaigns. It may have been legitimate for unbelievers to complain that the examiners included "spiritual sympathizers," but it was partisan if not hypocritical to question the girls' innocence *because* they allowed themselves to be subject to these inspections.

321 *The Fox sisters were followed by "trance lecturers"*: Ann Braude, *Radical Spirits: Spiritualism and Women's Rights in Nineteenth-Century America* (Boston: Beacon Press, 1989), p. 85.

322 *"a profusion of sunny ringlets"*: Earl Wesley Fornell, *The Unhappy Medium: Spiritualism and the Life of Margaret Fox* (Austin: University of Texas Press, 1964), pp. 79–80.

322 *fictionalized as the charismatic Verena Tarrant*: Barbara Goldsmith, *Other Powers: The Age of Suffrage, Spiritualism, and the Scandalous Victoria Woodhull* (New York: Knopf, 1998), p. 381.

322 *described as a "domestic hell" by intimates*: Greeley's wife, Mary, was a well-educated and sensitive young woman, who had been a teacher at a New York City school before her marriage to the young newspaperman in 1836. Her descent into depression began the following year, after the death of the couple's first child. When her next two pregnancies ended in miscarriages, Greeley described his wife's difficulties: "Mary is terribly ill and downhearted—a miscarriage of the worst kind and a great danger of the loss of her eyesight. She is now unable to bear the light and must be kept so all

winter without reading or doing anything, but I think she ultimately will recover." Ibid., p. 57. As he threw himself into work on his paper, the *New York Tribune*, launched in 1844 (the same year Pickie was born), Mary retreated, often on voyages to Europe. Greeley began to call his deserted home Castle Doleful, and after Pickie died, his mother turned to spiritualism. She summoned Kate Fox to live with her, to enable her to communicate with her dead son through a medium's intervention more or less at will. Ibid., p. 58.

322 *"The one sunburst of joy"*: Weisberg, *Talking to the Dead*, p. 108.

322 *communing with their dead*: Ibid., p. 121.

322 *Kate Fox was hired by the publisher*: Ibid., p. 170.

323 *rival sects erupted*: Ibid., p. 177.

324 *" 'Tis sweet to call to mind the loved"*: Goldsmith, *Other Powers*, pp. 68–69. See also Robert Cox, *Body and Soul: A Sympathetic History of American Spiritualism* (Charlottesville: University of Virginia Press, 2003), p. 71.

324 *"disease of a starved heart"*: Cox, *Body and Soul*, pp. 71–72.

324 *including two of the three Fox sisters*: Ibid., p. 124.

324 *"While some are crying against it"*: Fornell, *Unhappy Medium*, p. 79.

325 *"when a sympathetic cord"*: Cox, *Body and Soul*, p. 111.

325 *"spiritual telegraph"*: In 1855, George Templeton Strong marveled, "What would I have said six years ago to anybody who predicted that before the enlightened nineteenth century ended that hundreds of thousands of people in this country would believe themselves able

to communicate with the ghosts of their grandfathers?" *Diary of George Templeton Strong,* ed. Allan Nevins and Milton Halsey Thomas (New York: Macmillan, 1952), vol. 2, pp. 244–45.

325 *"The gentle whispers"*: Carroll, *Spiritualism,* p. 91.

325 *"the only religious sect in the world"*: And this resulted in a number of spiritualists being involved in feminist enterprises, such as the Woman's Medical College of Philadelphia. See Braude, *Radical Spirits,* p. 149. Susan B. Anthony, on a speaking tour of New York State in 1855, confessed that she envied trance speakers: "If the spirits would only just make me a trance medium and put the right thing into my mouth. . . . You can't think how earnestly I have prayed to be made a speaking medium for a whole week. If they would only come to me thus, I'd give them a hearty welcome." Braude, *Radical Spirits,* p. 96.

325 *"the medium may be man or woman"*: Braude, *Radical Spirits,* p. 23.

326 *"site designed for the contemplation"*: Ibid., p. 53.

326 *Harriet Beecher Stowe*: Barbara White, *The Beecher Sisters* (New Haven, CT: Yale University Press, 2003), p. 237.

326 *more than two million Americans subscribing to spiritualist beliefs*: Ibid., p. 233.

327 *"women who had lost their blue-eyed children"*: Fornell, *Unhappy Medium,* p. 120.

327 *Colchester lingered in Washington*: Ross, *President's Wife,* p. 185. See also: *Lincoln Lore,* no. 1497: Henry "discovered Colchester's secret. While traveling on a train

he happened to sit by a young passenger, and in making friendly conversation his traveling companion mentioned that he was a manufacturer of telegraph instruments. He even volunteered the information that 'I also made them for spiritualists.' The young man explained how his device would fit around the bicep [sic], whereby the medium, by expanding his muscle, could produce sharp clicks like a telegraph key. Henry's young friend said, 'Have you heard of that Colchester fellow? He uses my equipment.' Lincoln was pleased to learn the secret."

328 *she had been in contact with her son Willie*: Fornell, *Unhappy Medium*, p. 119. Mrs. Lincoln also told Browning that her husband would have to dismiss his Cabinet to win the war; apparently the spirits became vocal with advice on how to conduct the war.

328 *reflected his extreme curiosity and courtesy rather than any affinity*: See "Sec. Chase and Mrs. Case to see President," in Miers, *Lincoln Day by Day*, Sept. 15, 1862.

329 *Her belief that her dead boys appeared*: Her niece Katherine Helm cited it as evidence of her famous aunt's mental imbalance in her biographical reminiscences about Mrs. Lincoln. Helm, *Wife of Lincoln*, pp. 226–27.

329 *made much of the friendships*: Pinsker, *Lincoln's Sanctuary*, p. 60.

329 *their "clatter of sabers and spurs"*: Packard, *Lincolns in the White House*, p. 140.

329 *Lincoln might pull rank*: Pinsker, *Lincoln's Sanctuary*, p. 61.

330 *Derickson quickly became a favorite*: Although C. A. Tripp characterized this relationship as the president's being smitten from the day they met, one among

Lincoln's allegedly multiple homoerotic liaisons, Matt Pinsker's characterization of this relationship is perhaps a more nuanced appreciation of this string of clues.

330 *often a dinner companion*: Scholars disagree over interpretation of the comments about Derickson and Lincoln. See especially books by David Donald, Doris Kearns Goodwin, Jean Baker, C. A. Tripp, and Matthew Pinsker listed in the bibliography.

330 *"making use of His Excellency's"*: Pinsker, *Lincoln's Sanctuary*, p. 84.

330 *"What stuff!"*: Virginia Fox Diary, Fox Papers, Manuscript Collections, Library of Congress.

331 *sharing her husband's bed*: see Pinsker, *Lincoln's Sanctuary*, pp. 106–107.

331 *"a friend and acquaintance"*: Ibid., p. 85.

332 *This practice did not indicate*: Americans today might visit Europe and see many young women walking together arm in arm, and even kissing warmly at greeting or departure, and mistakenly read this intimacy as indicative of an erotic lesbian relationship—instead of being a cultural norm in another country, just as sharing a bed was a country custom of the American past, not necessarily proving a sexual relationship.

332 *She even attempted*: Pinsker, *Lincoln's Sanctuary*, p. 85.

334 *"If my good, patient Husband, were here*: Turner and Turner, *Mary Todd Lincoln*, p. 138.

334 *"a spy was discovered"*: "Spy Lurking Around the President's House," *National Republican*, Sept. 24, 1862.

335 *"I suppose you have heard"*: T. J. Barnett to S. L. M. Barlow, Oct. 27, 1862, Washington, Barlow Papers, Huntington Library, San Marino, CA.

336 *the President's Military Guard*: Pinsker, *Lincoln's Sanctuary*, p. 86.

336 *volunteered to sign a petition*: Turner and Turner, *Mary Todd Lincoln*, p. 142.

336 *"I have waited in vain"*: Ibid., p. 139.

337 *Mary's absence had allowed*: Pinsker, *Lincoln's Sanctuary*, p. 83.

337 *put him on the road to revolution*: See Eric Foner, "Lincoln, Slavery and Colonization," Cunliffe Lecture, Mar. 5, 2008, University of Sussex (publication forthcoming).

338 *"Fellow citizens"*: Randall, *Mary Lincoln*, p. 320.

TWELVE : GLOOMY ANNIVERSARIES

339 *"From this time until spring"*: Turner and Turner, *Mary Todd Lincoln*, p. 143.

340 *"I may not have made as great a President"*: Donald, *Lincoln*, p. 406.

340 *"traitoress Mrs. Lincoln"*: Ibid., p. 399.

341 *"a cheer that was almost deafening"*: James McPherson, *Marching Toward Freedom: The Negro in the Civil War, 1861–1865* (New York: Knopf, 1968), pp. 25–26.

342 *General Saxton in South Carolina organized*: J. Matthew Gallman, *The Civil War Chronicle* (New York: Gramercy Books, 2003), pp. 213–14.

342 *"Day by day her carriage is seen"*: National Republican, Jan. 26, 1863.

342 *"If she were worldly-wise"*: Burlingame, *Inside the White House*, pp. 87–88.

343 *"it was a day & occasion never to be forgotten"*: Eliza S. Quincy to Mary Lincoln, Jan. 2, 1863, Abraham Lincoln Papers, Library of Congress.

343 *"there was not a member of the Cabinet"*: Laas, *Wartime Washington*, p. 231.

343 *attending a patriotic reading*: Miers, *Lincoln Day by Day*, Jan. 19, 1863.

344 *"My notions of duty"*: Elizabeth Keckly suggested that Robert's lofty soul prevented him from stooping "to all the follies and absurdities of the ephemeral current of fashionable life." He would disdain such common fraternizing most of his adult life. Keckley, *Behind the Scenes*, p. 124.

344 *"wedding finery"*: The papers would report a wedding gift of a gorgeous set of Chinese fire screens, richly inlaid with gold, silver, and pearl, from the Lincolns. See Sandburg, *War Years*, p. 291.

344 *"tall, graceful, her small Greek head"*: Sandburg, *War Years*, p. 292.

344 *Virginia Fox called on Mary*: Virginia Fox Diary, Fox Papers, Manuscript Collections, Library of Congress.

345 *"Our heavy bereavement"*: MTL to Benjamin French, Mar. 10, 1863, Louise and Barry Taper Collection.

345 *"My dear Mrs. Welles"*: Turner and Turner, *Mary Todd Lincoln*, p. 147.

345 *"the affability for which she is distinguished"*: New York Herald, Feb. 22, 1863.

346 *"he had done nothing to make any human being remember"*: Miller, *Lincoln's Virtues*, p. 68.

346 *enlisting African Americans*: CWL, vol. 6, p. 158.

346 *Mrs. Lincoln looked particularly regal*: Washington *Chronicle*, Mar. 2, 1863, and French, *Witness*, p. 417.

346 *continued her hospital work*: MTL to Peter Watson, Mar. 10, 1863, Turner and Turner, *Mary Todd Lincoln*, p. 148. See also: MTL to George Harrington, Mar. 20, 1863, p. 149, and Telegram to Gustav Gumpert, p. 150. On April 22, Mrs. Lincoln requested a place for a Miss Coburn, who may have been the spiritualist who brought Mary such relief through séances after Willie's death. Turner and Turner, *Mary Todd Lincoln*, p. 150. On June 16, she wrote to the husband of her good friend Mary Jane Welles, Secretary of the Navy Gideon Welles, begging him to give "the bearer of this note" a post as assistant paymaster, confiding, "I am sure you will not refuse me this trifling courtesy." MTL to Gideon Welles, in Schwartz and Bauer, "Unpublished Mary Todd Lincoln," p. 5.

346 *Swisshelm was a radical abolitionist*: See Sylvia Hoffert, *Jane Grey Swisshelm: An Unconventional Life* (Chapel Hill: University of North Carolina, 2004).

347 *preserve the glove*: Randall, *Mary Lincoln*, pp. 321–22.

347 *"I recognized Mrs. L as a loyal"*: Ibid., p. 322.

347 *"to Emancipation, as a matter of right"*: Ibid., p. 323.

348 *"whatever aid or counsel"*: Ibid., p. 322.

348 *"Her expressions were strong"*: Ibid., p. 323.

348 *"often late in the evening"*: Ibid., p. 319.

349 *She married a titled European*: Prince Felix zu Salm-Salm left Europe because of his scandalous love affairs, duels, and enormous gambling debts, and came to America in 1861 to take part in the Civil War. He met General Blenker at Hotel Willard. The general wanted

to make Felix a colonel and his chief of staff, and after a short meeting with Felix, Lincoln gave his approval.

350 *"kissed him three times"*: Michael Burlingame, *Lincoln Observed: Civil War Dispatches of Noah Brooks* (Baltimore: Johns Hopkins University Press, 1998), p. 35.

350 *"I never knew until last night"*: Julia L. Butterfield, *A Biographical Memorial of General Daniel Butterfield* (New York: Grafton Press, 1904), p. 162. Butterfield's account does demonstrate Mary Lincoln's dictatorial attitude toward her husband's attentions and her vindictiveness toward those who might cross her by allowing the president to be exposed for any time to adoring female company. Yet this anecdote shows several different aspects of the relationship between the Lincolns.

351 *"Mrs. Lincoln's attire"*: *New York Herald*, Apr. 10, 1863.

351 *"They are fighting at the front"*: Nettie Coburn Maynard, *Was Abraham Lincoln a Spiritualist? or, Curious Revelations from the Life of a Trance Medium* (Whitefish, MT: Kessinger, 2003), p. 100. Coburn goes on to suggest that Mary Lincoln insisted upon an impromptu séance to "see if we can get anything from 'beyond?'" and after their session, was considerably soothed by the results.

352 *"Think you better put Tad's pistol away"*: *CWL*, vol. 6, p. 256.

352 *"I am very well"*: Lincoln Papers, Illinois State Historical Library.

352 *his brother-in-law Ninian Edwards*: *CWL*, vol. 6, pp. 275–76.

352 *"I do not think the raid into Pennsylvania"*: Lincoln Papers, Illinois State Historical Library.

353 *She organized the move*: "The afternoon papers say that President Lincoln and family would leave the city to-day for their summer retreat, the Soldier's Home." *New York Herald*, June 23, 1863.

353 *Mary was thrown to the ground*: "Mrs. Lincoln nearly met with a fatal accident on the 2nd July in consequence of her horses taking fright. She threw herself out of her carriage. Fortunately no bones were broken, and after some restoratives was taken to her residence." *Frank Leslie's Illustrated Newspaper*, July 18, 1863.

353 *"The driver's seat became detached"*: *New York Herald*, July 4, 1863.

354 *Mrs. Lincoln was attended by a doctor*: "Surgeons from the Carver Hospital came to her aid and she was then taken to the hospital. No bones broken. Dr. Judson brought his carriage to the door of hospital and Mrs. Lincoln taken home to White House." She was actually taken to the Soldiers' Home, which was closer. See Boyden, *Echoes from Hospital*, p. 139.

355 *"Anything else I can do?"*: Burlingame, *Inside the White House*, pp. 117–18. The details of this conversation would indicate that it took place before the head injury.

355 *"Can anybody measure the fierce intoxication"*: Ibid.

355 *Because Mary's accident*: "Don't be uneasy. Your mother very slightly hurt by her fall. A.L." Helm, *Wife of Lincoln*, July 3, 1863, p. 211.

356 *Draft Riots broke out*: Miers, *Lincoln Day by Day*, July 2–3, 1863.

356 *"every brute in the drove was pure Celtic"*: Gallman, *Civil War Chronicle*, p. 334.

356 *"Wherever Negroes were discovered"*: Salm-Salm, *Ten Years of My Life*, p. 56.

356 *"Why do I hear no more of you"*: *CWL*, vol. 6, p. 327.

357 *He penned letters of rebuke*: Goodwin, *Team of Rivals*, p. 536.

357 *"in tears with his head bowed"*: John S. Goff, *Robert Todd Lincoln* (Manchester, VT: Friends of Hildene, 1969), p. 52.

357 *Under Pomroy's vigilant care*: Mr. Lincoln called her "one of the best women I ever knew." Randall, *Mary Lincoln*, p. 325.

357 *"We are not above the tide-water here"*: Burlingame, *Inside the White House*, pp. 123–24.

358 *The newspapers reported that both men and horses*: John Hay to John Nicolay, Aug. 13, 1863, Burlingame, *At Lincoln's Side*, p. 50.

358 *to flee the capital*: Apparently Mary's elder son might have had something else to flee. John Hay commented that the wedding of a "much admired lady" had Robert Lincoln "shattered." Ibid.

358 *"Mrs. Lincoln was delighted"*: New York Daily Tribune, Aug. 22, 1863.

359 *"Tell dear Tad, poor 'Nanny Goat' is lost"*: See D. P. Bacon to Abraham Lincoln, Apr. 25, 1864, *CWL*, vol. 6, p. 372.

359 *"The Tycoon is in fine whack"*: Burlingame, *At Lincoln's Side*, p. 49. Some might use this to suggest that Lincoln was happiest without Mary around, but this would be a

stretch and a misreading of the complicated couple, each of whom was happiest believing the other was well and happy—by one another's side, or while absent.

359 *her various perambulations*: See the *New York Herald*, Sept. 1, 1863, p. 1; Sept. 17, 1863, p. 7; and Sept. 26, 1863, p. 3.

359 *"I neither see nor hear anything"*: AL to MTL, Sept. 20 and 21, 1863, Abraham Lincoln Papers, Library of Congress.

360 *"Mrs. Cuthbert did not correctly understand"*: Ibid., Sept. 22.

360 *Mary responded*: Mary Lincoln to Edward McManus, Sept. 21, 1863; and Mary Lincoln to Abraham Lincoln, Sept. 22, 1863: both in Turner and Turner, *Mary Todd Lincoln*, p. 157.

361 *"I feel as David of old did"*: Helm, *Wife of Lincoln*, p. 217.

361 *"She knew that a single tear shed"*: Ibid.

361 *to help secure a pass for travel*: John Helm to Betsy Humphreys Todd, Oct. 11, 1863, ibid., p. 219.

361 *Lincoln gave written permission*: Oct. 15, 1863, Abraham Lincoln to Lyman Todd, Abraham Lincoln Papers, Library of Congress.

361 *Mary was beset by worries*: *Lincoln Lore*, no. 155.

362 *the engagement ring, a solitaire worth $4,000*: Sandburg, *War Years*, p. 148.

362 *"take the cuss off the meagerness"*: Burlingame, *Lincoln Observed*, p. 90.

363 *"So many ladies here, all dressed in mourning"*: Faust, *Republic of Suffering*, p. 149.

364 *interring nearly one hundred Union corpses*: See Eileen F. Conklin, *Women of Gettysburg* (Gettysburg, PA: Thomas Publications, 1993).

364 *town to become a national shrine*: See James McPherson's story about Russian visitors to the United States during the Bicentennial, in James M. McPherson, *Hallowed Ground: A Walk at Gettysburg* (New York: Crown, 2003), p. 16.

365 *"Mrs. Lincoln informs me"*: Sandburg, *War Years*, p. 465.

367 *"I should be glad if I could flatter"*: Ibid., p. 466.

367 *lavished even more praise*: "Edward Everett," *Lincoln Lore*, no. 410.

367 *"may be credited to the influence"*: Helm, *Wife of Lincoln*, p. 221.

THIRTEEN : DIVIDED HOUSES

369 *"Send her to me, A. Lincoln"*: Helm, *Wife of Lincoln*, p. 222.

369 *"the scapegoat for both North and South"*: Helm, *Wife of Lincoln*, p. 225.

369 *Her mail was often so hateful*: Helen Nicolay, *Lincoln's Secretary: A Biography of John G. Nicolay* (1949; repr. Westport, CT: Greenwood Press, 1971), p. 192.

369 *Republican ally Orville Browning suggested*: 14 December 1863, *Diary of Orville Hickman Browning*, Dec. 14, 1863. The very same day, Lincoln signed an order indicating that Emilie claimed to "own some cotton at Jackson, Mississippi and also some in Georgia; and

I shall be glad, upon either place being brought within our lines, for her to be afforded the proper facilities to show her ownership, and take her property." Abraham Lincoln Papers, Library of Congress. He was trying to assist his destitute relative, but still insisted that she take the oath to have her property restored.

370 *when he came to her chamber*: Helm, *Wife of Lincoln*, p. 226.

370 *"He lives, Emilie!"*: Ibid., p. 227.

371 *"it is good for her"*: Ibid., p. 225.

374 *"My wife and I are in the habit"*: Ibid., p. 231.

374 *"my being here is more or less"*: Ibid.

374 *"Sister and I cannot"*: Ibid., p. 224.

374 *Lincoln expressed the hope*: Ibid., p. 233.

374 *Martha Todd White, created a scandal*: Martha Todd White to Abraham Lincoln, Dec. 19, 1863, Abraham Lincoln Papers, Library of Congress. See also: *National Intelligencer*, Mar. 29, 1864, and *New York Herald*, Apr. 2, 1864. I am indebted to Amy Murell Taylor for her unpublished essay on White's involvement with the Lincolns.

375 *"Mrs. J. Todd White"*: Berry, *House of Abraham*, p. 159.

376 *Cabinet heard details from the president*: *Diary of Gideon Welles, Secretary of the Navy Under Lincoln and Johnson*, ed. Howard K. Beale (New York: Norton, 1960), vol. 2, p. 20, Jan. 29, 1864. Welles reported that Mrs. White asked that her trunks not be searched, and the president refused. Lincoln had also warned that if she did not leave, "she might expect to find herself within twenty-four hours in the Old Capitol Prison."

376 *these rumors were damaging*: The story appeared in papers in Washington, New York, Chicago, Houston, Cincinnati, Detroit, Memphis, and St. Louis. See Berry, *House of Abraham*, pp. 159–60.

376 *"from the first without truth or foundation"*: Ibid., p. 60.

376 *asked her not to defend the First Lady*: Mrs. H. C. Ingersoll, "Memories of Mrs. Lincoln," *Springfield Daily Republican*, June 7, 1875.

377 *By Christmas, Lincoln felt fully recovered*: Lincoln to Stanton, Dec. 26, 1863, Miers, *Lincoln Day by Day*.

378 *while Mary Lincoln had abandoned*: Noah Brooks, *Mr. Lincoln's Washington: Selections from the Writings of Noah Brooks, Civil War Correspondent*, ed. P. J. Staudenraus (South Brunswick, NJ: T. Yoseloff, 1967), p. 273.

378 *inaugurating the winter season*: *New York Times*, Jan. 13, 1864.

379 *"if he thought he could make anything"*: Keckley, *Behind the Scenes*, p. 129.

379 *observed that Chase's conduct was so disgraceful*: Anson G. Henry to Isaac Newton, April 21, 1864, Abraham Lincoln Papers, Library of Congress.

380 *White House stables were consumed by the flames*: Miers, *Lincoln Day by Day*, Feb. 20, 1864.

380 *Varina Davis, seven months pregnant*: Cashin, *Varina Davis*, p. 148.

380 *Her personal losses*: Ibid., p. 150.

380 *complaining that he was a butcher*: Turner and Turner, *Mary Todd Lincoln*, p. 161.

381 *knew how important it would become*: Indeed as scholars have emphasized, it was the Union soldiers' votes

(true citizen-soldiers during the Civil War) who would tip the balance in Lincoln's favor. See especially Jennifer L. Weber, *Copperheads: The Rise and Fall of Lincoln's Opponents in the North* (New York: Oxford University Press, 2006).

381 *"I will be clever to them"*: Keckley, *Behind the Scenes*, pp. 145–46.

381 *became a shrewd adviser and go-between*: Letters between Lincoln and Wakeman in July and August 1864 confirm this.

382 *John Hay asserted this hearsay as fact*: Burlingame and Ettlinger, *Inside Lincoln's White House*, Feb. 13, 1867.

383 *"an anti-Lincoln man"*: Harry Carman and Reinhard H. Luthin, *Lincoln and the Patronage* (New York: Columbia University Press, 1943), p. 280.

383 *"Please say not a word"*: Turner and Turner, *Mary Todd Lincoln*, p. 181.

384 *"Mrs. Lincoln ransacked"*: *New York Herald*, May 2, 1864.

384 *"While her sister women scraped lint"*: Mary Clemmer, *Ten Years in Washington: Life and Scenes in the National Capital, as a Woman Sees Them* (Hartford, CT: A. D. Worthington, 1873), p. 237.

384 *ladies were agreeing not to use*: *New York Times*, May 2, 1864.

385 *She rationalized: "I must have more money"*: Keckley, *Behind the Scenes*, pp. 149–50.

386 *"hundreds of them are getting immensely rich"*: Ibid., pp. 149–50.

386 *Robert Lincoln would later confide*: Helm, *Wife of Lincoln*, p. 267.

387 *Mrs. Lincoln would "shed tears"*: Tyler Dennett, ed., *Lincoln and the Civil War in the Diaries and Letters of John Hay* (New York: Dodd, Mead, 1939), p. 274.

387 *"My life in College"*: See "Big Names for DKE History," http://www.dke.org/Heritage/Big_Names .htm.

388 *since she had redecorated it*: Packard, *Lincolns in the White House*, p. 188.

388 *"None were more kind or showed a nobler spirit"*: Virginia Brown, *Through Lincoln's Door* (Springfield, IL: Li-Co Art & Letter Service, 1952), p. 22.

389 *"My dear Mrs. Agen—"*: See Turner and Turner, *Mary Todd Lincoln*, p. 179.

390 *she complained that most days*: Turner and Turner, *Mary Todd Lincoln*, p. 187.

390 *bittersweet for Mary Lincoln*: Goodwin, *Team of Rivals*, p. 704. Nicolay wanted to give up his fourteen-hour days at the White House and to marry. He was eager for the diplomatic posting to the Paris consulate, and its salary of $5,000. Hay was equally delighted to take a post as secretary to a U.S. legation in the French capital.

391 *wanted to bring six hundred bales of her family's cotton*: Lowell Harrison, *Lincoln of Kentucky* (Lexington: University of Kentucky Press, 2000), p. 219.

391 *She wanted an exemption*: See Berry, *House of Abraham*, p. 173.

392 *Lincoln allowed Mrs. Hunt to claim her own property*: Ross, *President's Wife*, p. 151.

393 *"your Minnie [Minié] bullets"*: Berry, *House of Abraham*, p. 174.

393 *"Excuse me, Mr. Lincoln"*: Statement of Henry Samuels, cited in Miller, *Lincoln's Virtues*, pp. 356–57.

394 *through his contact with Frederick Douglass*: For a careful look at Lincoln and Douglass, see James Oakes, *The Radical and the Republican: Frederick Douglass, Abraham Lincoln, and the Triumph of Antislavery Politics* (New York: Norton, 2007).

394 *Lincoln was concerned*: Mary Edwards Raymond, ed., *Some Incidents in the Life of Mrs. Benjamin S. Edwards* (privately published, 1909), p. 16.

395 *"Poor Mr. Lincoln is looking so broken-hearted"*: Keckley, *Behind the Scenes*, p. 157.

FOURTEEN : WITH VICTORY IN VIEW

397 *deftly argued that slavery's end*: Chandra Manning, *What This Cruel War Was Over: Soldiers, Slavery, and the Civil War* (New York: Knopf, 2007).

398 *"The oppressed will shout the name"*: Donald, *Lincoln*, p. 541.

398 *Lincoln feared emancipation*: Lincoln asked his Cabinet to allow a financial settlement: $400 million appropriated to the former Confederates to end slavery and declare peace. His conciliatory attitude was roundly rejected.

399 *This appointment silenced a large chorus of Lincoln's*: Donald, *Lincoln*, pp. 550–51; Goodwin, *Team of Rivals*, pp. 676–81.

400 *Scandalous reports began to circulate*: See MTL to Abram Wakeman, Feb. 18 and 20, 1865, in Turner and Turner, *Mary Todd Lincoln*, pp. 201–202.

400 *Mrs. Lincoln was willing to use her powers of persuasion*: On March 12, 1865, she writes about Mr. and Mrs. James Gordon Bennett, in New York, "Weed may try to sour them—all this is between ourselves." Turner and Turner, *Mary Todd Lincoln*, p. 204.

400 *"sharing secrets with her other Abraham"*: Baker, *Mary Todd Lincoln*, p. 234.

400 *Mary destroyed Wakeman's letters*: She discusses "espionage" and is clearly trying to assess damage, and asks Wakeman, "Please burn this note." See Turner and Turner, *Mary Todd Lincoln*, p. 202.

401 *Elizabeth Cady Stanton reported* : Ross, *President's wife*, p. 216.

403 *Lincoln wrote out an actual reprieve*: See http://www.whitehouse.gov/holiday/thanksgiving. See also: Gerald Prokopowicz, *Did Lincoln Own Slaves?* (New York: Pantheon, 2007), p. 116.

403 *"the kindest, brightest, most considerate*: Randall, *Lincoln's Sons*, p. 107.

403 *"the tyrant of the White House"*: Ibid., p. 109.

404 *consoled himself with his father*: On Tad's role, see Randall, *Lincoln's Sons*, pp. 140–42.

404 *"There was no hilarity excepting"*: Harold Holzer, ed., *Lincoln as I Knew Him: Gossip, Tributes, and Revelations from His Best Friends and Worst Enemies* (Chapel Hill, NC: Algonquin Books of Chapel Hill, 1999), p. 244.

405 *not an unusual experience for them*: Randall, *Lincoln's Sons*, p. 178.

406 *Lincoln reportedly saw one hundred performances*: Goodwin, *Team of Rivals*, p. 309.

406 *He also set up a table*: Randall, *Lincoln's Sons*, pp. 193–94.

407 *Tad was leaning his head on his father's shoulder*: Auguste Laugel, *The United States During the Civil War*, ed. Allan Nevins (Bloomington: Indiana University Press, 1961), pp. 315–16.

407 *including Dr. Anson Henry*: Dr. Anson G. Henry was with Mrs. Lincoln on Feb. 8, 1865, and escorted her to a joint session of both houses of Congress. "President Lincoln went to the theatre with Dr. Anson G. Henry 'a few times just for a good hearty laugh, and because I felt that the President needed the relaxation it furnished, and which could not be had any other way.'" Dr. A. G. Henry to his wife, Washington, DC, May 8, 1865, Henry Papers, Illinois State Historical Library.

408 *"I want to give him all the toys"*: Randall, *Lincoln's Sons*, p. 180.

408 *"a little gun"*: Ibid., p. 179.

409 *and banished him* : Ibid., p. 108.

410 *was found guilty once again*: Ibid., pp. 114–15.

410 *once Tad was tucked away in bed*: Ibid., pp. 150–51.

410 *Tad was finally given his own personal firearm*: Ibid., pp. 160–61.

410 *"see something of the war"*: AL to Ulysses S. Grant, Jan. 19, 1865, *CWL*, vol. 8, p. 223.

411 *"I will be most happy"*: Goodwin, *Team of Rivals*, p. 684.

412 *sent along one of her bouquets*: Randall, *Lincoln's Sons*, p. 174.

412 *amused by his wife's attachment*: Helm, *Wife of Lincoln*, p. 274.

412 *about "our Iowa friend"*: Randall, *Lincoln's Son's*, p. 172.

414 *Mrs. Lincoln was resplendent*: Her gown alone was reportedly worth $2,000. Ross, *President's Wife*, p. 213.

414 *"greater advantage than usual"*: New York Herald, Mar. 5, 1865, p. 1.

416 *the First Lady suffered miserably*: See letters to Simeon Draper, Abram Wakeman, and others, Jan. 26–Mar. 20, 1865, Turner and Turner, *Mary Todd Lincoln*, pp. 199–206.

417 *She could not conceal*: Ross, *President's Wife*, p. 224.

419 *would end her days as the Countess*: Ibid., p. 226.

419 *"with an expression of pain and sadness"*: Adam Badeau, *Grant in Peace: From Appomattox to Mount McGregor, a Personal Memoir* (Hartford, CT: S. S. Scranton, 1887), pp. 356–65.

421 *"Will you not dine with us"*: Turner and Turner, *Mary Todd Lincoln*, pp. 211–12.

422 *"The President is looking much better"*: New York Herald, Apr. 9, 1865.

422 *Stretched out on Seward's sickbed*: Goodwin, *Team of Rivals*, p. 724.

423 *"If possible, this is a happier day"*: Turner and Turner, *Mary Todd Lincoln*, p. 216.

424 *"It is also unsatisfactory to some"*: CWL, vol. 8, p. 404.

424 *He vowed in anger*: See James Swanson, *Manhunt*, and Michael Kauffman, *American Brutus*.

425 *the whole town turned out*: *Illinois State Journal*, Apr. 11, 1865, p. 3.

427 *"We must* both *be more cheerful"*: F. B. Carpenter, *Six Months at the White House* (New York: Hurd & Houghton, 1867), p. 293.

427 *as Lincoln ever came to reprimanding*: Donald, *Lincoln*, p. 593.

428 *had turned the president's face "radiant"*: Randall, *Mary Lincoln*, p. 342.

428 *emblazoned in popular culture*: Timothy Good, ed., *We Saw Lincoln Shot: One Hundred Eyewitness Accounts* (Jackson: University Press of Mississippi, 1995).

428 *stimulated a daunting amount of literature*: There are over five hundred books listed on Amazon in connection with this topic, including at least two dozen scholarly books in print on the topic. The assassination's impact on Mary Lincoln is not underestimated, and yet there has been no corresponding scholarship.

432 *"Now he belongs to the ages"*: On whether Stanton said, "Now he belongs to the ages" or "Now he belongs to the angels," see Adam Gopnik's "Angels and Ages," *New Yorker*, May 28, 2007.

FIFTEEN : WAKING NIGHTMARES

435 *her care over to Elizabeth Keckly*: Keckley, *Behind the Scenes*, p. 191.

435 *Reverend Phineas Gurley presided*: To view his sermon, access http://showcase.netins.net/web/creative/lincoln/speeches/gurley.htm.

436 *a reverse of the seventeen-hundred-mile route*: With the exception of Cincinnati, which was bypassed on the return.

437 *Nearly all the copies of this last photo*: Stanton ordered all copies of the photograph destroyed, but one survived and was discovered in 1952.

439 *the coffin was taken to the state capitol for a final viewing*: Lincoln's body, displayed in an open coffin throughout the twenty-day journey, was powdered and changed daily by an undertaker who traveled from D.C.

442 *She donated scores of items*: Keckley, *Behind the Scenes*, p. 208.

442 *The dispersal of goods remains*: I was approached in February 2008 by the PBS program *The History Detectives* to discuss the provenance of some materials that may or may not have been handed down from the Lincoln White House to the Lincolns' coachman.

442 *threatened to bury her husband*: Thomas Craughwell, *Stealing Lincoln's Body* (Cambridge, MA: Harvard University Press, 2007), pp. 23–25.

443 *"Mary, you are younger than I"*: Tarbell, *Abraham Lincoln*, vol. 2, pp. 231–32.

445 *"Dear Madam"*: Turner and Turner, *Mary Todd Lincoln*, pp. 230–31. Her reply to Queen Victoria is on the same pages, where she suggests only the queen can "appreciate the *intense grief* I now endure."

446 *Mary penned a note*: May 22, 1865, Mary Lincoln to Madame Berghmans, Mary Todd Lincoln Collection, Lincoln Museum Collection, formerly held in Fort Wayne, Indiana.

448 *"buy an entire bolt of yard goods"*: Lloyd Ostendorf and Walter Olesky, eds., *Lincoln's Unknown Private Life: An Oral History by His Black Housekeeper, Mariah Vance, 1850–1860* (Mamaroneck, NY: Hastings House, 1995), p. 54.

449 *After several false starts*: Announcements about her impending departure appeared in the papers. On May 12, the *New York Herald* reported: "Mrs. Lincoln has nearly recovered. It is now announced that she will leave Washington for Chicago on Wednesday of next week May 17." *New York Herald*, May 12, 1865.

449 *"There was scarcely a friend"*: Keckley, *Behind the Scenes*, p. 208.

449 *she seemed in a stupor*: Margarita Spalding Gerry, ed., *Through Five Administrations: Reminiscences of Colonel William Crook, Bodyguard to President Lincoln* (New York: Harper & Brothers, 1910).

450 *"as soon be dead as be compelled"*: Keckley, *Behind the Scenes*, p. 212.

453 *regretted that appeals for the family*: Indeed, despite several public schemes, including one launched by Horace Greeley, only $10,000 was raised, which may have been a gift from a single donor.

455 *"She places her hand upon her forehead"*: Keckley, *Behind the Scenes*, pp. 311–12.

457 *gift giving*: See Kenneth Greenberg, *Honor and Slavery: Lies, Duels, Noses, Masks, Dressing as a Woman, Gifts, Strangers, Humanitarianism, Death, Slave Rebellions, the Proslavery Argument, Baseball, Hunting, and Gambling in the Old South* (Princeton, NJ: Princeton University Press, 1996).

457 "*sufficient to account for all*": See Clemmer, *Ten Years in Washington*, p. 241.

458 "from whence, *we have just come*": MTL to David Davis, June 27, 1865, Turner and Turner, *Mary Todd Lincoln*, p. 255.

458 *she felt she would like to slip*: MTL to Elizabeth Blair Lee, July 11, 1865, Turner and Turner, *Mary Todd Lincoln*, p. 258.

460 "*Roving generals have elegant mansions*": MTL to Alexander Williamson, Aug. 17, 1865, ibid., p. 265.

460 "*The family of the man*": MTL to Sally Orne, Jan. 10, 1866, ibid., p. 324.

460 "*Please do not mention that Sen. Sumner*": MTL to Alexander Williamson, Aug. 19, 1866, ibid., p. 383.

463 *She also made reference to other rich and beautiful materials*: MTL to Sally Orne, Aug. 31, 1865, ibid., p. 270.

464 *seven dozen pairs of gloves while claiming "rigid economy"*: See MTL to David Davis, Sept. 12, 1865, ibid., p. 275, n. 7.

465 *she wrote daily bulletins*: On Williamson's role, see Turner and Turner, *Mary Todd Lincoln*, pp. 247–48.

465 "*The family of the Martyred President*": Dec. 30, 1865, MTL to Sally Orne, ibid., p. 318.

466 *writing a letter to her favorite Washington jeweler*: MTL to Alexander Williamson, Jan. 3, 1866, ibid., p. 321.

466 *she did take them personally*: MTL to Sally Orne, Jan. 13, 1866, ibid., p. 326.

469 *Herndon's oral histories of Lincoln*: See Donald, *Lincoln*, and Wilson and Davis, *Herndon's Informants*.

471 *Herndon's campaign was particularly traumatizing*: Emerson, *Madness of Mary Lincoln*, p. 26. Although

I may have disagreements on several points, including substantial differences about Robert Lincoln's role and motives, I generally agree with many of Emerson's subtler suggestions about Mary Lincoln. Just because I choose not to diagnose the dead, I do not wholly condemn those who do.

471 *"the only one, he had ever thought of"*: MTL to Mary Jane Welles, Dec. 6, 1865, Turner and Turner, *Mary Todd Lincoln*, p. 296.

471 *no one should be compelled*: Emerson, *Madness of Mary Lincoln*, pp. 26–27.

472 *"There are hours of each day"*: MTL to James Smith, Dec. 17, 1866, Turner and Turner, *Mary Todd Lincoln*, p. 400.

SIXTEEN : WIDOW OF THE MARTYR PRESIDENT

476 *a robe that she had sewn for Mrs. Davis*: Fleischner, *Mrs. Lincoln and Mrs. Keckly*, p. 304.

476 *tenure as First Lady*: For a fascinating look at Mrs. Lincoln's clothing, see Donna D. McCreary, *Fashionable First Lady: The Victorian Wardrobe of Mary Lincoln* (Charlestown, IN: Lincoln Presentations, 2007). See especially the color plates between pp. 118–19.

476 *"Lizzie, I want to ask a favor of you"*: Excerpt from letter, MTL to Elizabeth Keckley [n.d., March 1867], Turner and Turner, *Mary Todd Lincoln*, p. 418.

477 *a pseudonym she had employed frequently*: Recently acquired telegrams by the Abraham Lincoln Presidential

Library demonstrate that Mrs. Lincoln used this name often in telegrams and to cloak any commercial dealings she wanted kept secret. See Thomas Schwartz, *Mary Todd Lincoln: First Lady of Controversy* (Springfield, IL: Abraham Lincoln Presidential Library and Museum, 2007).

480 *"I consider you my best living friend"*: MTL to Elizabeth Keckley, Oct. 6, 1867, Turner and Turner, *Mary Todd Lincoln*, p. 440.

481 *Two men seated next to her even discussed*: Keckley, *Behind the Scenes*, pp. 298–99.

481 *"pale as the tablecloth"*: Ibid., p. 300.

481 *"came up last evening like a maniac"*: MTL to Elizabeth Keckley, 6 October 1867, Turner & Turner, *Mary Todd Lincoln*, p. 440.

482 *"Mrs. Lincoln, the widow of the murdered President"*: Reprinted from the *New York World*, Oct. 7, 1867.

482 *"Will not somebody, for very shame sake"*: Fleischner, *Mrs. Lincoln and Mrs. Keckly*, p. 311.

482 *"her repugnant individuality"*: Turner and Turner, *Mary Todd Lincoln*, pp. 432–33.

483 *"Mrs. Lincoln invites and justifies"*: Ibid. A reporter from the *Pittsburg Gazette* asks what happened to the $10,000 and $5,000 raised from contributors for her support by citizens who sent their money to "B.B. Sherman of New York." See *New York Commercial Advertiser*, Oct. 15, 1867. Questions about money, scams, blackmail, and attacks upon character began to snake their way into press reports and resulted in nastiness—just as recent campaigns to fund families of those

killed on September 11, 2001, or solicitations for victims of the 2005 Katrina disaster have caused similar consternation in the public imagination.

484 *"If I had committed murder"*: MTL to Elizabeth Keckley, Oct. 9, 1867, Turner and Turner, *Mary Todd Lincoln*, p. 441.

484 *she wrote to a friend*: See MTL to Rhoda White, Oct. 18, 1867, ibid., pp. 444–45.

484 *"The simple truth, which I cannot tell to anyone not personally interested"*: Berry, *House of Abraham*, p. 186. I had suggested in earlier drafts of my study that Mrs. Lincoln suffered from "shopper's anorexia," but defer to Stephen Berry's vivid and apt characterization, which appeared in print shortly before this book went to press.

485 *his friend Edgar Welles offered him the chance*: MTL to Elizabeth Keckly, Turner and Turner, *Mary Todd Lincoln*, p. 449.

485 *"Those who swindled our Government"*: MTL to Alexander Williamson, October 29, 1867, ibid., p. 446.

485 *"to cross my purposes"*: MTL to Elizabeth Keckley, ibid., p. 449.

486 *Charity from African Americans*: Fleischner, *Mrs. Lincoln and Mrs. Keckly*, pp. 312–14.

487 *her clothes sale was proof of her diminished capacity*: Emerson, *Madness of Mary Lincoln*, pp. 28–29.

487 *Mary was aggrieved it had taken so long*: Davis had taken the roughly $75,000 from 1865 and parlayed it into an estate of more than $100,000, for which Mrs. Lincoln expressed her personal gratitude. In addition, Davis charged no personal expenses and did not take the

6 percent fee allowed by law. See Turner and Turner, *Mary Todd Lincoln*, p. 457.

488 *as he was "named for your husband"*: Dec. 19, 1867, MTL to Sarah Bush Lincoln, ibid., p. 465.

491 *The publisher, Carleton and Company, hawked the book*: Fleischner, *Mrs. Lincoln and Mrs. Keckly*, pp. 316–17.

492 *engineered by Redpath, apparently without Keckly's permission*: Fleischner, *Mrs. Lincoln and Mrs. Keckly*, pp. 316–17.

492 *"back stairs gossip of negro servant girls"*: Fleischner, *Mrs. Lincoln and Mrs. Keckly*, p. 317.

493 *dismissing her as "the colored historian"*: Fleischner, *Mrs. Lincoln and Mrs. Keckly*, p. 318.

493 *"says there is no doubt Mrs. L.—is deranged"*: MTL to Elizabeth Keckley, Oct. 9, 1867, Turner and Turner, *Mary Todd Lincoln*, p. 442.

494 *"only sunbeam in my sad future"*: Aug. 27, 1865, MTL to Rhoda White, ibid., p. 482.

494 *In the fall of 1867, she sold*: Evans, *Mrs. Lincoln*, p. 201.

494 *the peace she regularly required*: Turner and Turner, *Mary Todd Lincoln*, pp. 489–91; Fleischner, *Mrs. Lincoln and Mrs. Keckly*, p. 407.

496 *His mother admired the two gold bracelets*: MTL to Eliza Slataper, Sept. 25, 1868, Turner and Turner, *Mary Todd Lincoln*, pp. 484–85.

497 *"If you love me"*: MTL to Eliza Slataper, Sept. 27, 1868, Turner and Turner, *Mary Todd Lincoln*, p. 485.

498 *"His dark loving eyes—watching over me"*: MTL to Sally Orne, Dec. 29, 1869, ibid., p. 538.

499 *money worries*: MTL to George Boutwell, Dec. 4, 1868, ibid., pp. 491–92.

499 *"Nor can I live in a style"*: MTL to United States Senate, Dec. 1868, ibid., p. 493.

500 *she resisted because of high prices*: MTL to David Davis, Dec. 15, 1868, ibid., p. 497.

501 *legislative limbo*: Ibid., p. 300.

501 *"In memory of my great and good husband"*: Ibid., p. 507, n.

502 *Poor Sumner played Mary's messenger boy*: Ibid.

502 *she was sure she would have been welcomed*: MTL to Rhoda White, August 30, 1869, ibid., p. 517.

503 *"appears to me that we saw every place"*: MTL to Eliza Slataper, Aug. 21, 1869, ibid., p. 512.

503 *"Beautiful, glorious Scotland"*: Bartlett's Familiar Quotations.

504 *He was also able to joke*: Randall, *Lincoln's Sons*, p. 265.

504 *"My husband was so richly blessed"*: MTL to Sally Orne, Dec. 12, 1869, Turner and Turner, *Mary Todd Lincoln*, p. 534. See also her lie to Rhoda White that she had Robert when she was twenty, when she was actually closer to twenty-five: Dec. 20, 1869, ibid., p. 536.

505 *Mary was deeply offended*: MTL to Sally Orne, Dec. 16, 1869, ibid., p. 535.

505 *"I am alas! Out in THE COLD"*: MTL to Sally Orne, Feb. 11, 1870, ibid., p. 546.

506 *"As, in the midst of all my wickedness & transgressions"*: Ibid.

506 *"The gossips of the town"*: Emerson, *Madness of Mary Lincoln*, p. 30.

507 *"Surely the honorable members of the Senate"*: Ross, *President's Wife*, p. 298.

510 *"My husband, so fondly indulgent"*: MTL to Rhoda White, May 23, 1871, Turner and Turner, *Mary Todd Lincoln*, p. 589.

510 *"dangerously ill"*: MTL to Rhoda White, June 8, 1871, ibid., p. 590.

512 *"There is no life for me"*: Ross, *President's Wife*, p. 304.

SEVENTEEN : RISING FROM THE ASHES

517 *Nearly $2 million had gone up in smoke*: Although the fire would usher in a building boom, Black Friday in September 1873 would cause a national economic panic, effectively halting all new construction. In July 1874, a small fire (nicknamed the Little Fire) would consume millions of dollars' worth of post–1871 construction, requiring stricter building codes in the future. When the recession lifted, architects and developers combined to spearhead an urban renaissance in a style that came to be known as the Chicago School. This movement culminated in the creation of the White City, Chicago's contribution to the Columbian Exposition in 1893, pioneered by architect Daniel Burnham. See Erik Larson, *The Devil in the White City: Murder, Magic, and Madness at the Fair That Changed America* (New York: Crown, 2003), and Carl S. Smith, *The Plan of Chicago: Daniel Burnham and the Remaking of the American City* (Chicago: University of Chicago Press, 2006).

517 *"It is for these I grieve"*: Anna E. Higginson letter, in *Chicago: A Chronological and Documentary History*, Howard D. Furer, ed. (New York: Oceana Press, 1974), p. 112. With thanks to Taylor Patterson for sharing her work in progress on women in nineteenth-century Chicago.

518 *she divided Tad's inheritance*: "I understand that by the law, I am entitled to two thirds of them, but I should prefer only the half." MTL to David Davis, Turner and Turner, *Mary Todd Lincoln*, p. 597.

518 *"My own strength was then used up"*: Randall, *Lincoln's Sons*, p. 275.

519 *"I often tell Tad I can scarcely flatter myself"*: MTL to Mary Harlan Lincoln, Nov. 1870, Turner and Turner, *Mary Todd Lincoln*, p. 581.

519 *The return to Chicago*: Nearly all biographers agree on Mrs. Lincoln's ability to manipulate matters to try to gain sympathy for herself. But Jason Emerson argues, "While there is no doubt she suffered from physical ailments, she seems to have been something of a hypochondriac and to have used her illness to garner attention and sympathy, even to create her own martyrdom to parallel that of her husband, perhaps in an attempt to continue, at some level, her association with him." Indeed, Mrs. Lincoln may have been a hypochondriac, but it is an understandable fault that she would try to continue her "association with him." (Emerson, *Madness of Mary Lincoln*, p. 35.) Indeed this was the core of her identity—Abraham Lincoln's widow—to the end of her life.

521 *Disputes followed, aired in such articles*: Turner and Turner, *Mary Todd Lincoln*, p. 601.

521 *"sensational falsehoods"*: MTL to James Knowlton, Aug. 2, 1872, ibid., p. 598.

521 *A lady who was heavily veiled*: Joel Martin and William J. Birnes, *The Haunting of the Presidents: A Paranormal History of the U.S. Presidency* (New York: Signet, 2003), p. 189.

522 *"suffered from periods of mild insanity"*: Evans, *Mrs. Lincoln*, p. 307.

523 *"The following news item"*: *Illinois State Journal*, Aug. 7, 1874, p. 2, col. 2.

524 *She wrote to a friend, "The sudden and unexpected news"*: MTL to Elizabeth Swing, Mar. 12, 1874, Illinois State Archives.

524 *Mrs. Lincoln had arrived in Florida*: See *Illinois State Journal*, Dec. 1, 1874, p. 1, col. 1, and the *Aurora Beacon*, Nov. 25, 1874, p. 1, col. 3.

525 *in light of the events"*: Indeed, it is important to distinguish between the period of institutionalization and the period of "insanity," which many scholars suggest it was. Insanity was a legal term, and Mary Lincoln was declared insane, but she was also released from her institutionalization within a matter of weeks, and then given back her legal rights and unfettered freedom exactly one year after her trial, which does shed light on the legal as opposed to the medical aspects of this case.

526 *because of a longstanding estrangement*: Turner and Turner, *Mary Todd Lincoln*, p. 611.

527 *Mary predicted that there would soon be another devastating fire:* Emerson, *Madness of Mary Lincoln,* p. 45.

527 *pulling wires from inside her head:* Mark E. Neely Jr. and Gerald R. McMurtry Jr., *The Insanity File: The Case of Mary Todd Lincoln* (Carbondale: Southern Illinois University Press, 1986), p. 81.

528 *mixing medications:* For new insight into this topic, please see Oliver Sacks, "Patterns," posted Feb. 13, 2008, on the *New York Times* blog. I am deeply indebted to Jody Block for this reference, and many other insights and clippings to assist with my research and writing.

530 *He conferred with a group of legal and medical experts:* Emerson, *Madness of Mary Lincoln,* p. 50.

532 *"unnecessary excitement":* Ibid., p. 56.

533 *"Can the word of the experts, all men":* Neely and McMurtry, *Insanity File,* p. 31.

533 *Emerson argues that all these men:* Emerson, *Madness of Mary Lincoln,* p. 59. Class trumps gender in this interpretation.

533 *"O Robert, to think that my son":* Chicago Inter Ocean, May 20, 1875.

535 *"discharged his delicate and unhappy task":* Emerson, *Madness of Mary Lincoln,* p. 63.

535 *"the unfavorable impressions created by her conduct":* Emerson, *Madness of Mary Lincoln,* p. 64. This is a very loaded sentiment, indicating that Mrs. Lincoln was a publicity problem and that incarceration in an asylum was perceived as resolving a *longstanding* issue of public perceptions.

535 *who might question his decision*: Although Hay might have had a low opinion of Mrs. Lincoln while they sparred in the White House, he was concerned about the former First Lady being put into an institution rather than being cared for at home, as were most Americans who read about her situation.

536 *"now that a court of justice"*: Emerson, *Madness of Mary Lincoln*, p. 64.

536 *drugs on the night of May 19*: See Norbert Hirschhorn, "Mary Lincoln's 'Suicide Attempt': A Physician Reconsiders the Evidence," *Lincoln Herald*, vol. 104, no. 3, Fall 2003. Also see Emerson, *Madness of Mary Lincoln*, pp. 67–70.

537 *scholars who dismiss chloral hydrate*: See Oliver Sacks, forthcoming work on migraines and his comments on chemically induced hallucinations: http://oliversacks.com/news/2008/02/14/new-york-times-migraine-blog/.

541 *in a trunk held by descendants of a Lincoln family lawyer*: See Jason Emerson, "The Madness of Mary Lincoln," *American Heritage*, July 2006. In his article, Emerson details his discovery of copies of letters, the story of the destruction of the originals by Mary Harlan Lincoln's representatives, and how and why this adds importantly to the saga.

542 *many doctors seem to want to diagnose without reliable evidence*: For a recent example of this, see appendix 3 in Jason Emerson's *The Madness of Mary Lincoln*. For a very judicious attempt at such diagnosis, see Hirschhorn, "Mary Lincoln's 'Suicide Attempt.' "

545 *"loose" behavior*: MTL to Mercy Conkling, Nov. 19, 1864, Turner and Turner, *Mary Todd Lincoln*, pp. 187–88.

545 *during the years leading up to the Civil War*: Berry, *House of Abraham*, p. 133.

546 *"made the world hate her"*: Wilson and Davis, *Herndon's Informants*, p. 444.

547 *"to print the news and raise hell"*: Emerson, *Madness of Mary Lincoln*, p. 79.

548 *"It does not appear that God is good, to have placed me here"*: Ibid., p. 163.

548 *"behind the grates and bars"*: Emerson, ibid., p. 83.

549 *"a high priestess in a gang of Spiritualists"*: Ibid., p. 85.

549 *"cut off absolutely all communication"*: Ibid., p. 92.

550 *"It has been stated that she"*: *Chicago Tribune*, Aug. 29, 1875, p. 16.

551 *"making herself talked about by everybody"*: Emerson, *Madness of Mary Lincoln*, p. 61.

551 *"I would be ashamed to put on paper"*: Neely and McMurtry, *Insanity File*, p. 37.

552 *allowed a "visit" to her sister*: There is debate whether or not Mrs. Lincoln stopped off in Chicago to visit with the Bradwells on her way to see her sister.

552 *"She has dined at Mrs. Smith's"*: Emerson, *Madness of Mary Lincoln*, p. 102.

553 *Carrying a gun might or might not*: See Emerson, *Madness of Mary Lincoln*, pp. 108–10.

553 *on the night he died*: Again, thanks to Louise Taper for reminding me of this.

554 *"that the said Mary Lincoln is restored to reason"*: Neely and McMurtry, *Insanity File*, p. 102.

555 *"a faithful, devoted son"*: Emerson, *Madness of Mary Lincoln*, p. 172.

555 *highlighting her fierce indignation*: Ibid., pp. 168–70.

555 *"None of my treasure in the way of rich"*: MTL to Myra Bradwell, July 7, 1876, ibid., p. 170.

555 *"your game of robbery"*: MTL to Robert Lincoln, June 19, 1876, Turner and Turner, *Mary Todd Lincoln*, pp. 615–16.

556 *called Robert the "monster of mankind"*: Emerson, *Madness of Mary Lincoln*, p. 107.

556 *blaming him for stirring up even more trouble*: Ibid., pp. 170–72.

556 *Frances Wallace, who had been widowed*: Susan Krause, Kelley A. Boston, and Daniel W. Stowell, *Now They Belong to the Ages: Abraham Lincoln and His Contemporaries in Oak Ridge Cemetery* (Springfield: Illinois Historic Preservation Agency, 2005), p. 59.

EIGHTEEN : SMOLDERING EMBERS

560 *to confirm the attractive climate*: One physician's book, Alexander Taylor's *A Comparative Enquiry as to the Preventive and Curative Influence of the Climate of Pau and of Montpellier, Hyeres, Nice, Rome, Pisa, Florence, Naples, Biarritz, etc. on Health and Disease, with a Description of the Watering Places of the Pyrenees and of the Virtues of Their Respective Mineral Sources* (London: John W. Parker and Son, 1856), contained elaborate meteorological tables from 1843–1856, providing temperatures twice a day, as well as recording rainfall.

560 *"a capful of wind"*: Taylor, *Climate of Pau*, p. 37.

560 *Pau escaped extreme temperatures*: Ibid., p. 82.

561 *"ladies find that their hair retains the curl"*: As to more substantive matters, Taylor wrote, "The population of Pau and its neighborhood has possessed a very marked exemption from those epidemic diseases which have at different periods raged in Europe." Ibid., p. 112.

561 *guests seeking all kinds of improvements*: One invalid arrived in Pau and within days found that his overly rapid pulse fell several beats per minute, permanently. Ibid., p. 108.

561 *"the price of substantial articles"*: Ibid., p. 40.

562 *"congestion of the brain"*: Ibid., p. 272.

562 *its accommodations were "cheaper"*: Ibid., pp. 300–301.

563 *"write at once & direct to Pau, France"*: MTL to Edward Lewis Baker Jr., Oct. 17, Turner and Turner, *Mary Todd Lincoln*, p. 618.

564 *writing from the Hôtel de la Paix*: Turner and Turner, *Mary Todd Lincoln*, Apr. 23, 1877.

564 *a pocket guide to Pau*: *Liste d'Appartements, Hôtels & Villas à Louer: Deuxième Partie du Guide de l'Étranger à Pau et aux Environs* (Pau: Imprimerie et Lithographie Veronese, 1883), p. 41.

567 *disturbing news arrived from Illinois*: Thomas Craughwell, *Stealing Lincoln's Body* (Cambridge, MA: Belknap Press of Harvard University Press, 2007), pp. 154–64.

569 *"I was so cruelly persecuted"*: MTL to Edward Lewis Baker Jr., Apr. 11, 1877, Turner and Turner, *Mary Todd Lincoln*, p. 633.

569 *The letters from Mary to Bunn*: For a nuanced analysis of this correspondence and its meaning, see Turner and Turner, *Mary Todd Lincoln*, especially pp. 619–20.

570 *appeared on the "Lunatic Record"*: Ibid., p. 611.

570 *A little over a year later*: MTL to Jacob Bunn, Dec. 12, 1876, ibid., p. 622–23.

571 *her sister Frances*: Krause, Boston, and Stowell, *Now They Belong to the Ages*, p. 59.

572 *a plea that echoed Mary's*: MTL to Edward Lewis Baker Jr., Jan. 16, 1880, Turner and Turner, *Mary Todd Lincoln*, p. 694.

572 *sent her into a swoon*: MTL to Edward Lewis Baker Jr., June 22, 1879, ibid., pp. 683–84.

573 *"somewhere in Europe"*: Emerson, *Madness of Mary Lincoln*, p. 128.

574 *A local guidebook described*: Pierre Tucoo-Chala, *Pau Ville Américaine* (Pau: Cairn Editions, 1997), p. 18. Also see full menu, p. 19.

575 *"with the exception of a very few, I detest them all"*: MTL to Edward Lewis Baker Jr., June 12, 1880, Turner and Turner, *Mary Todd Lincoln*, p. 699.

576 *"In ill health & sadness"*: MTL to Edward Lewis Baker Jr., June 12, 1880, ibid., p. 700.

576 *actress Sarah Bernhardt*: Ibid., p. 704.

577 *"Rely, upon my passage"*: MTL to Edward Lewis Baker Jr., Oct. 7, 1880, ibid., p 703.

577 *"The widow of ex-President Lincoln"*: New York Herald, Oct. 28, 1880, p. 6, col. 6.

579 *Granddaughter Mamie came with Robert*: Emerson, *Madness of Mary Lincoln*, p. 130.

580 *"obliged to guard her health"*: *New York Daily Tribune*, Oct. 10, 1881, p. 4., col. 6.

583 *sisters from a local convent*: See Steven Spearie, "Final Comfort: Mary Todd Lincoln May Have Relied on the Help of Local Nuns in Her Last Days," *State Journal-Register* (Springfield, IL), July 14, 2006.

586 *Robert qualified for his Arlington burial plot*: This was painfully ironic in that Mary had been so opposed to his military service, and now this service qualified him to be buried away from her and his family, which would have displeased his mother no end.

587 *the family name died out*: Michael Beschloss, "Last of the Lincolns," *New Yorker*, Feb. 28, 1994, p. 54.

588 *was scheming, criminal*: "Michael Burlingame: King of Lincoln Researchers," interview by Gerald Prokopowicz, *Civil War Talk Radio*, Nov. 3, 2006.

588 *stemmed from the effects of syphilis*: Indeed, one author has speculated that Lincoln broke off his engagement to Mary because he was suffering from full-blown symptoms of tertiary syphilis in the winter of 1840–1841. See Deborah Hayden, *Pox: Genius, Madness, and the Mysteries of Syphilis* (New York: Basic Books, 2003), ch. 12. However, this fanciful claim is not supported by evidence or scholarship.

590 *scholarship has been deeply affected*: Open hostilities broke out a little more than a year after Herndon began collecting information, when he gave a series of lectures to capitalize on his research and interviews. More balanced and sympathetic accounts of Mary Lincoln would counter those charges. Besides the Morrow and Helm

volumes, there are valuable studies by William Evans and Carl Sandburg (1932), Ruth Painter Randall (1953), Justin and Linda Turner (1972), Ishbel Ross (1973), and Jean Baker (1987).

590 *Herndon's stock seemingly on the rise*: During the late twentieth century, historian Michael Burlingame has replaced Herndon as Mary Lincoln's harshest critic. Burlingame places blame for the marriage's failure (a given in his estimation) on Mary. He has detailed her shortcomings and their disastrous effects on her husband. Burlingame's one-sided perspective is showcased in his collected essays, *The Inner World of Abraham Lincoln* (1994), and will doubtless play a role in his forthcoming Lincoln biography. He directly challenges the work of biographers Baker and Randall, accusing them of cherry-picking the evidence about the Lincoln marriage, selectively providing quotes to support their theses. Perhaps this is essentially what *all* scholars might be accused of— but some do it more skillfully than others. In the twenty-first century, C. A. Tripp, author of *The Intimate World of Abraham Lincoln* (2005), surpassed Burlingame to become Mary Lincoln's most rabid detractor. As Jean Baker notes in her preface for Tripp's volume, "Tripp compares [her] to psychopaths like Hitler." Tripp proclaims that the marriage of Abraham and Mary ranks as "one of the worst marital misfortunes in recorded history" and that Mary was Lincoln's "cross to bear." Tripp's death has in many ways curtailed productive debate that his posthumously published work might have generated about Lincoln's private life. However, with

Tripp's Hitler analogy and with Burlingame's repeated assertions that Lincoln was a "battered husband," Mary Lincoln's reputation remains controversial.

590 *a vigil maintained by his widow*: See Jason Emerson, "The Madness of Mary Lincoln," *American Heritage*, July 2006.

591 *Mary Lincoln has been subjected*: See, for example, *The Trial of Mary Lincoln*, an opera by Thomas Pasatieri (1972); Andrew Holleran, *Grief* (New York: Hyperion, 2006); the off-Broadway musical, *Asylum: The Strange Case of Mary Lincoln*, book by June Bingham, music and lyrics by Carmel Owen (2006); and "A Short Reprise for Mary Todd, Who Went Insane, but for Very Good Reasons," by Sufjan Stevens. Several new operas featuring Mary Lincoln have appeared, roughly coinciding with the Abraham Lincoln bicentennial, including Philip Glass's *Appomattox* (2007) and John Shoptaw and Eric Sawyer's *Our American Cousin* (2007).

591 *a punch line in popular culture*: LexisNexis the phrase "Mrs. Lincoln" to get the long list of hits and the full impact of "Other than that, how did you enjoy the play, Mrs. Lincoln?" The term *shopaholic* is likewise linked with Mary Todd Lincoln in contemporary commentary, and in her own day *shopping mania* was a phrase applied to her.

Bibliography

PRIMARY SOURCES

Basler, Roy P., ed. See Lincoln, Abraham, *Collected Works.*

Boyden, Anna L. *Echoes from Hospital and White House: A Record of Mrs. Rebecca R. Pomroy's Experience in War-Times.* Boston: D. Lothrop, 1884.

Brooks, Noah. *Dispatches.* See Burlingame, Michael, ed., *Lincoln Observed.*

———. *Mr. Lincoln's Washington: Selections from the Writings of Noah Brooks, Civil War Correspondent,* ed. P. J. Staudenraus. South Brunswick, NJ: Thomas Yoseloff, 1967.

———. *Washington, D.C., in Lincoln's Time,* ed. Herbert Mitgang. Chicago: Quadrangle Books, 1971.

Browning, Orville. *The Diary of Orville Hickman Browning,* ed. Theodore Calvin Pease and James G. Randall. Springfield, IL: Illinois State Historical Library, 1925–1933.

Burlingame, Michael, ed. *At Lincoln's Side: John Hay's Civil War Correspondence and Selected Writings.* Carbondale: Southern Illinois University Press, 2000.

———. *Inside the White House in War Times: Memoirs and Reports of Lincoln's Secretary, William O. Stoddard.* Lincoln: University of Nebraska Press, 2000.

———. *Lincoln's Journalist: John Hay's Anonymous Writing for the Press, 1860–1864.* Carbondale: Southern Illinois University Press, 1998.

———. *Lincoln Observed: Civil War Dispatches of Noah Brooks.* Baltimore: Johns Hopkins University Press, 1998.

———. *An Oral History of Abraham Lincoln: John G. Nicolay's Interviews and Essays.* Carbondale: Southern Illinois University Press, 1996.

———. *With Lincoln in the White House: Letters, Memoranda, and Other Writings of John G. Nicolay, 1860–1865.* Carbondale: Southern Illinois University Press, 2000.

Burlingame, Michael, and John R. Turner Ettlinger, eds. *Inside Lincoln's White House: The Complete Civil War Diary of John Hay.* Carbondale: Southern Illinois University Press, 1997.

Carpenter, F. B. *Six Months at the White House.* New York: Hurd & Houghton, 1867.

Clemmer, Mary. *Ten Years in Washington: Life and Scenes in the National Capital, as a Woman Sees Them.* Hartford, CT: A. D. Worthington, 1873.

Dennett, Tyler, ed. *Lincoln and the Civil War in the Diaries and Letters of John Hay.* New York: Dodd, Mead, 1939.

Donald, David H., and Harold Holzer, eds. *Lincoln in the Times: The Life of Abraham Lincoln, as Originally Reported in the* New York Times. New York: St. Martin's Press, 2005.

Ellet, Mrs. E. F. *The Court Circles of the Republic.* Hartford, CT: Hartford Pub. Co., 1869.

French, Benjamin Brown. *Witness to the Young Republic: A Yankee's Journal, 1828–1870,* ed. Donald B. Cole and John J. McDonough. Hanover, NH: University Press of New England, 1989.

Gerry, Margarita Spalding, ed. *Through Five Administrations: Reminiscences of Colonel William Crook, Bodyguard to President Lincoln.* New York: Harper & Bros., 1910.

Good, Timothy, ed. *We Saw Lincoln Shot: One Hundred Eyewitness Accounts.* Jackson: University Press of Mississippi, 1995.

Grimsley, Elizabeth Todd. "Six Months in the White House." *Journal of the Illinois State Historical Society,* vol. 19, no. 3.

Hay, John. *Anonymous Writing for the Press.* See Burlingame, Michael, ed., *Lincoln's Journalist.*

———. *Correspondence.* See Burlingame, Michael, ed., *At Lincoln's Side.*

———. *Diaries and Letters.* See Dennett, Tyler, ed., *Lincoln and the Civil War.*

———. *Diary.* See Burlingame, Michael, and John R. Turner Ettlinger, eds., *Inside Lincoln's White House.*

Holzer, Harold, ed. *Lincoln as I knew Him: Gossip, Tributes, and Revelations from His Best Friends and*

Worst Enemies. Chapel Hill, NC: Algonquin Books of Chapel Hill, 1999.

Keckley, Elizabeth. *Behind the Scenes; or, Thirty Years a Slave, and Four Years in the White House.* New York: G. W. Carleton, 1868; repr. New York: Oxford University Press, 1988.

Laas, Virginia Jeans. *Wartime Washington: The Civil War Letters of Elizabeth Blair Lee.* Urbana: University of Illinois Press, 1991.

Lincoln, Abraham. *The Collected Works of Abraham Lincoln* (abbreviated *CWL* in notes), ed. Roy P. Basler. New Brunswick, NJ: Rutgers University Press, 1953.

McClure, J. B. *Lincoln's Stories.* Chicago: Rhodes & McClure, 1879.

Nicolay, John G., *Interviews and Essays.* See Burlingame, Michael, ed. *An Oral History of Abraham Lincoln.*

————. *Letters.* See Burlingame, Michael, ed., *With Lincoln in the White House.*

Rankin, Henry B. *Personal Recollections of Abraham Lincoln.* New York: Putnam's, 1916.

Raymond, Mary Edwards, ed. *Some Incidents in the Life of Mrs. Benjamin S. Edwards.* Privately published, 1909.

Rice, Allen T., ed. *Reminiscences of Abraham Lincoln by Distinguished Men of Our Time.* New York: North American Pub. Co., 1886.

Russell, William Howard. *My Diary, North and South,* ed. Fletcher Pratt. New York: Harper & Bros, 1954.

Salm-Salm, Agnes Elisabeth Winona Leclerq Joy, Princess zu. *Ten Years of My Life.* Detroit: Bedford Brothers, 1878.

Speed, Joshua F. *Reminiscences of Abraham Lincoln & Notes of a Visit to California.* Louisville, KY: John Morton, 1884.

Still, William. *The Underground Railroad.* Philadelphia: Porter & Coates, 1872.

Stoddard, William O. *Memoirs.* See Burlingame, Michael, ed., *Inside the White House in War Times.*

Turner, Justin G., and Linda Levitt Turner. *Mary Todd Lincoln: Her Life and Letters.* New York: Knopf, 1972.

Villard, Henry. *Lincoln on the Eve of '61: A Journalist's Story,* ed. Harold G. Villard and Oswald Garrison Villard. New York: Knopf, 1941.

Weik, Jesse W. *The Real Lincoln: A Portrait,* ed. Michael Burlingame. Lincoln: University of Nebraska Press, 2002.

Wilson, Douglas L., and Rodney O. Davis: *Herndon's Informants: Letters, Interviews and Statements About Abraham Lincoln.* Urbana: University of Illinois Press, 1998.

SECONDARY SOURCES

Abelson, Elaine S. *When Ladies Go A-Thieving: Middle-Class Shoplifters in the Victorian Department Store.* New York: Oxford University Press, 1989.

Applegate, Debby. *The Most Famous Man in America: The Biography of Henry Ward Beecher.* New York: Doubleday, 2006.

Badeau, Adam. *Grant in Peace: From Appomattox to Mount McGregor, a Personal Memoir.* Hartford, CT: S. S. Scranton, 1887.

Baker, Jean. *Affairs of Party: The Political Culture of Northern Democrats in the Mid-Nineteenth Century.* Ithaca, NY: Cornell University Press, 1983.

————. *Mary Todd Lincoln: A Biography.* New York: Norton, 1987.

Bates, Edward. *The Diary of Edward Bates, 1859–1866,* ed. Howard K. Beale. Washington, DC: U.S. Government Printing Office, 1933.

Bayne, Julia Taft. *Tad Lincoln's Father.* Lincoln: University of Nebraska Press, 2001.

Bearss, Edwin C. *Historic Structure Report: Lincoln Home National Historic Site.* Denver: Denver Service Center, Department of the Interior, 1973.

Berry, Stephen. *House of Abraham: Lincoln and the Todds, a Family Divided by War.* Boston: Houghton Mifflin, 2007.

Bleser, Carol K., ed. *In Joy and Sorrow: Women, Family, and Marriage in the Victorian South, 1830–1900.* New York: Oxford University Press, 1991.

Blight, David W. *Race and Reunion: The Civil War in American Memory.* Cambridge, MA: Belknap Press of Harvard University Press, 2001.

————, ed. *Passages to Freedom: The Underground Railroad in History and Memory.* Washington, DC: Smithsonian Books, 2004.

Braude, Ann. *Radical Spirits: Spiritualism and Women's Rights in Nineteenth-Century America*. Beacon Press: Boston, 1989.

Brown, Virginia Stuart. *Through Lincoln's Door*. Springfield, IL: Li-Co Art & Letter Service, 1952.

Burlingame, Michael. *The Inner World of Abraham Lincoln*. Urbana: University of Illinois Press, 1994.

Butterfield, Julia. *A Biographical Memorial of General Daniel Butterfield*. New York: Grafton Press, 1904.

Carman, Harry J., and Reinhard H. Luthin. *Lincoln and the Patronage*. New York: Columbia University Press, 1943.

Carroll, Bret E., ed. *Spiritualism in Antebellum America*. Bloomington: Indiana University Press, 1997.

Carwardine, Richard. *Lincoln: A Life of Purpose and Power*. New York: Knopf, 2006.

Cashin, Joan. *First Lady of the Confederacy: Varina Davis's Civil War*. Cambridge, MA: Harvard University Press, 2006.

Conklin, Eileen F. *Women of Gettysburg*. Gettysburg, PA: Thomas Publications, 1993.

Cox, Robert S. *Body and Soul: A Sympathetic History of American Spiritualism*. Charlottesville: University of Virginia Press, 2003.

Current, Richard Nelson. *The Lincoln Nobody Knows*. New York: Hill and Wang, 1963.

———. *Speaking of Abraham Lincoln: The Man and His Meaning for Our Times*. Urbana: University of Illinois Press, 1983.

Cuthbert, Norma B., ed. *Lincoln and the Baltimore Plot, 1861*. San Marino, CA: Huntington Library, 1949.

Detzer, David. *Dissonance: The Turbulent Days Between Fort Sumter and Bull Run.* New York: Harcourt, 2006.

Donald, David Herbert. *Charles Sumner and the Coming of the Civil War.* New York: Knopf, 1961.

————. *Lincoln.* New York: Simon & Schuster, 1995.

————. *Lincoln at Home: Two Glimpses of Abraham Lincoln's Family Life.* New York: Simon & Schuster, 2000.

————. *Lincoln Reconsidered: Essays on the Civil War Era.* New York: Knopf, 1956; Vintage, 2001.

Doty, William Kavanaugh. *The Confectionary of Monsieur Giron.* Charlottesville, VA: Michie Company, 1915.

Doyle, Arthur Conan. *The History of Spiritualism.* New York: George H. Doran, 1926.

Dusinberre, William. *Slavemaster President: The Double Career of James Polk.* New York: Oxford University Press, 2007.

Ehrmann, Bess. *The Missing Chapter in the Life of Abraham Lincoln.* Chicago: Walter M. Hill, 1938.

Emerson, Jason. *The Madness of Mary Lincoln.* Carbondale: Southern Illinois University Press, 2007.

Epstein, Daniel Mark. *The Lincolns: Portrait of a Marriage.* New York: Ballantine, 2008.

Evans, William Augustus. *Mrs. Abraham Lincoln: A Study of Her Personality and Her Influence on Lincoln.* New York: Knopf, 1932.

Faust, Drew Gilpin. *This Republic of Suffering: Death and the American Civil War.* New York: Knopf, 2008.

Ferguson, Andrew. *Land of Lincoln: Adventures in Abe's America.* New York: Atlantic Monthly Press, 2007.

Ferguson, Ernest G. *Freedom Rising: Washington in the Civil War.* New York: Knopf, 2004.

Fleischner, Jennifer. *Mrs. Lincoln and Mrs. Keckly: The Remarkable Story of the Friendship Between a First Lady and a Former Slave.* New York: Broadway Books, 2003.

Foner, Eric. *Freedom's Lawmakers: A Directory of Black Officeholders During Reconstruction.* Baton Rouge: Louisiana State University Press, 1996.

———. *Nothing But Freedom: Emancipation and Its Legacy.* Baton Rouge: Louisiana State University Press, 1983.

———. *Politics and Ideology in the Age of the Civil War.* New York: Oxford University Press, 1981.

———. *Reconstruction: America's Unfinished Revolution, 1863–1877.* New York: Harper & Row, 1988.

Forgie, George B. *Patricide in the House Divided: A Psychological Interpretation of Lincoln and His Age.* New York: Norton, 1979.

Fornell, Earl Wesley. *The Unhappy Medium: Spiritualism and the Life of Margaret Fox.* Austin: University of Texas Press, 1964.

Frederickson, George. *Big Enough to Be Inconsistent: Abraham Lincoln Confronts Slavery and Race.* Cambridge, MA: Harvard University Press, 2008.

Gallman, J. Matthew. *America's Joan of Arc: The Life of Anna Elizabeth Dickinson.* New York: Oxford University Press, 2006.

————. *The Civil War Chronicle: The Only Day-by-Day Portrait of America's Tragic Conflict as Told by Soldiers, Journalists, Politicians, Farmers, Nurses, Slaves, and Other Eyewitnesses.* New York: Gramercy Books, 2003.

Gernon, Blaine Brooks. *The Lincolns in Chicago.* Chicago: Ancarthe Publishers, 1934.

Goldsmith, Barbara. *Other Powers: The Age of Suffrage, Spiritualism, and the Scandalous Victoria Woodhull.* New York: Knopf, 1998.

Goodwin, Doris Kearns. *Team of Rivals: The Political Genius of Abraham Lincoln* (New York: Simon & Schuster, 2005).

Gray, Ralph. *Following in Lincoln's Footsteps: A Complete Annotated Reference to Hundreds of Historical Sites Visited by Abraham Lincoln (Illinois).* New York: Carroll & Graf, 2001.

Greenberg, Kenneth S. *Honor and Slavery: Lies, Duels, Noses, Masks, Dressing as a Woman, Gifts, Strangers, Humanitarianism, Death, Slave Rebellions, the Proslavery Argument, Baseball, Hunting, and Gambling in the Old South.* Princeton, NJ: Princeton University Press, 1996.

Guelzo, Allen C. *Lincoln's Emancipation Proclamation: The End of Slavery in America.* New York: Simon & Schuster, 2004.

Harris, William C. *Lincoln's Last Months.* Cambridge, MA: Harvard University Press, 2004.

Harrison, Lowell H. *Lincoln of Kentucky.* Lexington: University of Kentucky Press, 2000.

Helm, Katherine. *The True Story of Mary, Wife of Lincoln: Containing the Recollections of Mary Lincoln's Sister Emilie (Mrs. Ben Hardin Helm), Extracts from Her War-time Diary, Numerous Letters and Other Documents.* New York: Harper & Bros., 1923.

Herndon, William Henry, and Jesse W. Weik. *Herndon's Lincoln: The True Story of a Great Life* (repr. ed. Douglas L. Wilson). Chicago: Belford, Clarke, 1889; repr. Urbana: University of Illinois Press, 2006.

Heyrman, Christine Leigh. *Southern Cross: The Beginnings of the Bible Belt.* New York: Knopf, 1997.

Hoffert, Sylvia. *Jane Grey Swisshelm: An Unconventional Life, 1815–1884.* Chapel Hill: University of North Carolina Press, 2004.

Holloway, Laura C. *The Ladies of the White House; or, in the Home of the Presidents, Being a Complete History of the Social and Domestic Lives of the Presidents from Washington to the Present Time, 1789–1881.* Philadelphia: Bradley, 1885.

Holzer, Harold. *Lincoln at Cooper Union: The Speech That Made Abraham Lincoln President.* New York: Simon & Schuster, 2004.

Holzer, Harold, and Sara Vaughn Gabbard, eds. *Lincoln and Freedom: Slavery, Emancipation, and the Thirteenth Amendment.* Carbondale: Southern Illinois University Press, 2007.

Johnson, Patricia. "Sensitivity and the Civil War: The Selected Diaries and Papers, 1858–66, of Frances

Adeline (Fanny) Seward." Ph.D. diss., University of Rochester, 1963.

King, C. J. *Four Marys and a Jessie: The Story of the Lincoln Women.* Manchester, VT: Hildene Library, 2005.

Krause, Susan, Kelley A. Boston, and Daniel W. Stowell. *Now They Belong to the Ages: Abraham Lincoln and His Contemporaries in Oak Ridge Cemetery.* Springfield: Illinois Historic Preservation Agency, 2005.

Kromm, Jane. *The Art of Frenzy: Public Madness in the Visual Culture of Europe, 1500–1850.* London: Continuum, 2002.

Larson, Erik. *The Devil in the White City: Murder, Magic, and Madness at the Fair That Changed America.* New York: Crown, 2003.

Laugel, Auguste. *The United States During the Civil War,* ed. Allan Nevins. Bloomington: Indiana University Press, 1961.

Leonard, Elizabeth D. *Yankee Women: Gender Battles in the Civil War.* New York: Norton, 1994.

Lystra, Karen. *Searching the Heart: Women, Men, and Romantic Love in Nineteenth-Century America.* New York: Oxford University Press, 1989.

Manning, Chandra. *What This Cruel War Was Over: Soldiers, Slavery, and the Civil War.* New York: Knopf, 2007.

Maynard, Nettie Coburn. *Was Abraham Lincoln a Spiritualist? or, Curious Revelations from the Life of a Trance Medium.* Whitefish, MT: Kessinger, 2003.

McCreary, Donna. *The Victorian Wardrobe of Mary Lincoln*. Charlestown, IN: Lincoln Presentations, 2007.

McPherson, James M. *Hallowed Ground: A Walk at Gettysburg*. New York: Crown, 2003.

———. *Marching Toward freedom: The Negro in the Civil War, 1861–1865*. New York: Knopf, 1968.

———. *What They Fought For, 1861–1865*. Baton Rouge: Louisiana State University Press, 1994.

Menand, Louis. *The Metaphysical Club: A Story of Ideas in America*. New York: Farrar, Straus and Giroux, 2001.

Miller, William Lee. *Lincoln's Virtues: An Ethical Biography*. New York: Knopf, 2002.

Mitgang, Herbert, ed. *Abraham Lincoln—A Press Portrait: His Life and Times from the Original Newspaper Documents of the Union, the Confederacy, and Europe*, 3rd ed. Athens: University of Georgia Press, 1989.

Neely, Mark E., Jr. *The Civil War and the Limits of Destruction*. Cambridge, MA: Harvard University Press, 2007.

———. *The Fate of Liberty: Abraham Lincoln and Civil Liberties*. New York: Oxford University Press, 1991.

———. *The Last Best Hope of Earth: Abraham Lincoln and the Promise of America*. New York: Oxford University Press, 1993.

Neely, Mark E., Jr., and Gerald R. McMurtry Jr. *The Insanity File: The Case of Mary Todd Lincoln*. Carbondale: Southern Illinois University Press, 1986.

Nicolay, Helen. *Lincoln's Secretary: A Biography of John G. Nicolay.* New York: Longmans, Green, 1949; repr. Westport, CT: Greenwood Press, 1971.

Oakes, James. *The Radical and the Republican: Frederick Douglass, Abraham Lincoln, and the Triumph of Antislavery Politics.* New York: Norton, 2007.

Ostendorf, Lloyd, and Walter Olesky, eds. *Lincoln's Unknown Private Life: An Oral History by His Black Housekeeper, Mariah Vance, 1850–1860.* Mamaroneck, NY: Hastings House, 1995.

Packard, Jerrold. *The Lincolns in the White House: Four Years That Shattered a Family.* New York: St. Martin's, 2005.

Peterson, Merrill D. *Lincoln in American Memory.* New York: Oxford University Press, 1994.

Pinsker, Matthew. *Lincoln's Sanctuary: Abraham Lincoln and the Soldiers' Home.* New York: Oxford University Press, 2005.

Pratt, Harry. *The Personal Finances of Abraham Lincoln.* Springfield, IL: Abraham Lincoln Association, 1943; repr. Ann Arbor: University of Michigan Library, 2006.

Prokopowicz, Gerald. *Did Lincoln Own Slaves? And Other Frequently Asked Questions About Abraham Lincoln.* New York: Pantheon, 2007.

Randall, Ruth Painter. *Colonel Elmer Ellsworth: A Biography of Lincoln's Friend and First Hero of the Civil War.* Boston: Little, Brown, 1960.

———. *Lincoln's Sons.* Boston: Little, Brown, 1955.

———. *Mary Lincoln: Biography of a Marriage.* Boston: Little, Brown, 1953.

Rose, Anne C. *Victorian America and the Civil War*. New York: Cambridge University Press, 1992.

Ross, Ishbel. *The President's Wife: Mary Todd Lincoln*. New York: Putnam's, 1973.

———. *Proud Kate: Portrait of an Ambitious Woman*. New York: Harper & Bros., 1953.

Sandburg, Carl. *Abraham Lincoln*, vol. 1, The *Prairie Years* (New York: Harcourt Brace, 1926).

———. *Abraham Lincoln*, vol. 2, *The War Years* (New York: Harcourt Brace, 1939).

———. *Mary Lincoln: Wife and Widow*. New York: Harcourt Brace, 1932; repr. Bedford, MA: Applewood Books, 1994.

Schlereth, Thomas J. *Victorian America: Transformations in Everyday Life, 1876–1915*. New York: Harper, 1991.

Schreiner, Samuel Agnew. *The Trials of Mrs. Lincoln: The Harrowing Never-Before-Told Story of Mary Todd Lincoln's Last and Finest Years*. New York: Donald I. Fine, 1987.

Schwartz, Thomas. *Mary Todd Lincoln: First Lady of Controversy*. Springfield: Abraham Lincoln Presidential Library and Museum, 2007.

Schwartz, Thomas, and Kim Bauer, eds. "Unpublished Mary Todd Lincoln." *Journal of the Abraham Lincoln Association 17* (summer 1996).

Shenk, Joshua Wolf. *Lincoln's Melancholy: How Depression Challenged a President and Fueled His Greatness*. New York: Houghton Mifflin, 2005.

Simon, John, Harold Holzer, and Dawn Vogel. *Lincoln Revisited: New Insights from the Lincoln Forum*. New York: Fordham University Press, 2007.

Simpson, Brooks. *The Reconstruction Presidents.* Lawrence: University Press of Kansas, 1998.

Singleton, Esther. *The Story of the White House.* New York: McClure, 1907.

Smith, Carl S. *The Plan of Chicago: Daniel Burnham and the Remaking of the American City.* Chicago: University of Chicago Press, 2006.

Speed, James. *James Speed: A Personality.* Louisville, KY: John P. Morton, 1914.

Stauffer, John. *Giants: The Parallel Lives of Frederick Douglass and Abraham Lincoln.* New York: Twelve, 2008.

Swanson, James. *Manhunt: The 12-Day Chase for Lincoln's Killer.* New York: William Morrow, 2006.

Tarbell, Ida M. *The Life of Abraham Lincoln,* 2 vols. New York: Doubleday, Page, 1895.

Townsend, William H. *Lincoln and His Wife's Home Town.* Indianapolis: Bobbs-Merrill, 1929.

————. *Lincoln and the Bluegrass: Slavery and Civil War in Kentucky.* Lexington: University of Kentucky Press, 1955.

Tripp, C. A. *The Intimate World of Abraham Lincoln.* New York: Free Press, 2005.

Vance, Mariah. *Lincoln's Unknown Private Life.* See Ostendorf and Olesky, eds.

Van der Heuvel, Gerry. *Crowns of Thorns and Glory: Mary Todd Lincoln and Varina Howell Davis, the Two First Ladies of the Civil War.* New York: Dutton, 1988.

Washington, John E. *They Knew Lincoln.* New York: Dutton, 1942.

Weber, Jennifer L. *Copperheads: The Rise and Fall of Lincoln's Opponents in the North.* New York: Oxford University Press, 2006.

Weik, Jesse W. *The Real Lincoln: A Portrait,* ed. Michael Burlingame. Lincoln: University of Nebraska Press, 2002.

Weisberg, Barbara. *Talking to the Dead: Kate and Maggie Fox and the Rise of Spiritualism.* San Francisco: HarperSanFrancisco, 2004.

Welles, Gideon. *Diary of Gideon Welles, Secretary of the Navy Under Lincoln and Johnson,* ed. Howard K. Beale. New York: Norton, 1960.

Whitcomb, John, and Claire Whitcomb. *Real Life at the White House: 200 Years of Daily Life at America's Most Famous Residence.* New York: Routledge, 2000.

White, Barbara A. *The Beecher Sisters.* New Haven, CT: Yale University Press, 2003.

White, Ronald C., Jr. *Lincoln's Greatest Speech: The Second Inaugural.* New York: Simon & Schuster, 2002.

———. *A. Lincoln: A Biography* (New York, Random House, 2009).

Wilson, Douglas L. *Honor's Voice: The Transformation of Abraham Lincoln.* New York: Knopf, 1998.

Winkler, H. Donald. *Lincoln's Ladies: The Women in the Life of the Sixteenth President.* Nashville, TN: Cumberland House, 2004.

List of Illustrations

E xcept as noted below, all illustrations are courtesy of the Abraham Lincoln Presidential Library and Museum.

Acknowledgments

The long road to completing *Mrs. Lincoln* has been littered with obstacles, bedeviled by detours, which, along with transatlantic frequent-flyer miles, have worn me down along the way. Although I cannot adequately thank all the people who assisted my journey, I wish to acknowledge several whose time and generosity made measurable improvements. Like all authors within the burgeoning field of Lincoln studies, I live in fear of making or repeating errors, but take responsibility for my mistakes and thank those who tried to reduce them in both number and scope.

First and foremost, I must thank David and Aida Donald, who have been patient and generous with me over the years, and I cannot begin to express my gratitude for their taking time from their own busy schedules as editors, researchers, and authors to offer me insight and counsel. Several kind and generous scholars have provided close readings and offered critical corrections for which I remain indebted, including Eric Foner, James McPherson, Joseph Ellis, and Carol Bleser. As always, I appreciate your support, especially during the sprint for this finish line.

Doris Kearns Goodwin remains a deeply loyal colleague who offered assistance at several key junctures. I am both awed and inspired by her appreciation of all things Lincoln, and salute her sharing of research and insight—the afternoons spent in her library in Concord and, most of all, her faith in this project, which carried me through some difficult days.

Louise Taper, with characteristic warmth, opened her home—and, most generously, her vault. Current and future Lincoln scholars, especially this one, remain in her debt.

Nearly two decades ago, Gabor Boritt kindly invited me to Gettysburg to lecture on women at his summertime Civil War Institute. His gentle prodding, along with Jim McPherson, eventually converted me—although I am not sure either of them wants to take the credit or the blame. Harold Holzer's indefatigability and generosity have been inspirational. The Gilder Lehrman Institute of American History has supported me on previous projects and funded so many valuable works from which I have benefited.

Again, I wish to acknowledge those whose friendships made crucial contributions to this project: Paul Ulhmann Jr., Paul Uhlmann III, Fran D'Ooge, Edna Medford Green, Frank Williams, Jennifer Fleischner, Jason Emerson, Christine Heyrman, Amy McCandless, Jim Downs, Anne Bacon, Tom Block, Wayne Temple, Daniel Mark Epstein, Andrew Ferguson, Julie Bowen, Carol Berkin, Susan O'Hara, Lizzy Rockwell, Taylor Patterson, Lesley Herrmann, and Mary Rose Taylor. I very much miss Larry Appel, who did not live to see the completion of this book, of which he was so supportive.

I do not know how Roy Ritchie expects to keep us inside the Huntington Library, surrounded by such beauty . . . but many thanks for trying. Craig D'Ooge continues to supply me with links to new discoveries from his perch at the Library of Congress.

In Springfield, Tom Schwartz has repeatedly proffered thoughtful advice. I would like to thank others at the Abraham Lincoln Presidential Library and Museum: Rick Beard and his entire ALPLM team are paving the way for the next generation of Lincoln scholars. Many thanks to James Cornelius, Mary Michals, Christine Juliat, Roger Schmitz, and Peggy

Dunn. Members of the team at the National Park Service's Lincoln Home Historic Site were generous with tours and information. I also need to thank the Abraham Lincoln Presidential Bicentennial Commission, the staff at Hildene in Manchester, Vermont, staff at President Lincoln's Cottage in Washington, D.C., as well as Joan Flinspach and her staff at the former Lincoln Museum, in Fort Wayne, Indiana. Archivists at the Library of Congress, the Illinois State Library, the Abraham Lincoln Legal Papers, the New-York Historical Society, the Chicago Historical Society, Special Collections at Allegheny College, and the Archives de la Communauté d'Agglomération Pau-Pyrénées, as well as the American Society for Psychical Research were most helpful.

I appreciate Sandy Brue and her team at the Abraham Lincoln Birthplace National Historic Site in Hodgenville, Kentucky, Gwen Thompson at the Mary Todd Lincoln Home in Lexington, Bill Marshall at the University of Kentucky, Mark Wetherington at the Filson Historical Society, and friends who have hosted me while I was doing research on Mary's early years: Joan Chan (of New York and Lexington), Tom Appleton (of Lexington), and Emily Bingham (of Louisville).

My agent, Kris Dahl, remains a stellar friend, who introduced me to my wonderful editor, Tim Duggan.

I thank them both for remaining voices of calm and reason whose contributions considerably eased the process.

This book could not have been completed without the support of beloved friends whose mentoring, messages, and mettle kept me going with the ebb and flow of this project's emotional tidal waves: Virginia Gould, Deborah Tatro, Michele Gillespie, and most especially Jody Block (whose suggestions on an early draft allowed crucial changes). Each of you has kept me afloat during critical junctures, and I remain in your debt.

I received a fellowship from the Organization of American Historians and the White House Historical Society, and I wish to thank Bill Bushong, John Riley, and Sally Stokes. Most recently, Queens University, Belfast, has supported this project with a generous research fund to allow me to complete the book on time.

I want to thank the other Marys who have kept me company at College Green House—Mary Blair and Mary O'Dowd, who have made Belfast more than just a place of residence, but a welcoming community.

And for Dan, Drew, and Ned Colbert, you know how much you mean to me.

Catherine Clinton

Belfast, Northern Ireland, 2008

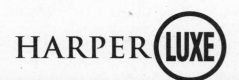